The Investigative Science Learning Environment: A Guide for Teacher Preparation and Professional Development

Online at: https://doi.org/10.1088/978-0-7503-5568-1

The Investigative Science Learning Environment: A Guide for Teacher Preparation and Professional Development

Eugenia Etkina
Department of Learning and Teaching, Graduate School of Education, Rutgers The State University of New Jersey, New Brunswick, NJ, United States of America

Gorazd Planinsic
Faculty of Mathematics and Phyics, University of Ljubljana, Ljubljana, Slovenia

IOP Publishing, Bristol, UK

© IOP Publishing Ltd 2024

All rights reserved. No part of this publication may be reproduced, stored in a retrieval system or transmitted in any form or by any means, electronic, mechanical, photocopying, recording or otherwise, without the prior permission of the publisher, or as expressly permitted by law or under terms agreed with the appropriate rights organization. Multiple copying is permitted in accordance with the terms of licences issued by the Copyright Licensing Agency, the Copyright Clearance Centre and other reproduction rights organizations.

Permission to make use of IOP Publishing content other than as set out above may be sought at permissions@ioppublishing.org.

Eugenia Etkina and Gorazd Planinsic have asserted their right to be identified as the authors of this work in accordance with sections 77 and 78 of the Copyright, Designs and Patents Act 1988.

ISBN 978-0-7503-5568-1 (ebook)
ISBN 978-0-7503-5566-7 (print)
ISBN 978-0-7503-5569-8 (myPrint)
ISBN 978-0-7503-5567-4 (mobi)

DOI 10.1088/978-0-7503-5568-1

Version: 20240501

IOP ebooks

British Library Cataloguing-in-Publication Data: A catalogue record for this book is available from the British Library.

Published by IOP Publishing, wholly owned by The Institute of Physics, London

IOP Publishing, No.2 The Distillery, Glassfields, Avon Street, Bristol, BS2 0GR, UK

US Office: IOP Publishing, Inc., 190 North Independence Mall West, Suite 601, Philadelphia, PA 19106, USA

This book is dedicated to teachers of all levels who found the courage, time, and persistence to implement the ISLE approach in their classrooms. It is dedicated to all our students who motivated us to improve our materials. It is also dedicated to all our colleagues who contributed to the development of the ISLE materials and who supported our work. Thank you!

Contents

Preface		xi
Acknowledgements		xii
Author biographies		xiii

1 Introduction — 1-1

1.1 Introduction to the introduction — 1-1
1.2 About the authors — 1-2
1.3 What can you learn about the ISLE approach in the first book, *Investigative Science Learning Environment: When Learning Physics Mirrors Doing Physics*? — 1-2
    1.3.1 Observational experiment and initial explanations — 1-7
    1.3.2 Observational experiment and models — 1-8
    1.3.3 Testing the models — 1-9
    1.3.4 Applying the model — 1-10
    1.3.5 Formative assessment — 1-11
1.4 What will you find and what will you not find in this book? — 1-12
1.5 Interlude: ISLE and multiple representations (by Eugenia and Gorazd) — 1-14
References — 1-19

2 Dispositions, knowledge, and skills of teachers and their relationship to teaching physics through the ISLE approach — 2-1

2.1 ISLE and dispositions — 2-1
2.2 ISLE and knowledge — 2-6
2.3 ISLE and skills — 2-19
2.4 *Interlude:* Habits which are the missing link to the implementation of the ISLE approach (by Eugenia) — 2-24
References — 2-25

3 Habits of physics teachers — 3-1

3.1 What are habits? — 3-1
3.2 Habits of mind — 3-5
3.3 Habits of practice — 3-8
3.4 Habits of maintenance and improvement — 3-14
3.5 *Interlude*: Eugenia and Gorazd on habits and routines — 3-15
References — 3-19

4 ISLE and the development of physics habits of mind   4-1

4.1 Treating physics as a process and not as a set of rules   4-1
     4.1.1 Physical phenomena and physical objects   4-2
     4.1.2 Models of phenomena, objects, systems, processes, and interactions   4-2
     4.1.3 Measuring instruments   4-8
     4.1.4 Physics devices   4-9
     4.1.5 Testing experiments   4-10
     4.1.6 Predictions   4-11
     4.1.7 Application experiments   4-12
     4.1.8 Assumptions   4-14
     4.1.9 How are the above elements connected to the physicist habits of mind?   4-14
     4.1.10 Inductive reasoning   4-15
     4.1.11 Spherical cow reasoning   4-19
     4.1.12 Analogical reasoning   4-20
     4.1.13 Hypothetico-deductive reasoning   4-21
     4.1.14 Theory   4-24
4.2 Noticing physics everywhere   4-25
4.3 Approaching problems as an expert   4-26
4.4 Treating mathematics in a physics way   4-31
4.5 *Interlude by Eugenia*: History of physics, physics habits of mind, physics teacher preparation, and professional development (why should physics teachers know where physics rules come from?)   4-35
     4.5.1 Historical examples of the development and use of the physics habits of mind   4-36
     4.5.2 Overview of the 'Development if Ideas in Physical Science'   4-38
     Appendix   4-39
     References   4-42

5 ISLE and the development of physics teacher habits of mind and practice   5-1

5.1 Development of inductive and hypothetico-deductive reasoning   5-1
     5.1.1 Cool glass   5-2
     5.1.2 Popping the balloon   5-6
5.2 Documenting physical phenomena in the outside world   5-6
5.3 Treating students' ideas as resources for the development of normative concepts, not as misconceptions that need to be weeded out   5-11

	5.3.1 How do we treat student ideas?	5-11
	5.3.2 Knowledge in pieces	5-12
	5.3.3 The ISLE approach and student ideas	5-15
	5.3.4 Unexpected treasure: student learning resources	5-16
	5.3.5 Developing teacher habits in recognizing and building on resources	5-18
5.4	Recognizing experimentally testable student ideas and knowing how to test student ideas	5-20
5.5	Applying systems reasoning to the analysis of physical phenomena	5-25
	5.5.1 Reason	5-27
5.6	Choosing the right language	5-38
5.7	*Interlude* Eugenia on motivation	5-42
	Appendix A	5-46
	Appendix B	5-56
	References	5-58

6 ISLE and development of routines — 6-1

6.1	Positioning students, group work, developing accountability	6-1
6.2	Setting up experimental work for the students	6-6
6.3	Reflecting	6-7
6.4	Questioning techniques	6-11
6.5	Responding (or not) to students' questions	6-16
6.6	Setting up assessment routines	6-18
6.7	Homework—to assign or not to assign?	6-29
6.8	Reading textbooks	6-32
6.9	*Interlude*: Gorazd about recognizing multiple levels of complexity in a specific physics experiment	6-35
	References	6-39

7 Organizing ISLE-based teacher preparation programs for the development of habits of mind and practice — 7-1

7.1	The importance of coherence and duration in the program	7-1
7.2	Course work	7-10
	7.2.1 'Development of Ideas in Physical Science' course	7-10
	7.2.2 'Teaching and Assessment in Physical Science' course	7-15
	7.2.3 'Multiple Representations in Physical Science' course	7-20

7.3	Clinical practice	7-23
	7.3.1 Observations of high school physics classes where students learn through the ISLE approach	7-24
	7.3.2 Teaching as lab or problem solving sessions' instructor in an ISLE-based introductory physics college course	7-26
	7.3.3 Microteaching in the physics methods courses where the students are peers in the program	7-27
	7.3.4 Full-time student teaching	7-33
7.4	Assessment	7-39
7.5	The role of a community in the development of habits	7-43
	7.5.1 In-the-program learning community	7-45
	7.5.2 After-graduation learning community	7-46
	7.5.3 Community and habits	7-48
7.6	*Interlude* by Bor Gregorcic: AI and the learning of physics	7-49
	7.6.1 Preparing physics teachers for a life with artificial intelligence	7-49
	7.6.2 What physics teachers should know about LLM-based technologies	7-50
	7.6.3 What opportunities LLM-based chatbots bring to physics teacher education	7-52
	7.6.4 The role of ISLE in a world with AI	7-54
	Appendix A	7-55
	Appendix B	7-62
	References	7-67
8	**Success stories in the development of habits**	**8-1**
8.1	Who are the authors of the stories?	8-1
8.2	June Lee's story	8-5
8.3	Allison Daubert's story	8-7
8.4	Danielle Buggé's story	8-9
8.5	Josh Rutberg's story	8-11
8.6	Reflection	8-12
9	**Summary**	**9-1**

Preface

This book is for those who read our first book, *Investigative Science Learning Environment: When Learning Physics Mirrors Doing Physics*, published by IOP publishing (Etkina, Brookes and Planinsic 2019), became interested in the ISLE approach and wish to learn how to develop the knowledge, skills, and mindset to implement it. It is also for those who have never heard of ISLE but wish their students to learn physics by practicing it, develop confidence, a growth mindset, and the ability to reason like a physicist. It is for those who love cool experiments and interesting problems, and wish to communicate this love to their students. It is for all those who tried an interactive engagement approach in their classes and felt that they needed more help. It is for all those who care about their students enjoying learning physics without fear. It is for all those who prepare such teachers. This book is for you!

Acknowledgements

Many people contributed to this book. We are grateful to Carolyn Sealfon, Andrew Yolleck, and Eugenio Tufino for reading every chapter, providing feedback, and helping to edit our work. We thank Bor Gregorcic for writing an interlude about the role of artificial intelligence in today's learning and teaching of physics. We are grateful to June Lee, Allison Daubert, Danielle Buggé, and Josh Rutberg who took time to write reflections on their own development of productive habits. We thank David Brookes for his work on the first ISLE book and all his contributions to the ISLE based materials that we have been using over the years. We thank Jane Jackson for her continuous support of our work. We also thank Caroline Mitchell for convincing us to write this book. Finally, we want to thank anonymous reviewers of our book proposal who provided invaluable advice before we even started this project.

Author biographies

Eugenia Etkina

Eugenia Etkina is an Emerita Distinguished Professor at Rutgers, the State University of New Jersey. She holds a PhD in physics education from Moscow State Pedagogical University and has more than 40 years of experience teaching physics. She is a recipient of the 2014 Millikan Medal, awarded to educators who have made significant contributions to teaching physics, and is a fellow of the AAPT. Professor Etkina designed and coordinated one of the largest programs in physics teacher preparation in the United States. She conducts professional development for high school and university physics instructors, and participates in reforms to undergraduate physics courses. In 1993 she developed a system in which students learn physics using processes that mirror scientific practice, the Investigative Science Learning Environment (ISLE) approach to learning and teaching physics. ISLE was described in the book *Investigative Science Learning Environment: When Learning Physics Mirrors Doing Physics*. She is the lead author of the textbook *College Physics: Explore and Apply* and the supporting *Active Learning Guide* and *Instructor Guide*. Professor Etkina has conducted over 200 workshops for physics instructors, and published over 100 peer-reviewed articles.

Gorazd Planinsic

Gorazd Planinsic is a Professor of Physics at the University of Ljubljana, Slovenia. He has a PhD in physics from the University of Ljubljana. Since 2000 he has led the Physics Education program, which prepares almost all high school physics teachers in the country of Slovenia. He started his career in MRI physics and later switched to physics education research. During the last ten years, his work has focused mostly on research into new experiments and how to use them more productively in teaching and learning physics. He is co-founder of the Slovenian hands-on science center House of Experiments. Professor Planinsic is co-author of more than 80 peer-reviewed research articles and more than 20 popular science articles, a co-author of *College Physics: Explore and Apply* and the *Active Learning Guide*, and the author of a Slovenian university textbook for future physics teachers. He is a recipient of the GIREP medal for his contributions in physics education and is a fellow of the IOP.

IOP Publishing

The Investigative Science Learning Environment: A Guide for Teacher Preparation and Professional Development

Eugenia Etkina and Gorazd Planinsic

Chapter 1

Introduction

1.1 Introduction to the introduction

This book, *The Investigative Science Learning Environment: A Guide for Teacher Preparation and Professional Development*, is the sequel to our first book *Investigative Science Learning Environment: When Learning Physics Mirrors Doing Physics*, published by IOP Publishing in 2019 (Etkina *et al* 2019a). In that book we described the Investigative Science Learning Environment (ISLE, pronounced as in the small island) approach to learning physics, including its philosophy and main components, and gave brief examples. In this book we systematically discuss how to use the ISLE approach to help teachers (prospective and current) develop productive habits of mind and practice that will allow them to prepare their students for success in the twenty-first century.

The ISLE approach engages all students in carefully organized collaborative explorations of the physical world with the subsequent construction of their own understanding of normative concepts. This approach develops epistemic knowledge and competencies such as experimental design, evaluation of multiple solutions, communication, and many others. What is important about the ISLE approach is that it is an intentional approach to curriculum design and learning activities (Macmillan and Garrison 1988). Intentionality means that the process through which learning occurs is as crucial for learning as the final outcome or learned content. We see two main intentionalities of the ISLE approach: the first intentionality is that the process through which students learn mirrors the practice of physics. This mirroring involves not only the process of the development of new ideas that is based on systematic patterns of physics experimentation and reasoning, but also the collaborative nature of science and its continuous opportunities to improve one's work. The latter connects the first intentionality to the second: the process through which students learn improves their motivation, confidence, and develops a growth mindset (Brookes *et al* 2020).

It takes a very special teacher to engage students in such a process of collaborative exploration and to help them grow. Teaching is a habitual activity. Teachers undertake many decisions and actions in the moment of teaching, e.g. in the 'fog of war'. Therefore, only those who can engage their students habitually in collaborative exploratory work can systematically 'not lecture' but facilitate active student learning. Such teachers must endure the pressures of teaching without reverting to the transmission mode of instruction. Several years ago, E Etkina together with B Gregorcic and S Vokos created a framework for the development of productive teaching habits as the foundation of high-quality teaching practice (Etkina *et al* 2017), called Development of Habits in A Community (DHAC). In this book we combine the ISLE approach with the DHAC framework to describe an approach to physics teacher preparation and professional development to develop productive teaching habits that engage students in physics practices, learning, and growth.

1.2 About the authors

The authors of this book have a combined experience of over 40 years of preparing physics teachers and running professional development programs. Eugenia Etkina (the founder of the ISLE approach) was trained as a physics teacher in the Soviet Union, taught there for 13 years, earned her PhD in physics education, and later became a professor at the Rutgers Graduate School of Education in the USA. At Rutgers, she led the physics teacher preparation program that has prepared over 150 physics teachers since 2002. She has conducted over 200 professional development workshops for physics educators of all levels. Her pioneering work on ISLE and physics teacher preparation was recognized with the Robert Millikan medal in 2014. The medal is awarded every year by the American Association of Physics Teachers (AAPT) to one educator who 'has made notable and intellectually creative contributions to the teaching of physics'.

Gorazd Planinsic was trained as a condensed matter experimentalist before he was given charge of preparing physics teachers at the University of Ljubljana, Slovenia, in 2000. Since then, he has been preparing almost all physics teachers in the country of Slovenia. His skills in experimental design are unmatched. In 2018 he received the Groupe International de Recherche sur l'Enseignement de la Physique (GIREP) Medal, awarded yearly to one educator who made 'special contributions to Physics Education'.

1.3 What can you learn about the ISLE approach in the first book, *Investigative Science Learning Environment: When Learning Physics Mirrors Doing Physics*?

The first book introduced the readers to the ISLE approach, and provided examples of how students learn new material through ISLE, develop problem solving expertise, and participate in new forms of assessment. Here we will briefly summarize the ISLE approach to the student construction of physics concepts, mathematical relationships, and problem solving.

Although the foundation of the ISLE approach was developed by E Etkina while she was teaching high school physics in Moscow, Russia, in the 1980s, ISLE in the present form is the result of collaboration and the contributions of many people. The most important is Alan Van Heuvelen who was the first person in the US to recognize the uniqueness of the ISLE philosophy and to try it with his students at the Ohio State University in 2000. The success of this first implementation was documented in the very first publication about ISLE in 2001 by E Etkina and A Van Heuvelen in the first *Proceedings of the Physics Education Research Conference* (PERC) (Etkina and Van Heuvelen 2001). Van Heuvelen not only wholeheartedly adopted the ISLE philosophy, but he also contributed greatly to its development by introducing reasoning tools—multiple representations—to help students identify patterns, devise and test explanations of physical phenomena, and make a bridge between abstract words and mathematical expressions.

The first complete description of the method was done in the chapter 'Investigative Science Learning Environment—a science process approach to learning physics' in the book edited by E Redish, *Research Based Reform of University Physics* (Etkina and Van Heuvelen 2007). In 2014 Etkina, Van Heuvelen, and Michael Gentile published the first edition of an algebra-based textbook where the ISLE approach was fully implemented in every chapter (Etkina *et al* 2014b). The textbook was accompanied by the *Active Learning Guide* (ALG) with student activities (Etkina *et al* 2014c) and the *Instructor Guide* (Etkina *et al* 2014a) that helped educators use the textbook and the ALG. Etkina and Van Heuvelen collaborated with a large group of people at Rutgers University (S Brahmia who coined the term ISLE and was the first to use the ISLE approach with students-at-risk at Rutgers, D T Brookes, M Gentile, A Karelina, S Murthy, D Rosengrant, M Ruibal Villasenhor, J Rutberg, A Warren, D Andres, J Pinheiro, D Jammula, S Ahmed, R Zisk) and at other universities (B Gregorcic, S Zou, C Sealfon, D Demaree, Y Lin, Y Young, M Selen). Gradually, high school teachers started contributing to the development and research of the ISLE methods (T Bartiromo, M Blackman, D Buggé, M J Finley, J Flakker, D Lee, H Lopez, E Resnick, J Santonasita, T Spero, and many others). Since 2012 Gorazd Planinsic has made the most significant contributions to the new ISLE-based materials. He not only contributed new types of problems, experiments, videos, and photos to the second edition of the ISLE-based introductory physics textbook *College Physics: Explore and Apply* (Etkina *et al* 2019c) but he also started using the ISLE approach in the physics teacher preparation program at the University of Ljubljana, Slovenia. Last year two of the introductory physics courses at the University of Ljubljana were transformed to adopt the ISLE approach under Planinsic's leadership. With the start of the COVID-19 pandemic, Etkina and Planinsic created the Online ALG (OALG) to be used when teaching remotely. This collaborative work has developed the ISLE approach into a premier student-centered method of teaching and learning physics in the world today. In the last ten years the ISLE approach has spread to Canada, Slovenia, Indonesia, the Netherlands, South Africa, Sweden, Italy, New Zealand,

Brazil, and many other countries on all the continents. As we apply iterative ISLE processes to develop and improve ISLE itself, high school and university physics educators who implement the ISLE approach also contribute to its development. The ISLE-based Facebook group called 'Exploring and Applying Physics' now over 2500 members, physics teachers of all levels from all over the world.

Other disciplines are starting to implement the ISLE approach. The graduates of the Rutgers Physics Teacher Preparation program are developing ISLE-based chemistry curriculum materials. Professor Julie Maybee and colleagues at Lehman College, City University of New York, applied it to teaching philosophy. We hope that, with time, the learner-centered benefits of ISLE ideas will spread to many other disciplines.

There are three key features of the ISLE approach, which mirror the features of a scientific inquiry environment while at the same time allowing students to develop traditionally valued physics knowledge (normative concepts).

1. Students develop normative physics concepts as their own ideas by repeatedly going through the following process:
    a. Observing pre-selected phenomena and looking for patterns (usually experiments but also could be simulations or previously collected data, photos, videos, etc) *without making any predictions about the outcomes in advance.*
    b. Developing explanations/models/mathematical relations for these patterns.
    c. Using these explanations/models/relations to make predictions about the outcomes of testing experiments that students propose.
    d. Deciding if the outcomes of the testing experiments match the predictions.
    e. Revising the models/relations if necessary and finally arriving at the normative physics knowledge.
    f. Applying this knowledge for practical purposes (solving problems, building devices, determining the values of physical quantities, etc). The process is depicted in figure 1.1.

Note that the ISLE process does not start with predictions but with observational experiments. The students collect data (qualitative or quantitative) and look for patterns without having any expectations about the patterns. This approach is radically different from the predict–observe–explain approach (White and Gunstone 1992), where students are asked first to make predictions of the outcome of the experiment using their intuition. The reasons we do not ask for predictions before observing new experiments are numerous and we discuss these in chapter 4. Here, we only cite *The Hitchhiker's Guide to the Galaxy* by Douglas Adams (Adams 1985). Adams described the ISLE process in book 4 of the trilogy (*So Long and Thanks for all the Fish*), towards the end of chapter 31 (the italics are ours):

When Wonko returned he was carrying something that stunned Arthur. Not the sandals, they were perfectly ordinary wooden-bottomed sandals. 'I just thought you'd like to see,' he said, 'what angels wear on their feet.

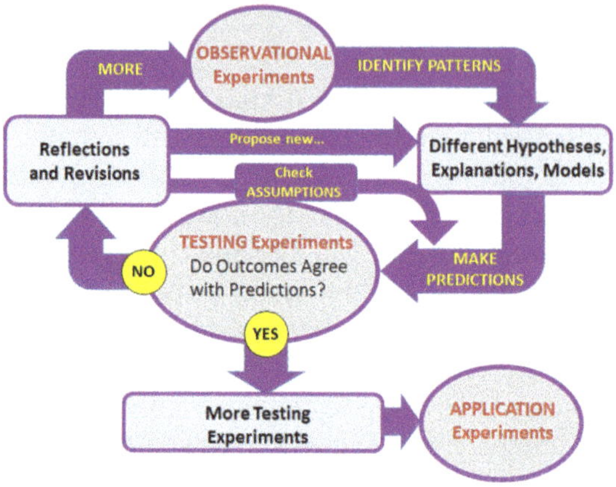

The Investigative Science Learning Environment (ISLE) approach

Figure 1.1. Logical progression of the ISLE process.

Just out of curiosity. I'm not trying to prove anything, by the way. I'm a scientist and I know what constitutes proof. But the reason I call myself by my childhood name is to remind myself that a scientist must also be absolutely like a child. If he sees a thing, he must say that he sees it, whether it was what he thought he was going to see or not. *See first, think later, then test. But always see first. Otherwise, you will only see what you were expecting.* Most scientists forget that. I'll show you something to demonstrate that later. (p 157)

2. While engaged in steps a–f, students represent physical processes in multiple ways to help them develop productive tools for qualitative reasoning and for problem solving.

    Examples of such representations are sketches, motion diagrams, graphs, force diagrams, energy bar charts, and mathematical representations. The graphical representations serve as bridges between words and mathematical representations. The following is an example of multiple representations used in kinematics (see figure 1.2). The consistency of the different representations is important as it allows students to evaluate their solutions. The interlude at the end of the chapter is dedicated to multiple representations.

3. While engaged in steps a–f, students work collaboratively in groups of 3–4 using whiteboards and then share their findings, designs, and solutions in a whole-class discussion.

The combination of these features applies to every conceptual unit in the ISLE learning system. However, there is more to the ISLE approach. We found that helping students to develop a growth mindset (Yeager and Dweck 2012) and to feel like a member of a learning community (Bielaczyc and Collins 1999) are crucial to ISLE's success. These additional goals affect how you set up your classroom and

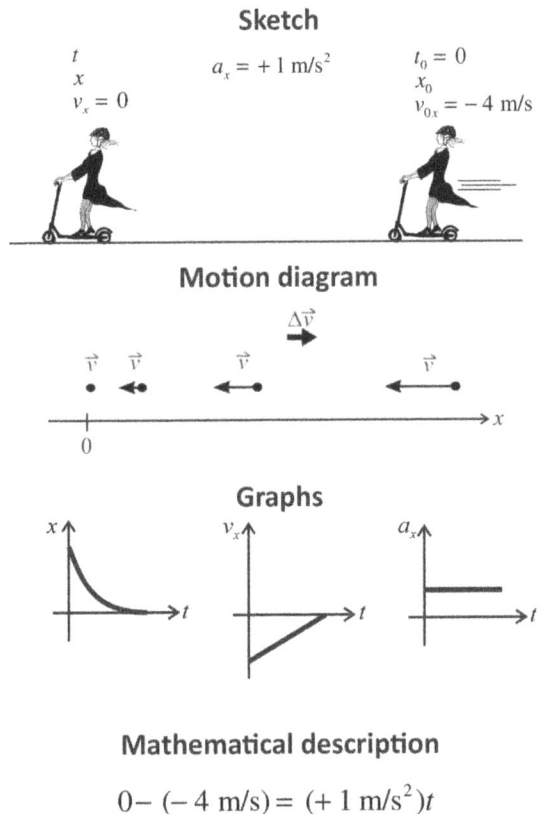

Figure 1.2. Multiple representations—an example from kinematics.

how you assess your students. Specifically, students should be able to collaborate with each other outside of class as well as in class, and to improve their work (homeworks, quizzes, lab reports, etc) without being punished for multiple attempts. It turns out that these two features (collaboration and improvement of work) are also important features of scientific work.

Below we present an example of how the ISLE approach works in the classroom to help students construct and apply the ray model of light emitted by an extended light source.

The first step in the ISLE process is to create the 'need to know' for the students (see chapter 3 for more details). Here we can do it by bringing the students into a completely dark room, let them sit there for a while (5 min) and then ask them if they could see anything. Alternatively, we can assign a homework activity before the lesson to find a completely dark room in the house (a closet in a dark room should work), stay there for about 5 min and then report the results in class the next day. Many students think that if we sit in a dark room for a long time, we will eventually be able to see. This idea is based on their prior experiences of being in dark rooms with some small sources of light and never experiencing 'true darkness'. However, after spending some time in a completely dark room, they all agree that they cannot see anything. This becomes their first observation. It raises a question: 1. What do we need to see objects? Students come up with an answer: we need a source of light. But do we see light itself? And 2. How does light travel? Notice that we do not start with a question(s); the questions to investigate come from observations.

To help students answer these two questions we put students in groups and each group receives a laser pointer, a water spray bottle, a ruler, and a few white cards. We tell students that a laser is a special light source that sends a very narrow parallel beam of light. They need to complete the activity below (adapted from the *Active Learning Guide*, chapter 22):

1.3.1 Observational experiment and initial explanations

Instructions for the students: Place a laser pointer in the center of a room and observe a bright spot on the wall toward which the laser points. Record your observations in the table below.

What path did the light follow to reach the wall?	
Why can't you see the beam of light itself but you can see the bright spot on the wall or on a piece of paper that intersects with the beam? Write possible explanations.	
Devise a way to see the beam of light using available equipment.	
Discuss the conditions needed for us to see the light beam and objects that do not emit light themselves.	

Every group in the room sees the bright spot but no one sees the beam (figure 1.3(a)). Students working in groups come up with the idea that they see the spot as light reflects off it (they sometimes say 'bounces off the wall') and travels to their eyes. To see the beam of light, students suggest spraying water between the laser and the screen. To answer the question about the conditions needed for us to see the beam of light and objects that do not emit light, the students quickly come up with an answer that they can see the path because light reflects off (bounces off) tiny droplets into our eyes (figure 1.3(b)) and thus to see any object there has to be light reflected off the object that reaches our eyes.

As the students see that the light beam forms a straight line, the instructor now can give this line a name: 'a ray'. A ray is a model of a very narrow light beam. The students are now also equipped with a key representational tool of geometrical optics: a light ray can be represented with a straight line drawn with a ruler. Using their knowledge of light propagation and the light ray representation, the students are ready to figure out how extended objects emit light.

1.3.2 Observational experiment and models

Each group of students receives a frosted bulb (any LED bulb would work). They turn it on in a dark room and observe that the walls and the ceiling are lit. How does the bulb emit light rays? Students discuss possible models in groups and usually come up with two models: each point of the bulb emits one ray pointing radially outward from the bulb (figure 1.4(a)), or each point of the bulb emits multiple rays in all directions (figure 1.4(b)). If the students do not come up with the model in figure 1.4(b) at this stage of the investigation, they will eventually devise it when their testing experiments reject the model in figure 1.4(a).

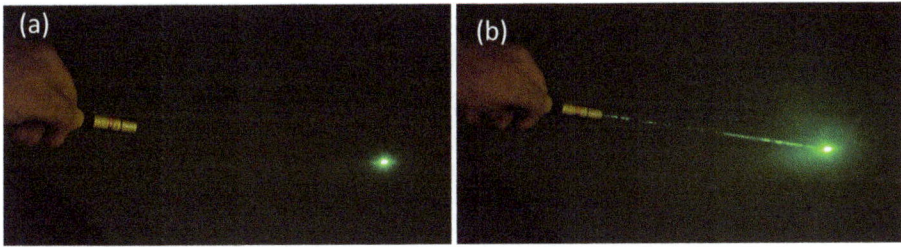

Figure 1.3. Observational experiments that lead to the concept of a ray as a model of a very narrow light beam; (a) initial observational experiment and (b) after spraying water droplets.

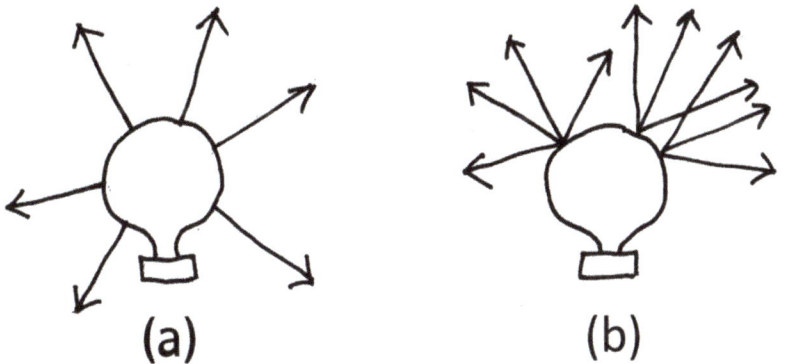

Figure 1.4. Two models of extended sources of light: (a) each point on the light source emits one ray and (b) each point on the light source emits multiple rays.

1.3.3 Testing the models

To test the models that they had invented, students are asked to design experiments whose outcomes they can predict using the models that they proposed. The progression of their thought (based on the observations of many student groups) is shown in table 1.1. The students use a hypothetico-deductive reasoning chain: *If* a model of light emitted by an extended source (their own model) is correct *and* I do the following experiment (the description of the experiment), *then* I will observe (the description of the outcome) *because* (the explanation of how the outcome follows from the model), *however* the outcome did not match the prediction, *therefore* I reject the model. Or *and* the outcome matched the prediction, *therefore* the model was not rejected.

Table 1.1 shows that the testing experiments reject the one-ray model and fail to reject the multiple-ray model. Through experimentation and reasoning, students have established a key idea of geometrical optics, that each point on the surface of an extended source of light behaves as a point source of light and sends an infinite number of rays in all directions.

Table 1.1. Testing two explanations for how an extended light source emits light.

	Testing experiment	Prediction base on		Outcome
		One-ray model	Multiple-ray model	
1	Cover the bulb with aluminum foil and poke a hole in the part of the foil facing the wall. Turn the bulb on.	Small spot on the wall directly in front of the hole.	Whole wall will be dimly lit.	The whole wall is dimly lit.
2	Turn on a lightbulb and place a pencil close to the wall between the bulb and the wall.	Dark, sharp shadow behind the pencil.	Dark sharp shadow behind the pencil.	Dark sharp shadow on the wall.
3	Turn on a lightbulb and place a pencil closer to the bulb between the bulb and the wall.	Dark, sharp shadow behind the pencil.	Light, fuzzy shadow on the wall.	Light, fuzzy shadow on the wall.

1.3.4 Applying the model

To apply the newly constructed model the students work on the following activity (similar activities can be found in the *Active Learning Guide* chapter 22). Each group receives an asymmetrical light source, a piece of thin cardboard with an opening, a whiteboard, and markers. A simple asymmetrical light source can be made by using six white LEDs arranged in an L shape (preferably use LEDs with a flat top) and two 9 V batteries all connected in series (see figure 1.5(a)).

Text for the activity:

Imagine that you are holding an L-shaped light source and place a piece of cardboard with a tiny hole between the light source and a nearby wall. Use the sketch (figure 1.5(b)) to draw a ray diagram to predict what you will see on the wall. Work with your group members to justify your prediction with a ray diagram. Then switch on the light source and turn off the room lights to observe the outcome of the experiment. Then revise your diagram if necessary.

While many students draw correct ray diagrams as in (figure 1.5(c)), it is difficult for them to predict that they would see the L shape upside down (figure 1.5(d)). The instructor might need to provide some help here asking for the details of the shape that the students expect to see.

In the above activity the students essentially used a pinhole camera. To design a real pinhole camera, the students can use a cardboard box. To reduce bright ambient light, they can use a cellphone camera to record images and videos that form on the screen inside the closed pinhole camera (figure 1.6(a)), note that the camera is pointing away from the object of interest. Watching the recorded upside-down image of the objects that students predicted earlier (figure 1.6(b)) creates a feeling of success and excitement, even if students have seen the pinhole camera before.

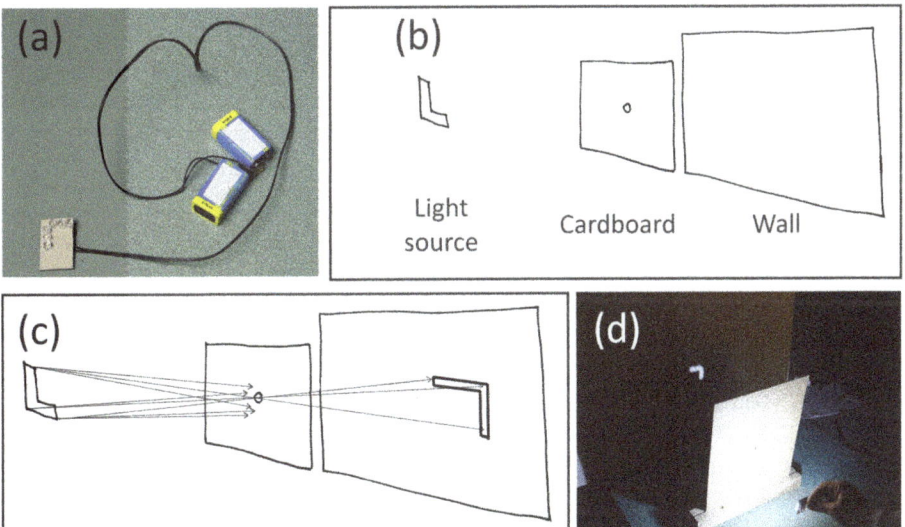

Figure 1.5. (a) L-shaped light source, (b) sketch of the set-up for student handout, (c) correct ray diagram, and (d) outcome of the experiment.

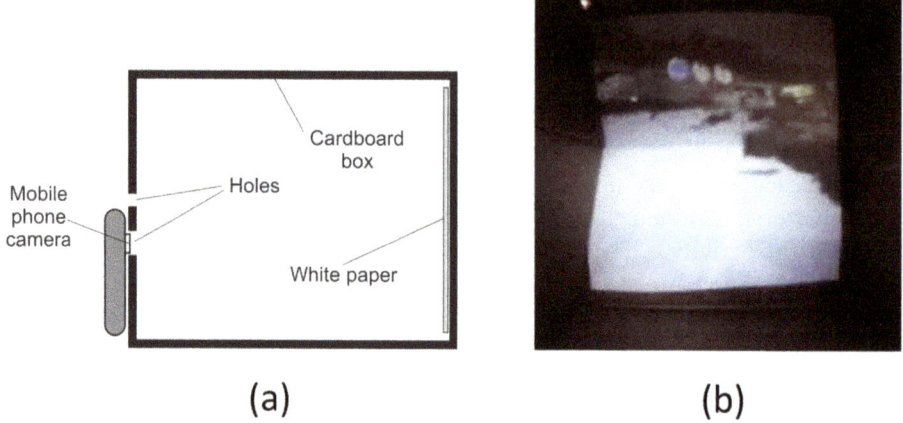

Figure 1.6. Pinhole camera—an application experiment: (a) set-up, (b) photograph of the image on the screen.

From this example we see that students construct a model of extended objects emitting light (crucial for understanding how mirrors and lenses form images) following a logical progression from observing simple experiments, to devising multiple models, explaining them, testing the models, and applying them for practical purposes. To be successful in this process the students use graphical representations when constructing models and when making predictions of the outcomes of the testing experiments. These graphical representations make critical reasoning tools visible.

1.3.5 Formative assessment

The following problems and questions can be used to formatively assess student newly constructed ideas:
1. Imagine that you have a candle on a table and set a piece of thin cardboard with an opening cut into it on the table between the candle and a nearby wall (see figure 1.7). Use a ray diagram to predict what will happen as you move the cardboard closer to the wall and then move it closer to the candle. Draw what you will see on the wall.
2. Your summer ecology research job involves documenting the growth of trees at an experimental site. One day you forget your tree-height-measuring instrument. How can you determine the height of trees without it? Provide a sketch for your method.

How can we make for ourselves a big picture of the ISLE process? David Brookes, one of the authors of the first ISLE book, describes his understanding of the ISLE process as follows:

> The way I think about the ISLE process now is that it lays out the 'rules' of an 'epistemic game' (Collins and Ferguson 1993) or a series of epistemic

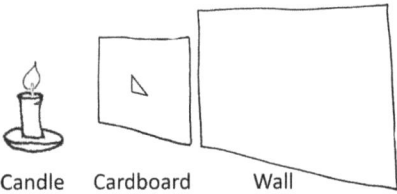

Figure 1.7. Experimental set-up.

questions that students should be asking over and over again as they do physics. For example, when someone suggests an idea, a teacher's response should not be 'no that's wrong,' but rather, 'how can we test this idea?' When results don't go as planned, possible questions could be, 'how can we explain that?' or 'what assumptions did we make?' When students have gathered data, questions might be 'how can we describe and represent these results?' or 'what is the pattern in these data?' That is what the ISLE process looks like in action. My goal in the classroom is to habituate students into asking those questions. When they do, the class almost runs itself because students are asking the questions they epistemologically ought to be asking (Macmillan and Garrison 1983). If students leave my classroom with these questions ingrained into how they think and reason about the world around them, I believe I have made a difference in the world, which is why I got into teaching in the first place (Etkina *et al* 2019a, pp 1–16).

1.4 What will you find and what will you not find in this book?

In the first book we not only described the ISLE approach and gave examples of how it works in the classroom, but we also suggested classroom set-ups, assessment techniques, and provided testimonials of the educators who use the ISLE approach in different environments—high school, college, university, teacher preparation programs, etc While being familiar with the first book is not mandatory to understand the content of this book, we strongly encourage the reader to familiarize themselves with the material in *Investigative Science Learning Environment: When Learning Physics Mirrors Doing Physics* (Etkina *et al* 2019a).

In the current book we connect the ISLE approach to physics teacher preparation and professional development, focusing on the development of productive teaching habits. Specifically, we will discuss:
- How the ISLE approach relates to the three pillars of teacher preparation (dispositions, knowledge, and skills—all explained in chapter 2).
- How the ISLE approach helps teachers develop productive habits of mind (such as teaching physics as a process not a set of rules, applying systems reasoning to the analysis of physical phenomena, and many others), and

productive habits of practice (such as recognizing multiple levels of complexity in a physics experiment, questioning techniques, and many others),
- How to organize your physics teacher preparation program or a professional development program around the development of these productive habits. There are eight chapters in the book and six of them end with interludes. These interludes connect specific teaching habits to the intricacies of teaching specific physics content. For example, we discuss how to create physics problems that develop student metacognition and epistemic cognition, how to develop new linguistic habits when working with forces and energy, and many others.

Overall, we hope that this book will help you set up your own program of physics teacher preparation or professional development that will allow you to prepare physics teachers who uncover the potential of every student and prepare them for success in any field of work they will choose. If you are yourself a physics teacher (of any level), we hope that this book will inspire and support you in your life-long journey of self-improvement and professional growth.

It is also important to note what you will not find in this book. We will not show how to help students develop normative physics knowledge (velocity, acceleration, Newton's laws, momentum, energy, electric charge, wave model of light, etc). To read about how to do this using the ISLE approach, please see the algebra-based textbook *College Physics: Explore and Apply* by Etkina *et al* (2019c) and supporting materials such as the *Active Learning Guide* (ALG; Etkina *et al* 2019b), the *Online Active Learning Guide* (OALG; Etkina *et al* 2020) and the *Instructor Guide* (IG; Etkina *et al* 2019d). The goal of this book is to help teacher educators use the existing resources and the theoretical framework of productive habits to prepare future and practicing teachers to teach physics through the ISLE approach. But what if you are preparing future teachers and you are not going to use the ISLE approach? Will this book be useful for you? The answer is affirmative. The habits that we wish ISLE-based teachers develop will be extremely useful for any approach to teaching physics.

Important note: in the rest of this book, we will be referring to examples of ISLE-based physics curriculum that are similar to the materials from the following sources:
1. *College Physics: Explore and Apply* by Etkina *et al* (2019c).
2. The *Active Learning Guide* (ALG; Etkina *et al* 2019b).
3. The *Online Active Learning Guide* (OALG; Etkina *et al* 2020).
4. The *Instructor Guide* (IG; Etkina *et al* 2019d).

To avoid using the full citations each time we mention one of those resources, we adopt the following notation that we will use in the rest of the chapters (the numbers below refer to the numbers for the resources given above):
1. CP: E&A
2. ALG
3. OALG
4. IG.

1.5 Interlude: ISLE and multiple representations (by Eugenia and Gorazd)

Every physicist and every physics teacher has had the same experience. When somebody asks you 'What do you do?' and you reply 'I do physics research' or 'I teach physics', the response is almost identical no matter where you are or who the person asking the question is. The response is usually full of respect: 'Wow, physics! You must be smart. It is so difficult! It has so much math! I never took physics.' Or, 'I took physics but I hated it.' There are minor variations in the responses but they all communicate the same fear of physics and the admiration for those who are brave enough to do it.

Why? What is it about physics that makes it universally considered to be one of the most difficult subjects accessible to only the brightest? One reason can be the way we teach it. We start by using abstract words and we attach abstract mathematical relations to those abstract words. However, the studies of the brain tell us that to learn something, one first needs to have an image of it (Dehaene 2021, Zull 2002). Imagine that we tell you that we are learning about *malum* today. A malum can be green, yellow, or red. It can be hard or soft. Some malums can be sweet and some can be sour. Will you know what malum is? Probably not. But if I show you a picture of one malum (see figure 1.8), you will know exactly what malum is (we used the Latin language here).

I (Eugenia) can only solve a physics problem if I can imagine what is going on. I can see moving pulleys and ice skaters doing their rotations but for the life of me, I cannot figure out the rules of baseball. So, any problem involving baseball is impossible for me to solve. Why? The quantities in the problem have no meaning for me. I cannot visualize them.

How does the ISLE approach deal with this problem of lacking concrete images to help students bridge abstract words and abstract symbolic representations used in the practice of physics? It links these abstract ways of describing the world to more concrete descriptions.

As we already said above, ISLE students learn to represent physical processes in multiple ways and learn to move from one representation to another in any

Figure 1.8. *Malum.*

direction. This moving back and forth between representations helps them make connections between concrete ways of representing a process (pictures and diagrams) that create images in their brains and more abstract ways of representing the same processes (graphs and equations). In fact, in CP: E&A the first half of every chapter does not have any mathematical representations. Instead, we focus on descriptions and videos of experiments, sketches, and graphical representations specific to physics (such as motion diagrams, force diagrams, energy bar charts, etc). This approach helps students construct their personal understanding and qualitative image of the concept and feel comfortable with it before using abstract mathematics. In a way, using representations other than mathematics not only addresses the first intentionality of the ISLE approach (help students learn physics by practicing it) but also the second intentionality (improve student motivation, self-confidence and persistence). Graphical representations in physics play a crucial role. We can think of the Feynman diagrams as an example. Quantum electrodynamics (QED) was largely inaccessible for many physicists before Richard Feynman invented a graphical representation that allowed the physicists to keep track of particles and interactions. Galileo took pages of his book *Two New Sciences* (1638) to explain how to derive the quadratic dependence of distance on time for the objects moving at constant acceleration, while if we use the graphical representation of mean speed theorem invented in the fourteenth century by the Calculators of the Merton College (a group of thinkers at Oxford) and proved by Nicolas Oresme, we can do it using one graph[1].

Representations also help students group information in bigger units (the process called chunking, see Gobet and Simon 1998) which allows them to process larger amounts of information at once. The mind can supposedly hold five to seven chunks of information. Experts with years of experience group many small ideas together in one of these chunks. Thus, their seven chunks are actually much bigger. Each chunk for a novice is small. This makes it difficult for novices to develop understanding about a whole process with these few small chunks stored in their mind. They must go back multiple times to the problem statement. It seems less fatiguing to solve the problem by finding an equation that seems appropriate and plugging the known information into that equation—the infamous plug-and-chug problem solving strategy.

Thus, constructing a sketch of the process, for example, allows a student to see the problem situation without having to rely on storing the information in their mind. They can then focus on using a more expert-like strategy to solve the problem. In such an expert strategy, graphical representations serve as bridges between the abstract words describing the problem situation and the mathematical equations that are needed to obtain the numerical answer.

In other words, the different representations serve as tools that allow us to solve problems and develop new knowledge more easily—similar to having tools in construction. We teach students what 'hammers' and 'saws' they can use to build their

---

[1] The mean speed theorem says that a uniformly accelerated object starting from zero initial velocity travels the same distance as the object with the constant velocity equal to the half of the final velocity of the accelerated object. Oresme proved it using a trapezoid with base being the time of travel and the sides being initial and final velocity.

'physics house'. Earlier we showed an example of different representations in kinematics (see figure 1.2). Here you can see another example of using graphical representations in dynamics and helping students learn how to check consistency of representations.

Examine figure 1.9(a). The teacher is catching a vertically falling medicine ball. The ball has a blinking LED attached to it and you can see the traces of light indicating that the ball's speed is decreasing after it comes in contact with the teacher's hands. Figure 1.9(b) shows the motion diagram for the ball during the contact with the hands. Figure 1.9(c) shows the force diagram for the ball. We choose the ball as our system of interest and put a dot on the right of the motion diagram to represent the system. We then find all objects with which the system object interacts (either directly touching or without the direct contact if there is a long-range force like the gravitational force that the Earth exerts on the object). Then we use this sketch and the identified system to construct a new representation that is called a force diagram or a free-body diagram (we actually prefer the word force diagram to the free-body diagram as when external forces are exerted on an object, it is not free). We then use the diagram to write Newton's second law in component form. (The method that we use is based on work by Heller and Reif and further developed by Etkina and Van Heuvelen in the first edition of *College Physics*.)

Now let's go back to the force diagram representing the forces exerted on the medicine ball as you are catching it. How long should the force arrows on the force diagram be? There are two very important ideas at this moment. First, the relative lengths of the force arrows should be approximately consistent with the magnitudes of the forces (if the information is known). And here is where the motion diagram helps. If the student constructs the motion diagram next to the force diagram, then the velocity

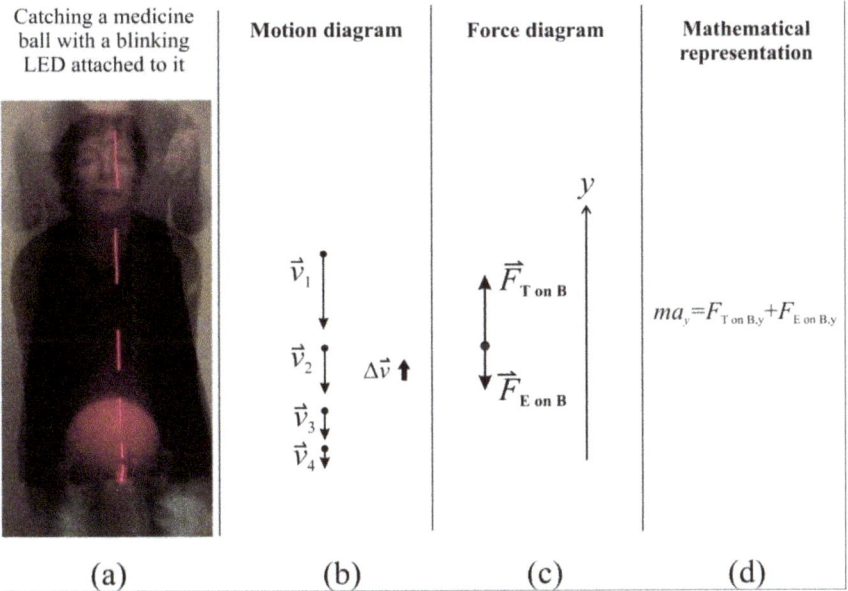

Figure 1.9. (a) A photo of Eugenia catching the medicine ball; (b) motion diagram for the ball; (c) force diagram; and (d) mathematical representation.

change arrow on the motion diagram will indicate the direction of the sum of the forces. The direction of the sum of the forces will help the students to adjust the lengths of the force arrows on the force diagram. This step helps students develop the habit of evaluating the consistency of the diagrams and also helps them decide what lengths the force arrows should be. Students need to practice constructing force diagrams and matching them with the motion diagrams before they start solving quantitative problems. We provide examples of such activities in CP: E&A, ALG, and OALG.

Next students use the force diagram to help write Newton's second law in component form. Having the force diagram as an intermediate reasoning step helps students avoid mistakes such as calculating the force that the hands exert on an accelerating medicine ball by setting ma_y equal to the force exerted by the teacher $F_{\text{Teacher on ball}}$ (or $F_{\text{T on B}}$) and neglecting the gravitational force that the Earth exerts on the ball.

Once the students have learned to construct these different representations, they need to learn to use them to evaluate their work. For example, is the velocity change arrow in the motion diagram in the same direction as the sum of the forces in the force diagram? Have all of the forces been included in the application of Newton's second law in component form? Students need to practice converting from one representation to another—for example, from equations to a force diagram or to a word description of a problem. In the textbook CP: E&A, multiple representations are integral to the problem solving strategy. The problem-solving strategy has four steps: sketch and translate, simplify and diagram, represent mathematically, and solve and evaluate. All four steps are related to different representations. See an example in table 1.2 to learn how we look for consistency among different presentations.

Table 1.2. An exemplary solution of a problem (see a similar problem in **CP: E&A**, chapter 3).

Sketch and translate	Initial	Final
• Sketch and visualize the process. • Decide what system you will use for analysis. • Decide on a coordinate system. • Use everything you know about the situation to label the sketch. • Decide what the unknown is and label it with a question mark on the sketch.	$m_H = 70$ kg $t_0 = 0$ $v_0 = -36$ m/s $y = 0$	$y_0 = +2.0$ m $v = 0$ $F_{\text{S-G on H}} = ?$

We start by sketching the process. We choose Holmes as the system (H). We need to determine the average force that the shrubbery and ground (S-G) exert on him from when he first touches the shrubbery to the moment when he stops. Let the y-axis point up. The origin is at the ground where Holmes finally stops.

Table 1.2. (*Continued*)

Simplify and diagram • Decide what simplifying assumptions you need to make. For example, can you neglect the size of the system? Can you assume that forces or acceleration is constant? • Represent the process with a motion diagram and/or force diagram(s). check for consistency of the diagrams with each other.	Let's model Holmes as a point-like object and for simplicity assume that the forces exerted on him are acceleration is also constant. We draw a motion diagram for his motion while stopping and the corresponding force diagram. To draw the force diagram, we first identify the objects interacting with Holmes as he slows down: the shrubbery and ground (combined as one interaction) and Earth. The shrubbery and ground exert an upward force $F_{\text{S-G on H}}$ on Holmes. Earth exerts downward gravitational force $F_{\text{E on H}}$. The force diagram is the same for all points of the motion diagram because the acceleration is constant. On the force diagram the arrow for $F_{\text{S-G on H}}$ must be longer to match the motion diagram, which shows the velocity change arrow pointing up. We can use the motion diagram and kinematics to find his acceleration while stopping. We can use the force diagram and Newton's second law to find the average force that the shrubbery and ground exerted on him while stopping him.	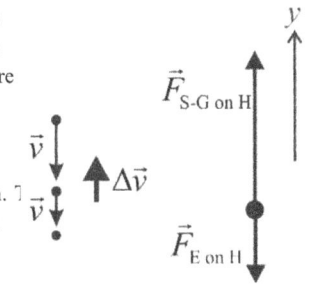

Represent mathematically

• Use these qualitative representations to write quantitative mathematical descriptions of the situation. Kinematics equations and Newton's second law for motion along the axis will be useful here. Decide on the signs of the force components in the equations. Add the force components (with either positive or negative signs) to find the sum of the forces.

We find the y-component of Holmes's average acceleration to be

$$a_y = \frac{v_y^2 - v_{y0}^2}{2(y - y_0)}$$

From the force diagram: the y-component of Newton's second law with the positive y-direction up is

$$a_y = \frac{\Sigma F_{\text{on H }y}}{m_H}$$

The y-component of the force exerted by the shrubbery-ground on Holmes is $F_{\text{S-G on H }y} = +F_{\text{S-G on H}}$ and the y-component of the force exerted by Earth is $F_{\text{E on H }y} = -F_{\text{E on H}} = -m_H g$ therefore,

$$a_y = \frac{F_{\text{S-G on H }y} + F_{\text{E on H }y}}{m_H} = \frac{(+F_{\text{S-G on H}}) + (-F_{\text{E on H}})}{m_H} =$$
$$\frac{+F_{\text{S-G on H}} - m_H g}{m_H} \rightarrow F_{\text{S-G on H}} = m_H a_y + m_H g$$

Solve and evaluate

• Substitute the known values into the mathematical expressions and solve for the unknowns.

• Evaluate your work to decide if it is reasonable (check units, limiting cases, and whether the answer has a reasonable

Holmes' average acceleration was

$$a_y = \frac{0^2 - (-36 \text{ m s}^{-2})}{2(0 - 2.0 \text{ m})} = +324 \text{ m s}^{-2}$$

Holmes's y- component of the initial velocity is negative. His initial position is +2.0 m at the top of the shrubbery, and his final position is zero at the ground. Note, that his velocity in the negative direction is decreasing, therefore the velocity change and the acceleration both point in the opposite direction (positive). The average magnitude of the force exerted by the shrubbery and ground on Holmes is

magnitude). Check whether all representations—mathematical, pictorial, and graphical—are consistent with each other.	$F_{\text{S-G on H}} = m_H a_y + m_H g = (70 \text{ kg})(324 \text{ m s}^{-2}) + (70 \text{ kg})(9.8 \text{ N/kg})$ $= 22.680 \text{ kg} \cdot \text{m s}^{-2} + 686 \text{ N} = 23{,}366 \text{ N} \approx 23{,}000 \text{ N}$ The result is consistent with the force diagram and the motion diagram as the force's magnitude is greater than the force exerted by Earth. The magnitude is huge (as we expected) and the units are correct. A limiting case for zero acceleration gives us a correct prediction—the force exerted on Holmes by the shrubbery and ground equals the force exerted by Earth.

Table 1.2 shows an exemplary solution for a problem that shows the multiple representation approach to problem solving. Notice the attention paid to different representations and their consistency. Also, notice the use of words as another representation to show the reasoning process inside the head of the solver. The problem statement says: Michael Holmes (70 kg) was moving downward at 36 m s^{-1} (80 mi h^{-1}) and was stopped by 2.0 m high shrubbery and the ground. Estimate the average force exerted by the shrubbery and ground on his body while stopping his fall. Note that the example is based on a real case. Michael Holmes survived the fall.

References

Adams D 1985 *So Long, and Thanks for All the Fish* (London: Pan)

Bielaczyc K and Collins A 1999 Learning communities in classrooms: a reconceptualization of educational practice *Instructional Design Theories and Models: A New Paradigm of Instructional Theory* ed C M Reigeluth vol 2 (Mahwah, NJ: Lawrence Erlbaum) pp 269–92

Brookes D, Etkina E and Planinsic G 2020 Implementing an epistemologically authentic approach to student-centered inquiry learning *Phys. Rev. Phys. Educ. Res.* **16** 020148

Collins A and Ferguson W 1993 Epistemic forms and epistemic games: structures and strategies to guide inquiry *Educ. Psychol.* **28** 25–42

Dehaene S 2021 *How We Learn: Why Brains Learn Better Than any Machine…for Now* (London: Penguin)

Etkina E, Brookes D and Van Heuvelen A 2014a *Instructor Guide for College Physics* (New York: Pearson)

Etkina E, Brookes D and Planinsic G 2019a *Investigative Science Learning Environment: When Learning Physics Mirrors Doing Physics* (IOP Concise Physics) (San Rafael, CA: Morgan and Claypool)

Etkina E, Brookes D, Planinsic G and Van Heuvelen A 2019b *Active Learning Guide (ALG) for College Physics: Explore and Apply* 2nd edn (New York: Pearson)

Etkina E, Brookes D, Planinsic G and Van Heuvelen A 2019d *Instructor Guide* 2nd edn (San Francisco, CA: Pearson)

Etkina E, Brookes D, Planinsic G and Van Heuvelen A 2020 *On-line Active Learning Guide (OALG) for College Physics: Explore and Apply* 2nd edn (New York: Pearson)

Etkina E, Gentile M and Van Heuvelen A 2014b *College Physics* (New York: Pearson)

Etkina E, Gentile M and Van Heuvelen A 2014c *The Physics Active Learning Guide* 2nd edn (New York: Pearson)

Etkina E, Gregorcic B and Vokos S 2017 Organizing physics teacher professional education around productive habit development: a way to meet reform challenges *Phys. Rev. Spec. Top. Phys. Educ. Res.* **13** 010107

Etkina E, Planinsic G and Van Heuvelen A 2019c *College Physics: Explore and Apply* 2nd edn (San Francisco, CA: Pearson)

Etkina E and Van Heuvelen A 2001 Investigative science learning environment: using the processes of science and cognitive strategies to learn physics *Proc. 2001 Physics Education Research Conf.* S Franklin, J Marx and K Cummings (Melville, NY: American Institute of Physics) pp 17–21

Etkina E and Van Heuvelen A 2007 Investigative science learning environment—a science process approach to learning physics *Research-Based Reform of University Physics* ed E F Redish and P J Cooney vol 1 (College Park, MD: American Association of Physics Teachers)

Gobet F and Simon H A 1998 Expert chess memory: revisiting the chunking hypothesis *Memory* **6** 225–55

Macmillan C J B and Garrison J W 1983 An erotetic concept of teaching *Edu. Theor.* **33** 157–66

MacMillan C J B and Garrison J W 1988 *A Logical Theory of Teaching: Erotetics and Intentionality* (Dordrecht: Kluwer)

White R and Gunstone R 1992 *Probing Understanding* (London: Falmer)

Yeager D S and Dweck C S 2012 Mindsets that promote resilience: when students believe that personal characteristics can be developed *Educ. Psychol.* **47** 302

Zull J E 2002 *The Art of Changing the Brain: Enriching Teaching by Exploring the Biology of Learning* 1st edn (Sterling, VA: Stylus)

IOP Publishing

The Investigative Science Learning Environment: A Guide for Teacher Preparation and Professional Development

Eugenia Etkina and Gorazd Planinsic

Chapter 2

Dispositions, knowledge, and skills of teachers and their relationship to teaching physics through the ISLE approach

In this chapter we will discuss research findings on teacher preparation and how these findings relate to the ISLE approach. Teacher preparation (and consequently, professional development) should attend to three important issues: teacher dispositions, teacher knowledge, and teaching skills (Pajares 1992, Ball and Cohen 1999, Darling-Hammond *et al* 2005, Hammerness *et al* 2005). We will examine these issues separately, connect them to the ISLE approach, and then combine them together to create an image of a teacher whose students learn physics through the ISLE approach. In some models of teacher knowledge, teacher disposition is called 'orientation towards teaching' (Magnusson *et al* 1999). We consider these terms synonymous[1].

2.1 ISLE and dispositions

Cambridge dictionary defines a disposition as 'a natural tendency to do something, or to have or develop something'[2]. This general definition does not help us understand teacher dispositions. We can conceive of a disposition as a *strong (often subconscious) belief or attitude related to some aspect of teaching, that in concert with other factors, shapes a teacher's thought and behavior* (see Korthagen and Lagerwerf 1996). Dispositions strongly affect how teachers analyse new information

---

[1] We remind the reader that abbreviation CP: E&A stands for the textbook *College Physics: Explore and Apply*, ALG stands for the *Active Learning Guide*, OALG stands for the *Online Active Learning Guide*, and IG stands for the *Instructor Guide*. Proper citations for these materials are in chapter 1.
[2] https://dictionary.cambridge.org/dictionary/english/disposition.

and teaching situations, and how they plan and act in the classroom (Fives and Buehl 2012).

How do pre-service teachers (PSTs) and in-service teachers (ISTs) develop their dispositions? We can speculate that such development starts very early when they observe their own teachers and interpret their actions. Before a person starts professional training, they usually have 12–16 years of formal schooling. These years undoubtedly form many memories and affect how a person conceptualizes the teaching profession. In fact, many think that so-called 'apprenticeship by observation' (Lortie 1975) usually affects PSTs' dispositions more than formal teacher education (Feiman-Nemser and Remillard 1996, Ball and Cohen 1999, Hammerness *et al* 2005). Practicing teachers solidify their dispositions based on their experiences with the schools where they teach, administration, parents, and, most importantly, administrative policies that affect teachers' professional lives.

What are examples of dispositions of physics teachers of all levels? The most important is probably the disposition towards students learning physics. Can all students learn physics or only very special, selected individuals? Teachers of these two different dispositions will run their classrooms (high school or college) very differently. A teacher who believes that all students can learn physics will create opportunities for the students to show their learning in multiple ways, improve their work without being punished for several tries, and develop a mistake-rich environment (Zull 2002). A teacher who believes that only selected students can learn physics will have one method of assessment (multiple choice exams, for example) and make their judgment of students' abilities to learn physics based on those assessments.

Another important disposition is towards learning itself—is learning an activity that a person does on their own, or it is a communal activity that requires collaboration and communication with other students? Think of how these two different dispositions affect, for example, the set-up of a classroom. When we walk into an amphitheater with the rows of individual seats all turned towards the stage where the teacher lectures to the audience, we see the implementation of a disposition that students learn individually by listening and studying their lecture notes. When we walk into a classroom with tables set for 3–4 students to face each other, with small whiteboards placed on those tables, we see the implementation of a disposition that students learn by working collaboratively and sharing their ideas.

An example of a disposition related directly to physics concerns the role of 'theory' in students' learning. Does theory come first or experiment? A teacher who believes in the former will start their lessons with the description of theoretical knowledge related to the topic of learning (introducing vocabulary, definitions and mathematical relations) and then showing experiments illustrating the 'theory', while the teacher who believes in the role of experiment as the driving force behind the construction of physics knowledge would begin their lessons with experiments that would help their students see where 'theory' comes from.

Below we list some of the teacher dispositions that will help them implement the ISLE approach in the classroom. Some of them apply for all teachers (marked with one asterisk), some for all physics teachers (marked with two asterisks), some for the

physics teachers who implement active learning techniques in their classrooms (three asterisks), and some for those who help their students learn physics through the ISLE approach (four asterisks). Note that for teachers implementing the ISLE approach all of the dispositions on the list below should be important. The order of the bullets does not represent the order of importance of the dispositions.
* Dispositions that a caring teacher of any subject should have.
** Dispositions related to physics learning.
*** Dispositions related to interactive engagement in physics learning (interactive engagement means that the students learn by actively and collaboratively participating in the process, not listening to a lecture, or watching a teacher solve problems on the board, or following cook-book instructions in a lab).
**** Dispositions related to learning physics through the ISLE approach.

Before you read on, reflect on your own dispositions. Make a list and as you continue reading, compare our list with yours.
- It is not about me. It is not personal.* Whatever happens in my classroom is about my students, not my emotions or my feelings (e.g. you might feel disrespected and upset if a student is late for class, but if you make it about the student, you might ask them what happened and you would discover that there was no disrespect meant but the student had a problem at home). This is the most important disposition for all teachers of all subjects and it means that the teacher believes that what happens in the classroom (student behavior) should always be about the students not the teacher's feelings about themselves.
- Learning is changing one's brain—literally (Zull 2002). My role is to help my students rewire their brains, not demonstrate the prowess of mine. Therefore, my role is to facilitate student learning, not to present the material.* This disposition again focuses on the students, not the teacher in the classroom but this time it relates to student learning, not their behavior.
- Learning is a social activity. My role is to create conditions in the classroom for the students to continuously collaborate and learn from each other in an atmosphere of mutual respect.* This disposition again focuses on the students and combines beliefs about their behaviors, classroom set-up and their learning.
- I am a physicist, and I am a teacher at the same time. When I think about physics, I think about it as a teacher; when I think about student learning, I bring in my inner physicist.** This disposition means that encountering any physics problem or learning some new physics, the teacher believes in the importance of thinking about how to make this content accessible and exciting for their students or what experiences their students might already have with this content. When encountering a problem or a situation related to some episode during a lesson, the teacher thinks of the observed phenomenon (what happened) and tries to develop multiple explanations for it, similar to how physicists construct new knowledge.

- I am a perpetual learner.** I continuously read and apply the results of physics education research based on decades of carefully studying how students learn the content and how to help students overcome difficulties. I also try to follow the recent developments of physics.
- Physics is all around me; I notice phenomena and apply physics to every (or almost every) one that I observe in my everyday life. And this is exactly what I wish my students to do.** This disposition relates to life of the teacher outside the classroom they are naturally excited to observe, explain, estimate, calculate, or wonder about the causes of the observations—let it be waves passing through two nearby stones that invoke diffraction patterns, a thunder delayed from the lightning that makes them calculate the distance to the storm, or a run to catch a bus that makes them wonder how many joules of their body's chemical energy were just converted into kinetic energy.
- All of my students would want to learn physics, given the right opportunities, and are capable of learning physics. They might need different time scales and different tools, but they all can learn.**** This disposition means believing in the growth mindset and therefore that students need different times, different supports, and different representations to learn physics and it is the duty of the teacher to help students master those different representations. It also means that the teacher believes that the students should be given multiple opportunities to demonstrate their learning.
- My students come from different backgrounds and cultures and bring different strengths and different life experiences into my classroom. I need to learn as much as I can about my students and make sure that their cultural backgrounds and life experiences are respected and provide the foundation for their learning.* This means that the teacher believes that individual and cultural diversity affect student learning and it is their responsibility to learn as much as possible about students' backgrounds and cultures.
- If some of my students are not learning, it is not because they cannot, it only means that they need more help.**** This disposition means that the teacher believes in their responsibility to find the ways to help struggling students.
- In order to help my students learn physics, I need to listen to my students and try to understand what they are saying instead of hunting for the right answer. *** This disposition means that the teacher believes that all student ideas can become productive, such as in a different context or expressed in a different language, as student ideas are based on their life experiences and it is their existing ideas that will determine own their future learning.
- Listening to my students is more important than talking myself.*** This disposition means the belief that learning is about students constructing their knowledge not about the transmission of the teacher's knowledge.
- Students' questions are the most important part of the lesson.*** This disposition means the belief that it is student curiosity and creativity that should be the goal and the driving force of their learning.

- Students talking to each other about physics is the most valuable part of their experience of the lesson.**** This disposition means that class time prioritizes the instances when students talk to each other about physics, as this when they learn and develop community, exactly the same way as physicists do.
- It is my role to motivate my students every day, not to expect them to always self-motivate.**** This disposition means the belief in the importance of the second intentionality of the ISLE approach.
- For a physicist, a mistake is a source of learning, and so it should be for my students. They should be able to improve their work and not feel punished for not getting it right on the first try.****
- Experiment (such as an observational experiment) is the start of learning in physics.****
- All student ideas are valuable and need to be carefully used to develop their knowledge further. Student ideas are not to be treated as misconceptions to hammer out but productive ideas to build on.****
- Students should learn by testing their ideas experimentally, not merely believing in authority.****
- Students are very talented. I believe in them. They are born scientists and my role is to help them feel that they belong in the physics world, that they can develop physics identity.****

As we said above, dispositions, in concert with other factors, shape a teacher's thought and behavior. The following example (see table 2.1) shows how thoughts and behaviors are shaped by dispositions. Imagine two teachers, teacher A and teacher B:

Where do these dispositions, these beliefs, often subconscious, come from? We can point to many different sources. We observe practicing teachers and instructors in teacher preparation programs and reflect on those observations. We can read books and papers about learning and teaching (Arons 1997, Dehaene 2021, Zull

Table 2.1. Connections among dispositions, thoughts, and behaviors.

	Teacher A	Teacher B
Disposition	Only selected students can learn physics.	Every student can learn physics.
Thought	Sees assessment as a one-time grading filter.	Sees assessment as a tool to help students grow.
Behavior	In the classroom, the students receive one grade for each submitted assignment.	The students are invited to revise and resubmit their assignments for improvement and receive a grade that they deserve after multiple tries. This grade is not lowered for multiple tries.

2002, 2011, Dweck 2016) as well as about physics practice (e.g. Einstein and Infeld 1967, Swartz 2003, Grimvall 2007, Butterworth 2017, Lemons 2017), in particular about experiments and easily observable phenomena (e.g. Sutton 1938, Erlich 1997, Walker 1977, Leim 1987, Minaert 1993). We connect our reading and observations to introspective reflections on one's own learning and classroom experiences, school administration feedback (for ISTs), and many others. All these sources of beliefs of PSTs and ISTs need to be coherent to form strong dispositions. Therefore, teacher preparation programs and professional development programs offer opportunities to contribute significantly to the formation of the dispositions. While dispositions formed by observing somebody's teaching can be called 'accidental' or 'random', the ones formed by professionals in teacher preparation programs can be considered 'purposeful'. How can physics teacher preparation programs achieve the coherence of the dispositions consistent with learning physics through the ISLE approach? We will come back to how we can achieve this coherency at the end of this chapter.

2.2 ISLE and knowledge

There are multiple theoretical frameworks for the knowledge that teachers need to have to help their students master the subject matter successfully. In this book we will adhere to the content knowledge for teaching (CKT) framework (Ball *et al* 2008), created to explain the nature of knowledge of mathematics teachers, and the application of this framework for physics (Etkina *et al* 2018).

While L Shulman (Shulman 1987) recognized that knowing the subject matter as an expert (a practicing physicist, for example) is not enough to help students master this subject and coined the term pedagogical content knowledge (PCK, the knowledge that allows a teacher make the subject matter comprehensible for the students) a long time ago, it was Ball and colleagues who conceptualized this specific knowledge as used in practice. What do teachers need to know to be able to successfully carry out complex tasks (called the tasks of teaching) that lead to student learning? Ball and colleagues made a list of the elements of such knowledge and grouped them into two big categories: subject matter knowledge (SMK) and pedagogical content knowledge—similar to Shulman's PCK.

Below we will analyse these types of knowledge and give examples from physics, specifically from the domain of energy. As you continue reading, ask yourself if you use these types of knowledge in your work with teachers or students, and if you do, where you learned it.

Subject matter knowledge consists of:
  a. *Common content knowledge*—this is the knowledge any person who has taken a physics course (a high school course or an introductory college course) should have.
     - Example: 'The total energy of an isolated system is constant'.
  b. *Horizon content knowledge*—this is the knowledge that goes beyond the knowledge of introductory physics and can be learned in advance courses.
     - Example: 'The law of energy conservation is the consequence of the symmetry of time'.

c. *Specialized content knowledge*—this is the knowledge that is necessary for a teacher to have to help students learn the topic of energy.
  - Example: 'Energy analysis should start with choosing your system. For different systems, the analyses can be different.'

Pedagogical content knowledge consists of:
a. *Knowledge of content and students*—this is the knowledge of student ideas related to the topic (based on the findings of physics education research (PER) and the teacher's personal experience).
  - Example: Research shows that visualizing the conversion of mechanical energy to internal energy is difficult for the students as internal energy is not easily perceptible (Daane *et al* 2015).
b. *Knowledge of content and teaching*—this is the knowledge of how to help students overcome difficulties and be successful learning this specific topic.
  - Example: If you consider the difficulty of visualizing mechanical energy conversion (described above), using an infrared camera attached to a cellphone would help students 'see' where the kinetic energy of a system consisting of a moving object sliding along a surface goes when the object stops (Planinsic *et al* 2022).
c. *Knowledge of content and curriculum*—this is knowledge of the place where the specific topic is in the 'big picture' of physics and in the specific curriculum, what concepts students need to learn before and what concepts are based on the one under current study.
  - Example: Energy is a cross-cutting concept in the US Next Generation Science Standards (NGSS Lead States 2013) and therefore there have to be explicit connections between the study of energy in mechanics and thermodynamics (connections between the mechanical and internal energies mentioned above).

The most important idea behind the CKT framework is that teaching knowledge is not just specific to physics as a discipline as was suggested by the initial models of PCK (Shulman 1987, Magnusson *et al* 1999), but it is specific to a narrow topic of study (kinematics, dynamics, conservation laws, etc). In other words, teachers need more than just general content knowledge in a discipline; they also need a deep understanding of how to teach specific topics within that discipline. Another important idea is that CKT is practical knowledge that can be assessed by observing classroom instruction, analysing teacher artifacts (lesson plans, assessment, etc) while PCK was hypothesized as something that a teacher knows. Being abstract and theoretical, PCK is not as easily observed or assessed through classroom instruction or artifacts. In their 2009 paper Ball and Forzani emphasized that

> ...practice must be at the core of teachers' preparation and that this entails close and detailed attention to the work of teaching and the development of ways to train people to do that work effectively, with

direct attention to fostering equitably the educational opportunities for which schools are responsible.

> By 'work of teaching,' we mean the core tasks that teachers must execute to help pupils learn…. Skillful teaching requires appropriately using and integrating specific moves and activities in particular cases and contexts, based on knowledge and understanding of one's pupils and on the application of professional judgment. This integration also depends on opportunities to practice and to measure one's performance against exemplars. Performing these activities effectively is intricate work. (Ball and Forzani 2009, p 497)

Building on the CKT ideas, a team of researchers lead by D Gitomer at Rutgers University created a theoretical framework for the content knowledge for teaching a specific physics domain, energy in the context of mechanics, for the first physics course that students take (called CKT-E; Etkina *et al* 2018).

This framework consists of three elements: tasks of teaching (ToT), student energy targets (SET), and horizon content knowledge (HCK). ToTs are productive activities in which all physics teachers engage, SETs are the list of disciplinary ideas, cross-cutting concepts, and scientific practices that the students should learn and be engaged in while learning about energy, and HCK is the knowledge beyond high school energy instruction that is not specified in detail in the framework. Below we list abbreviated ToT as they are relevant for the whole program of physics teacher preparation. The complete list of both ToTs and SETs and the results of the study involving the framework was originally published by Etkina *et al* (2018). A detailed statistical analysis of the study can be found in Phelps *et al* (2020).

As you are reading the list and the examples, we ask you to reflect on whether you engage in these tasks of teaching while teaching students who are learning physics or those who are preparing to be physics teachers. In your opinion, which of these ToTs are especially important for those teachers who wish to create an ISLE-based classroom? As teaching physics through the ISLE approach requires students to communally participate in the practice of physics while constructing and applying new knowledge using multiple representations, some of the ToTs are crucial for the implementation of the ISLE approach. As above, in table 2.2 we have marked with different numbers of asterisks ToTs that are important for teachers of all subjects, teachers of physics, interactive engagement teachers of physics, and the ToTs which are unique for ISLE-implementing teachers of physics. ISLE-implementing teachers engage in all of the ToTs listed in table 2.2 with special emphasis on the tasks with four asterisks.

While pre-service teachers develop some of the knowledge necessary to carry out the ToTs marked with one asterisk in general education courses, and some of the tasks marked with two asterisks in their physics courses, to learn how to engage in

Table 2.2. Tasks of teaching (stated as in Etkina *et al* 2018).

Task of teaching	Description	Specific tasks
I. Anticipating student thinking around science ideas	While planning and implementing instruction, teachers anticipate particular patterns in student thinking and challenges in developing an understanding of science concepts and mathematical models. Teachers are also familiar with student interests and background knowledge and enact instruction accordingly.	Teachers: I. a) Anticipate specific student challenges related to constructing scientific concepts, conceptual and quantitative reasoning, experimentation, and the application of science processes.*** I. b) Anticipate likely partial conceptions and alternative conceptions, including partial quantitative understanding about particular science content and processes.*** I. c) Recognize student interest and motivation around particular science content and practices.** I. d) Understand how students' background knowledge both in physics and mathematics can interact with new science content.**
II. Designing, selecting, and sequencing learning experiences and activities	Classroom learning experiences and activities are designed around learning goals and involve key science ideas, key experiments, and mathematical models relevant to the development of ideas and practices. Learning experiences reflect an awareness of student learning trajectories and support both individual and collective knowledge generation on the part of students.	Teachers: II. a) Design or select and sequence learning experiences that focus on sense-making around important science concepts and practices, including productive representations, mathematical models, and experiments in science that are connected to students' initial and developing ideas.**** II. b) Include key practices of science such as experimentation, reasoning based on collected evidence, experimental testing of hypotheses, mathematical modeling, representational consistency, and argumentation.**** II. c) Address projected learning trajectories that include both long-term and short-term goals and are based on evidence of actual student learning trajectories.***

(*Continued*)

Table 2.2. (*Continued*)

Task of teaching	Description	Specific tasks
		II. d) Address learners' actual learning trajectories by building on productive elements and addressing problematic ones.***
		II. e) Integrate, synthesize, and use multiple strategies and involve students in making decisions.****
		II. f) Prompt students to collectively generate and validate knowledge with others.****
		II. g) Help students draw on multiple types of knowledge, including declarative (definitions), procedural (science practices), schematic (multiple representations), and strategic (problem solving strategy).***
		II. h) Elicit student understanding and help them express their thinking via multiple modes of representation.****
		II. i) Help students consider multiple hypotheses, approaches, or solutions, including those that could be later considered to be incorrect.***
III. Monitoring, interpreting, and acting on student thinking	Teachers understand and recognize challenges and difficulties students experience in developing an understanding of key science concepts; understanding and applying mathematical models and manipulating equations; designing and conducting experiments, etc. Teachers also recognize productive developing ideas and know how to leverage them. Teachers engage in an ongoing and multifaceted assessment, using a variety of tools.	Teachers: III. a) Employ multiple strategies and tools to make student thinking visible.*** III. b) Interpret productive and problematic aspects of student thinking and mathematical reasoning.** III. c) Identify specific cognitive and experiential needs or patterns of needs and build upon them through instruction.* III. d) Use interpretations of student thinking to support instructional choices both in lesson design and during the course of classroom instruction.* III. e) Provide students with descriptive timely feedback.*

		III. f) Engage students in metacognition and epistemic cognition.*
		III. g) Devise assessment activities that match the goals of instruction.*
IV. Scaffolding meaningful engagement in a science learning community	Productive classroom learning environments are community-centered. Teachers engage all students as full and active classroom participants. Knowledge is constructed both individually and collectively, with an emphasis on coming to know through the practices of science. The values of the classroom community include evidence-based reasoning, the pursuit of multiple or alternative approaches or solutions, and the respectful challenging of ideas.	Teachers: IV. a) Engage all students to express their thinking about key science ideas and encourage students to take responsibility for building their understanding, including knowing how they know.**** IV. b) Develop a climate of respect for scientific inquiry and encourage students' productive deep questions and rich student discourse.**** IV. c) Establish and maintain a 'culture of physics learning' (including growth mindset) that scaffolds productive and supportive interactions between and among learners.*** IV. d) Encourage broad participation to ensure that no individual students or groups are marginalized in the classroom.** IV. e) Promote negotiation of shared understanding of forms, concepts, mathematical models, experiments, etc, within the class.*** IV. f) Model and scaffold goal behaviors, values, and practices aligned with those of scientific communities.**** IV. g) Make explicit distinctions between science practices and those of everyday informal reasoning as well as between scientific expression and everyday language and terms.**** IV. h) Help students make connections between their collective thinking and that of scientists and science communities.****

(Continued)

Table 2.2. (*Continued*)

Task of teaching	Description	Specific tasks
		IV. i) Scaffold learner flexibility and the development of independence.****
		IV. j) Create opportunities for students to use science ideas and practices to engage real-world problems in their own contexts.**
V. Explaining and using examples, models, representations, and arguments to support students' scientific understanding	Teachers support and scaffold students' ability to use models, examples, and representations to develop explanations and arguments. Mathematical models are included as a key aspect of physics understanding.	Teachers: V. a) Use representations, examples, and models that are consistent with each other and with the theoretical approach to the concept that they want students to learn.** V. b) Help students understand the purpose of a particular representation, example, or model and how to integrate new representations, examples, or models with those they already know.** V. c) Encourage students to invent and develop hypotheses, models, and representations that account for relevant observations support relevant learning goals.**** V. d) Encourage students to explain features of representations and models (their own and others') and to identify/evaluate both strengths and limitations.** V. e) Encourage students to create, critique, and shift between representations and models with the goal of seeking consistency between and among different representations and models.**** V. f) Provide examples that allow students to analyse situations from different frameworks such as energy, forces, momentum, and fields.**

VI. Using experiments to construct, test, and apply concepts	Teachers provide timely and meaningful opportunities throughout instruction for students to design and analyse experiments to help students develop, test, and apply particular concepts. Experiments are an integral part of student construction of physics concepts and are used as part of scientific inquiry in contrast with simple verification.	Teachers: VI. a) Provide opportunities for students to analyse quantitative and qualitative experimental data to identify patterns and construct explanations/models/hypotheses.**** VI. b) Provide opportunities for students to design and analyse experiments using particular frameworks such as energy, forces, momentum, field, etc.**** VI. c) Provide opportunities for students to test experimentally or apply particular ideas in multiple contexts.**** VI. d) Provide opportunities for students to pose their own questions and investigate them experimentally.**** VI. e) Use questioning, discussion, and other methods to draw student attention during experiments to key aspects needed for subsequent learning, including the limitations of the models used to explain a particular experiment.** VI. f) Help students draw connections between classroom experiments, their own ideas, and key science ideas.** VI. g) Encourage students to draw on experiments as evidence to support explanations and claims** and to test explanations and claims by designing experiments to rule them out.****

*ToTs of any subject teachers.
**ToTs of all physics teachers.
***ToTs of physics teachers implementing interactive engagement.
****ToTs of learning physics through the ISLE approach.

the tasks marked with three and four asterisks (in their application to physics), pre-service physics teachers need physics-specific teaching methods courses with educators skilled in student-centered instruction and in the ISLE approach. If we believe that students learn physics by constructing their own ideas, then this suggests that teachers also learn to teach by constructing their own ideas. Therefore, to help pre-service teachers develop the knowledge of these tasks of teaching, they need to engage in and practice these ToTs during their preparation.

The general tasks of teaching shown above are implemented in the context of each specific topic. For the ISLE approach, the physics applications of these tasks are articulated in the IG and ALG for the textbook CP: E&A. The textbook provides possible observational and testing experiments, paths to concept development, multiple representations as reasoning tools, connections to real life, etc, for each topic of a general physics course. The ALG turns this content into activities for students to do in groups and the OALG does it for learning online. We recommend having these resources handy as you are going through this book. For example, if you wish your teachers to learn how to implement the ISLE approach to learning kinematics (chapter 2 in the textbook and supporting materials), to prepare your lessons or professional development sessions, we recommend reading the textbook and the IG with the lens of ToTs and then examine activities in the ALG. Below, we will analyse an example from the textbook (see figure 2.1) using the ToT lens. The example is a modified observational experiment table from chapter 3 of the textbook, Newton's laws. Students previously developed the representations of force and motion diagrams. Specifically, students need to recognize that the velocity change arrow on the motion diagram is in the same direction as the sum of the forces exerted on the object. In other words, the objects change their motion in the direction of the sum of the forces.

While you are reading this analysis, think of what elements of the text and the activity itself implement specific ToTs as applied to Newtonian dynamics. After analysing this question, we will show activities from the ALG that help students develop the same ideas.

Here we can see several tasks of teaching. Specifically:
- II. Designing, selecting, and sequencing learning experiences and activities (II: a, b, g, i).
- IV. Scaffolding meaningful engagement in a science learning community (IV: a, h).
- V. Explaining and using examples, models, representations, and arguments to support students' scientific understanding (V: a, c, e).
- VI. Using experiments to construct, test, and apply concepts (VI: a, g).

Observational experiment	Motion diagrams	Force diagrams at any moment
Experiment 1: A bowling ball rolls on a horizontal linoleum floor. The ball is moving at constant speed.  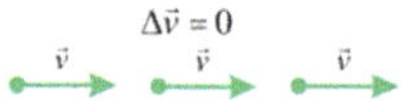 	$\Delta \vec{v} = 0$	$\vec{F}_{S\,on\,B}$ (up), $\vec{F}_{E\,on\,B}$ (down)
Experiment 2: A person using a ruler lightly pushes the rolling ball in the direction of its motion. The ball moves faster and faster. 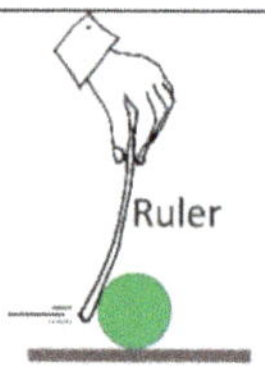 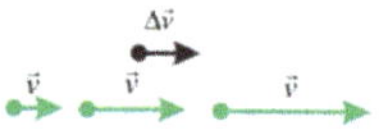 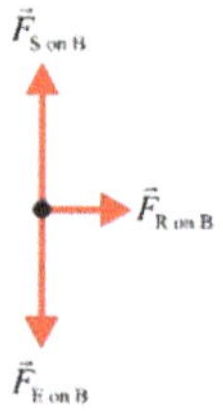	$\Delta \vec{v}$ (right)	$\vec{F}_{S\,on\,B}$ (up), $\vec{F}_{R\,on\,B}$ (right), $\vec{F}_{E\,on\,B}$ (down)
Experiment 3: The person using a ruler lightly pushes the rolling ball in the direction opposite the ball's motion. The ball moves slower and slower. 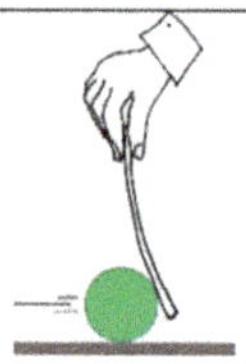 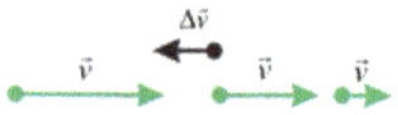 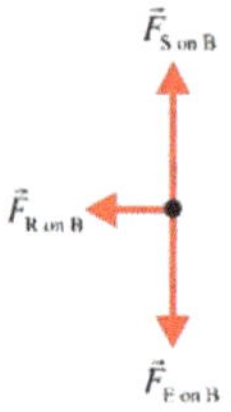	$\Delta \vec{v}$ (left)	$\vec{F}_{S\,on\,B}$ (up), $\vec{F}_{R\,on\,B}$ (left), $\vec{F}_{E\,on\,B}$ (down)

PATTERNS:

- In all the experiments, the forces perpendicular to the floor add to zero and cancel each other. We consider only forces exerted on the ball in the direction parallel to the floor.
- In the first experiment, the sum of the forces exerted on the ball is zero; the ball's velocity remains constant.
- In the second and third experiments, the velocity change arrow ($\Delta \vec{v}$ arrow) points in the same direction as the sum of the forces.

Summary: In all experiments, the $\Delta \vec{v}$ arrow points in the same direction as the sum of the forces. There is no pattern relating the *direction* of the velocity $\vec{v}$ to the direction of the sum of the forces. In Experiment 3, the velocity and the sum of the forces are in opposite directions, but in Experiment 2, they are in the same direction.

Figure 2.1. How are motion and forces related? Identifying ToTs. (To see the original text consult CP: E&A, chapter 3). See the video at https://youtu.be/H-9Q41VTECk.

The page in figure 2.1 provides a teacher with multiple opportunities to engage in the ToTs consistent with the ISLE approach. However, reading the textbook does not help the teacher envision the activities that the students will do in the classroom to construct the relationship. The help comes in the ALG (chapter 3) which provides a specific activity:

Observe and find a pattern
Equipment per group: 1 bowling ball, multiple sugar packets, stopwatch (alternative: cellphone app.), meter stick.
Perform the experiments in the left column (see table below) and analyse the observations using motion and force diagrams. All experiments need to be performed on a smooth, hard floor.

Observational experiment	Analysis	
	Motion diagram	Force diagram
Experiment 1. One group member will use both hands to push very hard but only once on a bowling ball on smooth floor so it rolls in a straight line. The second member will count seconds (or use a metronome). The third group member drops sugar packets on the floor next to the bowling ball on every count *after* it has started rolling. Then the group records the locations of the sugar packets.		
Experiment 2. Repeat experiment 1, only now the fourth group member pushes the moving bowling ball very lightly in the direction opposite to the direction of the ball's motion, trying to exert a constant push (it is easier to do if you use a ruler and keep the same bend). Other group members repeat the same procedure. The sugar packets need to be put on the floor after the ruler touches the ball.		
Experiment 3. Repeat experiment 1, only now the first group member continuously pushes the bowling ball trying to exert a constant push (it is easier to do if you use a ruler and keep the same bend). Other group members repeat the same procedure. The sugar packets need to be put on the floor after the ruler touches the ball.		

Patterns
Discuss with your group members the patterns in the direction of the $\Delta \vec{v}$ arrow on the motion diagram and the vector sum of the forces exerted on the ball. Make sure that all your forces are labeled with two subscripts.

Here we see the students collaboratively working on performing experiments, recording data, analysing the patterns and devising a hypothesis. They are encouraged to discuss the patterns with each other and receive guidance on how to identify the forces (two subscripts represent two interacting objects; see chapter 1 of this book for the example of a force diagram representation). In addition to the ToTs identified above, the teacher will engage in the following ToTs facilitating the activity (see table 2.2):
  I. Anticipating student thinking around science ideas (I: a).
  II. Designing, selecting, and sequencing learning experiences and activities (II: f; IV: a, b, e, f).
  III. Explaining and using examples, models, representations, and arguments to support students' scientific understanding (all ToTs in item V) and Using experiments to construct, test, and apply concepts (VI: e).

The follow up activity for the students to test their pattern will demand teacher engagements in all of the ToTs in section VI.

Finally, we show a short example in which we analyse the support for ToTs in a paragraph from the IG (chapter 3, p 3-2). We repeat the relevant ToTs below and then include the ToTs numbers directly in the text.
  I.a Anticipate specific student challenges related to constructing scientific concepts, conceptual and quantitative reasoning, experimentation, and the application of science processes.
  I.b Anticipate likely partial conceptions and alternate conceptions, including partial quantitative understanding about particular science content and processes.
  II.c Address projected learning trajectories that include both long-term and short-term goals and are based on evidence of actual student learning trajectories.
  II.d Address learners' actual learning trajectories by building on productive elements and addressing problematic ones.
  IV.a Engage all students to express their thinking about key science ideas and encourage students to take responsibility for building their understanding, including knowing how they know.
  IV.b Develop a climate of respect for scientific inquiry and encourage students' productive deep questions and rich student discourse.
  V.e Encourage students to create, critique, and shift between representations and models with the goal of seeking consistency between and among different representations and models.
  VI.a Provide opportunities for students to analyse quantitative and qualitative experimental data to identify patterns and construct explanations/models/hypotheses.
  VI.b Provide opportunities for students to design and analyse experiments using particular frameworks such as energy, forces, momentum, field, etc.

At the beginning of their study, many students believe that an object's velocity is in the same direction as the sum of the forces exerted on it because most of them think of force as a property of motion, not the cause of motion (I: a, b). Section 3.3 addresses this difficulty. We suggest that students first do the ALG activities 3.3.1–3.3.3 (VI: a, b), and then read the textbook. If students start putting a 'force of motion' label on their force diagrams, ask them to include two subscripts to identify the two interacting objects (II: c, d). The key idea that students need to learn is that if they can't identify the object exerting the force, that force shouldn't be included on the diagram (II: d). (V: e) Without doing qualitative activities such as those described in sections 3.3 and 3.4, students may have the same ideas at the end of their study as they had at the beginning (II: c; V: e).

The above example shows how the ToTs framework allows us to analyse the ISLE curriculum materials. Such analysis is important for teacher educators. Once we know what tasks our future teachers will need to perform, we can create an educational environment to help them practice these tasks.

Concerning the content knowledge for teaching each specific physics topic, the IG provides a list of learning goals for the students at the beginning of every chapter. For example, for chapter 3 in the textbook (from which the above activities were taken), the IG lists the following content goals (goal 4 was addressed in the activities above) on its pages 3-1–3-2:

Students should be able to:
1. Identify a system for analysis and objects interacting with the system.
2. 'Read and write' with force diagrams, labeling forces with two subscripts.
3. Find force components along chosen axes (in one dimension).
4. Find consistency between a motion diagram and a force diagram for a system (recognize the relationship between $\sum \vec{F}$ and $\Delta \vec{v}$. Explain how we know that objects do not move in the direction of the sum of the forces exerted on them (by referring to the experiments).
5. Explain the role of inertial reference frames for using Newton's laws to analyse motion and thus the role of Newton's first law in the set of laws.
6. Describe the experiments from which they developed Newton's laws and use the laws to predict the outcomes of simple one-dimensional processes.
7. Write Newton's second law in component form for a system using a force diagram.
8. Compare and contrast Newton's second and third laws.
9. Explain the difference between an operational definition of acceleration $\vec{a} = \frac{\Delta \vec{v}}{\Delta t}$ and a cause-effect relationship $\vec{a} = \frac{\sum \vec{F}}{m}$.
10. Apply Newton's laws to solve problems.
11. Explain why objects have the same free-fall acceleration on Earth. (IG, pp 3-1–3-2)

While the above knowledge is important, we want to emphasize our practice-oriented approach to teacher preparation as advocated by Ball and colleagues and as implemented in the CKT-E framework by Etkina *et al*:

To make practice the core of the curriculum of teacher education requires a shift from a focus on what teachers know and believe to a greater focus on what teachers do. This does not mean that knowledge and beliefs do not matter but, rather, that the knowledge that counts for practice is that entailed by the work. A practice-based theory of knowledge for teaching (Ball and Bass 2003) is derived from the tasks and demands of practice and includes know-how as well as declarative knowledge. But a practice-focused curriculum for learning teaching would include significant attention not just to the knowledge demands of teaching but to the actual tasks and activities involved in the work. (Ball and Forzani 2009, p 503).

2.3 ISLE and skills

For our purposes, we define a skill as 'a precompiled procedure that one deploys automatically without consciously thinking about it.' (Etkina *et al* 2017, p 010107-4). We group the skills into mental, technical, and emotional skills. Related to the example discussed above, examples of mental skills include drawing motion and force diagrams; examples of emotional skills include the skill of 'keeping cool' when students who have supposedly learned those skills already cannot demonstrate them; examples of technical skills include pushing the bowling ball with a constant force (this is not as easy as it might seem). Below we make a list of skills that are useful to develop for those who wish to implement the ISLE approach (see table 2.3). These skills are derived from the list of ToTs relevant to the ISLE approach (marked with four stars in table 2.2). Some of the skills repeat as they are used in enacting several ToTs.

Table 2.3. Connections between ISLE-specific ToTs and skills.

Task of teaching	Tasks of teaching relevant for the implementation of the ISLE approach	Skills
II. Designing, selecting, and sequencing learning experiences and activities	Teachers: II. a) Design or select and sequence learning experiences that focus on sense-making around important science concepts and practices, including productive representations, mathematical models, and experiments in science that are connected to students' initial and developing ideas. ****	The skills involve: To recognize and understand key results from PER and how to integrate them into one's instructional design. To handle and repair equipment, analyse data, using EXCEL or some other software to analyse data, conduct video analysis, data collection using phone apps, use hypothetico-deductive

(Continued)

Table 2.3. (*Continued*)

Task of teaching	Tasks of teaching relevant for the implementation of the ISLE approach	Skills
	II. b) Include key practices of science including experimentation, reasoning based on collected evidence, experimental testing of hypotheses, mathematical modeling, representational consistency, and argumentation. **** II. e) Integrate, synthesize, and use multiple strategies and involve students in making decisions. **** II. f) Prompt students to collectively generate and validate knowledge with others. **** II. h) Elicit student understanding and help them express their thinking via multiple modes of representation. ****	reasoning (see chapter 3 for details) etc. To monitor group work without interfering when the students are engaged in productive discussions. To run whole class discussion, to withhold validation. To help students articulate their thinking and choose appropriate representations to facilitate such thinking.
IV. Scaffolding meaningful engagement in a science learning community	Teachers: IV. a) Engage all students to express their thinking about key science ideas and encourage students to take responsibility for building their understanding, including knowing how they know. **** IV. b) Develop a climate of respect for scientific inquiry and encourage students' productive deep questions and rich student discourse. **** IV. f) Model and scaffold goal behaviors, values, and practices aligned with those of scientific communities. **** IV. g) Make explicit distinctions between science	The skills involve: To monitor group work without interfering, to run whole class discussion, to withhold validation. To facilitate cognitive reflection. To demonstrate personal behavior of respect to student ideas. To experimentally test student ideas. To develop and maintain classroom management rules mirroring a scientific community. To manage multiple resubmissions of student work. To use appropriate language when talking about physics (see chapter 5 for more).

	practices and those of everyday informal reasoning as well as between scientific expression and everyday language and terms. ****	To integrate historical ideas physicists had and how they discarded some (to avoid student learning of only correct ideas).
	IV. h) Help students make connections between their collective thinking and that of scientists and science communities. ****	
	IV. i) Scaffold learner flexibility and the development of independence. ****	
V. Explaining and using examples, models, representations, and arguments to support students' scientific understanding	Teachers:	The skills involve:
	V. c) Encourage students to invent and develop hypotheses, models, and representations that account for relevant observations support relevant learning goals. ****	To use simple experiments that allow for multiple explanations.
		To monitor group work without interfering, to run whole class discussion, to withhold validation.
	V. e) Encourage students to create, critique, and shift between representations and models with the goal of seeking consistency between and among different representations and models. ****	To inspire students to devise multiple explanations.
		To use different representations of the same phenomenon and seek consistency (for example consistency between motions and force diagrams discussed above).
VI. Using experiments to construct, test, and apply concepts	Teachers:	The skills involve:
	VI. c) Provide opportunities for students to analyse quantitative and qualitative experimental data to identify patterns and construct explanations/models/ hypotheses. ****	To use simple experiments that allow for multiple explanations.
		To monitor group work without interfering, to run whole class discussion.
		To withhold validation.
	VI. b) Provide opportunities for students to design and analyse experiments using particular frameworks such as energy, forces, momentum, field, etc. ****	To conduct experimental design and analysis.
		To design experiments that can test hypotheses.
		To use hypothetico-deductive reasoning to make

(*Continued*)

Table 2.3. (*Continued*)

Task of teaching	Tasks of teaching relevant for the implementation of the ISLE approach	Skills
	VI. c) Provide opportunities for students to test experimentally or apply particular ideas in multiple contexts. ****	predictions about the outcomes of testing experiments using hypotheses under test. To develop positive emotional response to the testing experiments that reject one or more of their ideas. To find appropriate equipment for the student investigations.
	VI. d) Provide opportunities for students to pose their own questions and investigate them experimentally. ****	
	VI. g) Encourage students to draw on experiments as evidence to support explanations and claims** and to test explanations and claims by designing experiments to rule them out. ****	To design experiments that can potentially test hypotheses, use hypothetico-deductive reasoning to make predictions about the outcomes of testing experiments using hypotheses under test.

While the list of skills in the table above might feel overwhelming, many skills repeat again and again. These are the skills that are the most important to develop to be able to implement the ISLE approach in your class. Before you read on, we encourage you to go through the table above to find those skills.

The most important skills involve monitoring group work without interfering, running whole class discussion, and withholding validation. These three skills are vital for the implementation of the ISLE approach. The students need time to work on the activities and when we interrupt them asking questions, they lose their train of thought and try to follow ours. Therefore, it is extremely important not to ask leading questions when the students are working collaboratively on the activities but carefully watch what is going on. If you feel that the students are stuck, try to ask a question that will help them fall back on something that they know, so that they make the conceptual leap themselves, thus using their brain connections and building on them.

While it is important to let students figure out things on their own, how do you keep the activities on time then? The time that each activity takes, depends on how much guidance/scaffolding the teacher provides. Therefore, sometimes it is trade-off between prompting the class with leading questions or cutting 'critical' activities. While our interruptions hurt student learning, it is a true challenge of trying to implement ISLE at scale.

During class discussions, there is always a moment when a student responds to another student, this is the most precious moment as it might give rise to the conversation among the students, which is our ultimate goal. If you withhold validation, and they figure out the issue themselves, this 'self-figuring' not only will help them better remember what happened but also will contribute to the development of self-efficacy—they did it themselves! Finally, these skills are followed by the skills of facilitating cognitive reflection. These reflections (what did you learn? How did you learn it? What remained unclear? What was the most difficult part? What helped you learn?) can be a part of a whole class discussion, group reflections or a homework assignment. Physics teachers need to reflect on what they learned as a physicist and what they learned as a physics teacher (see more in chapter 6).

The following two skills can be combined together: searching through physics education publications (e.g. articles from journals such as *The Physics Teacher* or *Physics Education*) to find simple observational experiments to help student construct new concepts, and searching historical literature to be familiar with what ideas physicists had and how they discarded some (to avoid student learning of only correct ideas). Those skills help the teacher learn how the ideas that they wish their students to learn were developed by scientists.

The next set of skills is comprised of the skills that are uniquely important for the implementation of the ISLE process using the lens of the first intentionality (helping students learn physics by engaging in the processes that mirror scientific practice): choosing/using simple experiments that allow for multiple explanations (available in the ISLE-based materials, such as the textbook CP: E&A and the ALG); inspiring students to devise multiple explanations, using different representations of the same phenomenon and seeking consistency (e.g. the consistency between the motion and force diagrams discussed above); designing experiments that can potentially test hypotheses, using hypothetico-deductive reasoning to make predictions about the outcomes of testing experiments using hypotheses under test. As you look at figure 1.1 in chapter 1, you will see that all skills in table 2.3 in this chapter are the skills necessary to help students progress from one step of the ISLE process to the next.

Teachers also need skills that address the second intentionality of the ISLE approach (improving student motivation, persistence, self-efficacy and developing community).

In 2.3 we can find the following skills: personal behavior of respect for student ideas, developing positive emotional response to testing experiments that reject one or more of their ideas, developing and maintaining classroom management rules mirroring a scientific community, and managing multiple resubmissions of student work.

Finally, some technical skills are necessary for the successful implementation of the ISLE process in real time. These skills involve handling and repairing equipment, analysing data, using EXCEL or some other software to analyse data, video analysis, data collection using phone apps, hypothetico-deductive reasoning (see chapter 4 for details); etc. The first five in this batch are not unique to ISLE but the last one is.

2.4 *Interlude:* Habits which are the missing link to the implementation of the ISLE approach (by Eugenia)

A long time ago, at the very beginning of my program of physics teacher preparation, one of my former students in her first year of teaching (who was an excellent pre-service physics teacher: diligent, focused, creative in course work and caring during her student teaching internship, but doing her internship with a traditional cooperating teacher) came to one of our regular community meetings and said: 'Eugenia, I hate to tell you, but I went to the Dark Side today. I spent the whole lesson lecturing to my students.' She was very upset about it and considered this experience a failure. I, on the other hand, saw it as an opportunity to talk to her and the rest of the meeting participants about the reasons for switching to the transmission mode of teaching. Lecturing is useful while summarizing what the students have figured out by themselves, or repeating an elegant solution that a student proposed, or explaining something that goes beyond the level of the current topic, or even sharing your solution to compare it with the students.

However, lecturing about curriculum material, with which the students have no experience yet, is not useful as it prevents the students from constructing the knowledge themselves. So, why would the teacher, who is skilled in the methods in active engagement, cares deeply about student learning, and has excellent physics knowledge, switch to a model of interaction of which they themselves do not approve?

This teacher had the dispositions for interactive engagement; she had knowledge of how to structure the lesson, what experiments to use, and what ideas students might have; she was skilled in all of the skills that I listed above. And yet, she spent 45 minutes of her class lecturing. Why?

Let me ask you a question: Is there something that you know is the right thing to do, you know how to do it, you are skilled in doing it, and you still do not do it? Think for example of flossing. We all believe that it is the right thing to do as we know that it saves the gums from bacteria, and we are all skilled in moving the thread between our teeth. But how many people do it habitually after every meal or even once a day every day? Other examples are drinking enough water, getting up and moving for 5 minutes for every hour of sitting, eating five servings of fruits and vegetables a day. These are all the behaviors that we believe are the right things to do, and we know why they are the right things, and we are skilled in doing them. However, how many of us do drink eight glasses of water every day or walk 10 000 steps? But those who do those things (floss, drink eight glasses, walk 10 000 steps) are usually the people who cannot *not* do them. Why? Because these things became their habits. Habits are indeed the powerful drivers of our behavior.

The examples above show that dispositions, knowledge and skills are not enough for us to do something when there are outside pressures. Without developed habits, we forget our bottle at home, we get submerged in a project, and all our plans to drink water remain unrealized. If we have the habits, we *cannot* leave the house without the water bottle, we *cannot* go to sleep if our step counter is at 9600. We will walk around the bedroom to get those 400 steps—this is how powerful the habits are.

The same is true for teaching. We can have the disposition to engage our students in learning physics in a specific way (in our case, through the ISLE approach), we can read papers and books and participate in a workshop about ISLE (so that we have the knowledge), and we may have even practiced individual components of the ISLE process and devised a method of evaluating resubmissions of our student work. But in a real classroom with the constraints of curriculum, time, equipment, technology, family issues, etc, it is very difficult not to switch to traditional modes of instruction with one-shot assessment strategies.

This switch is less likely to occur for a person who developed the habits of ISLE-based learning and teaching. What are those habits and how to develop them? What are the conditions under which habits thrive and what interrupts the habits? The rest of the book is dedicated to the answers to these questions.

There is one thing that I would like to add here. We all know that it is easier to develop a habit than to break one and develop a new one instead. Therefore, developing ISLE-based learning and teaching habits is much easier if you have not developed habits *inconsistent* with this style of teaching. Does it mean that the education of pre-service teacher should focus on the development of such habits while it is too late for in-service teachers used to traditional teaching to develop them? Although it is definitely more difficult for the latter group, I do not think that it is impossible. Watching many of my colleagues and friends switch and enjoy this new style of learning and teaching, I would say that it is possible. But it is not easy. The proof is in the story with which the interlude started. The teacher with all the dispositions, knowledge and skills did not develop the habits of ISLE-based learning, and in a difficult moment, reverted to lecturing.

References

Arons B A 1997 *Teaching Introductory Physics* (New York: Wiley)
Ball D L and Bass H 2003 Making mathematics reasonable in school *A Research Companion to Principles and Standards for School Mathematics* (National Council of Teachers of Mathematics) pp 27–44
Ball D L and Cohen D K 1999 Developing practice, developing practitioners: toward a practice-based theory of professional education *Teaching as the Learning Profession: Handbook of Policy and Practice* ed G Sykes and L Darling-Hammond (San Francisco, CA: Jossey-Bass) pp 3–32
Ball D L and Forzani F M 2009 The work of teaching and the challenge for teacher education *J. Teach. Educ.* **60** 497–511
Ball D L, Hoover Thames M and Phelps G 2008 Content knowledge for teaching: what makes it special? *J. Teach. Educ.* **59** 389–407
Butterworth J 2017 *Atom Land* (New York: The Experiment)
Daane A R, McKagan S B, Vokos S and Scherr R E 2015 Energy conservation in dissipative processes: Teacher expectations and strategies associated with imperceptible thermal energy *Phys. Rev. Spec. Top.-Phys. Educ. Res.* **11** 010109
Darling-Hammond L, Hammerness K, Grossman P, Rust F and Shulman L 2005 The design of teacher education programs *Preparing Teachers for a Changing World* ed L Darling-Hammond and J Bransford (San Francisco, CA: Jossey-Bass) pp 390–441

Dehaene S 2021 *How We Learn: Why Brains Learn Better Than any Machine…for Now* (London: Penguin)

Dweck C 2016 What having a 'growth mindset' actually means *Harv. Bus. Rev.* **13** 2–5

Einstein A and Infeld L 1967 *The Evolution of Physics* (New York: Touchstone)

Erlich R 1997 *Why Toast Lands Jelly-Side Down* (Princeton, NJ: Princeton University Press)

Etkina E, Gitomer D, Iconangelo C, Phelps G, Seeley L and Vokos S 2018 Design of an assessment to probe teachers' Content Knowledge for Teaching: an example from energy in HS physics *Phys. Rev. Phys. Educ. Res.* **14** 010127

Etkina E, Gregorcic B and Vokos S 2017 Organizing physics teacher professional education around productive habit development: a way to meet reform challenges *Phys. Rev. Spec. Top. Phys. Educ. Res.* **13** 010107

Feiman-Nemser S and Remillard J 1996 Perspectives on learning to teach *The Teacher Educator's Handbook: Building a Knowledge Base for the Preparation of Teachers* ed F B Murray (San Francisco, CA: Jossey-Bass) 1st edn pp 63–91

Fives H and Buehl M M 2012 Spring cleaning for the 'messy' construct of teachers' beliefs: What are they? Which have been examined? What can they tell us? *APA Educational Psychology Handbook, Volume 2. Individual Differences and Cultural and Contextual Factors* ed K R Harris, S Graham, T Urdan, S Graham, J M Royer and M Zeidner (American Psychological Association) pp 471–99

Grimvall G 2007 *Brainteaser Physics* (Baltimore, MD: Johns Hopkins University Press)

Hammerness K, Darling-Hammond L, Bransford J, Berliner D, Cochran-Smith M, McDonald M and Zeichner K 2005 How teachers learn and develop *Preparing Teachers for a Changing World* ed L Darling-Hammond and J D Bransford (San Francisco, CA: Jossey-Bass) pp 358–89

Korthagen F and Lagerwerf B 1996 Refraining the relationship between teacher thinking and teacher behaviour: Levels in learning about teaching *Teachers Teaching* **2** 161–90

Leim T L 1987 *Invitation to Science Inquiry* (Science Inquiry Enterprises)

Lemons D S 2017 *Drawing Physics* (Cambridge, MA: MIT Press)

Lortie D C 1975 *Schoolteacher: A Sociological Study* (Chicago, IL: University of Chicago Press)

Magnusson S, Krajcik J and Borko H 1999 Nature, sources, and development of pedagogical content knowledge for science teaching *Examining Pedagogical Content Knowledge: The Construct and its Implications for Science Education* vol 31 ed J Gess-Newsome and N G Lederman (Dordrecht: Kluwer Academic) pp 95–133

Minaert M G J 1993 *Light and Color in the Outdoors* (Berlin: Springer)

NGSS Lead States 2013 *Next Generation Science Standards: For States, by States* (Washington, DC: The National Academies Press)

Pajares M 1992 Teachers' beliefs and educational research: cleaning up a messy construct *Rev. Educ. Res.* **62** 307

Phelps G, Gitomer D H, Iaconangelo C J, Etkina E, Seeley L and Vokos S 2020 Developing assessments of content knowledge for teaching using evidence-centered design *Educ. Assess.* **25** 91–111

Planinsic G, Nered U and Etkina E 2022 An infrared camera: multiple ways to use a modern device in introductory physics courses *Thermal Cameras in Science Education* (Innovations in Science Education and Technology vol 26) ed J Haglund, F Jeppsson and K Schönborn (Cham: Springer) pp 147–67

Shulman L 1987 Knowledge and teaching: foundations of the new reform *Harv. Educ. Rev.* **57** 1–23

Sutton R M 1938 *Demonstration Experiments in Physics* (New York: McGraw-Hill)
Swartz C 2003 *Back-of-the-Envelope Physics* (Baltimore, MD: Johns Hopkins University Press)
Walker J 1977 *The Flying Circus of Physics* (New York: Wiley)
Zull J E 2002 *The Art of Changing the Brain: Enriching Teaching by Exploring the Biology of Learning* 1st edn (Sterling, VA: Stylus)
Zull J E 2011 *From Brain to Mind: Using Neuroscience to Guide Change in Education* (Sterling, VA: Stylus)

Chapter 3

Habits of physics teachers

3.1 What are habits?

In chapter 2, we discussed the roles of dispositions, knowledge, and skills in the development of a physics teacher and how those of teachers who implement the ISLE approach need to be consistent with the intentionalities of ISLE. We also noted that dispositions, knowledge, and even skills are not sufficient for the successful implementation of the ISLE approach. We need to be able to use the skills and knowledge habitually, so that in the spur of the moment we do not revert to traditional transmissive teaching[1].

However, we have not defined habits yet. What is a habit? According to Lally *et al* (2010), one can think of habits as *spontaneous* responses to situational cues that take time to develop and tend to cement when environmental conditions do not change. In her recent book (Wood 2019), Wendy Wood defines a habit in more detail:

> A habit is a mental association between a context cue and a response that develops as we repeat an action in that context for a reward. But a shorthand definition is this: automaticity in lieu of conscious motivation. A habit turns the world around you into a context, a trigger to act. However, a habit is not equal to automaticity, it is connected to conscious decisions. Habits are foundations of persistent behaviors.

---

[1] We remind the reader that the abbreviation CP: E&A stands for the textbook *College Physics: Explore and Apply*, ALG stands for the *Active Learning Guide*, OALG stands for the *Online Active Learning Guide*, and IG stands for the *Instructor Guide*. Proper citations for these materials are in chapter 1.

In her book, Wood provides many examples of habits and conditions for their development. Those conditions are:
  a. The context that triggers the habit (this serves as a situational cue).
  b. The rewards that come with practicing the habit (this leads to the release of dopamine in the brain, which makes us want to repeat the action).
  c. Multiple repetitions of the actions involved in the habit (this is required to achieve automaticity).
  d. Reducing difficulties (called reducing 'friction') or obstacles that are making us stop and think or do something extra before enacting the habit (reducing such obstacles avoids engaging cognitive functions of our brain which moves the habit out of an automatic action realm into the cognitive realm).

Think about the development of the habit of brushing your teeth in the morning. Let's imagine that your parents wanted you to develop a habit of brushing your teeth after breakfast and before going to bed. Here we see two clearly defined contexts that were needed to trigger this habit—finishing your breakfast in the morning and putting your pajamas on at night. The rewards for practicing those habits were probably the nice taste of kids' toothpaste, parental praise, clean breath, some gifts at the dentist during regular checkups, and the lack of cavities. You repeat brushing your teeth twice a day, every day—lots of repetition! Finally, as a child, you would always find the toothpaste and toothbrush in the same place. You did not need to stop and search for them around the house or run to the store to buy a new one. This is what your parents or your guardians did in order to help you develop a habit of brushing your teeth twice a day. They reduced 'friction'. Now you do it without thinking. And this is the key to the importance of the habits: habits free our mind to think about problems that require solutions, while stuff that we do every day goes smoothly without us noticing. Those who have healthy eating habits do not think about food much—they buy food on the perimeters of supermarkets automatically. Those of us who run every day do not contemplate every day about what to wear for a run and where to run as their clothes and routes are picked habitually. There are lots of examples of productive habits that make our life easier, healthier, and fuller. There are also bad habits—smoking, unhealthy snacks, being late, and so forth. In each of these habits you can find all of the four conditions described above —context, rewards, repetition, and reduction of difficulties.

You must be thinking—what does it have to do with teaching? Teaching is a creative activity that requires decision making throughout every second of the process. It cannot be habitual. But is this true? If you read the interlude in the previous chapter, you probably already see the connection between habits and teaching. All teachers develop habits with practical experience as many of their thoughts and actions repeat every day. Think of productive habits that you have developed (or wish to develop) in your teaching life. You habitually prepare lessons for the next day (or if you developed another habit, you might plan the upcoming week's lessons on the weekend, this is what I did—EE), you habitually start each lesson by making eye contact with every student in your class, you habitually toss their questions back to the class for everyone to think instead of answering directly, you habitually return or reset all equipment after it has been used in today's lesson. These are great habits to have as they make you a successful teacher.

Also think about unproductive habits that some teachers could have developed: coming to school in the morning without clear lesson plans for today, checking attendance at the beginning of class instead of getting students to work right away, answering students' questions without giving them an opportunity to think about them, leaving equipment on your desk after class so that it is hard to find anything the next day. I could go on and on about habitual things that we do as teachers.

Teachers' knowledge, belief structures, and practical work affect the development of their habits. With the development of habits many of the tasks of teaching that we described in the previous chapter become automated and thus allow teachers to focus on other, unexpected situations that come up during a lesson or during lesson preparation.

We all know that the first years of teaching are often spent in a 'survival mode'. The teacher responds to a continuous stream of demands put on them by the students, administration, parents, curriculum, time and space constraints, equipment, etc. Therefore, it is crucial that a beginning teacher starts their professional life with a set of habits, that will help them focus their practice on the productive tasks of teaching instead of merely surviving (or worse, developing bad habits!). If we wish to prepare a physics teacher who is going to implement the ISLE approach in their classroom, then we need to focus on the development of ISLE-specific teaching habits during their time in a teacher preparation program. However, there is more to the development and maintenance of the habits. It is one's environment—supportive or not—that affects the habits and, most importantly, it is one's community that helps maintain the habits. Think of the groups that people form to help them develop and maintain habits. The first that comes to mind is AA. This community helps people abstain from alcohol. Another example is exercise classes. It is easier to continue workouts when you are surrounded by like-minded people who support your efforts. This is true for both AA and exercise classes. Therefore, when we think about the development of habits, we need to keep in mind that the development of a supportive community is a part of habit development.

Dewey pointed to the importance of habits a long time ago. For Dewey, they are more than just the tools that a teacher possesses. According to Dewey, the habits themselves direct the teacher's dispositions, thoughts, and actions. They are the tools that carry some degree of agency—they also direct one's thoughts and actions (Dewey 1922).

Although reflection on practice is one of the characteristics of a successful teacher (Schön 1983), many of the decisions and actions of a teacher are routine[2] (Schoenfeld 1998, 2010). This combination of routines and continuous reflection is what makes a successful teacher. If we take Dewean conceptualization of habits and combine it with the reflective nature of teaching and its routines, we see that habits help us to shape spontaneous behavior responsive to a specific teaching/learning situation.

---

[2] A routine is a series of behaviors that repeat frequently. We all have a habit of brushing our teeth in the morning (hopefully) but the routines of this brushing are different. Some use mechanical brushes, some use electric brushes. Some squeeze the toothpaste by rolling it bottom up, and some just squeeze it.

Seeing habits as mental associations between a context cue and a response helps us connect them to the tasks of teaching of a teacher who implements the ISLE approach and does not revert to traditional transmissive teaching. In the ISLE approach, learning contexts repeat again and again. Students work in groups every day. When developing new concepts, they proceed from observing simple experiments, to finding patterns, to explaining them, and then to testing their explanations in new experiments. Then they apply the explanations that have not been rejected to solve practical problems. In doing all these activities, they continuously use multiple representations. As this happens, it is natural to think about these contexts as cues for the development of productive habits. Because the habits form as responses to situational cues, it is important that new teachers experience situations where they practice the ISLE-consistent way of teaching long enough to form specific habits. Therefore, one of the major foci of physics teacher preparation programs should be on the development of habits.

According to the conditions necessary for the development of habits, they form when a person spends sufficient time in an environment that does not change. Therefore, if we wish for our future teachers to develop specific habits and phase out some unproductive habits that they have developed prior to starting a program, habit formation should be one of the major goals of any physics teacher preparation program. It follows from this goal that programs need to set up conditions that will encourage the development of productive habits and the phasing out of unproductive habits. These conditions need to be in place for a significant amount of time. In addition, if the program is set up with the goal of forming specific habits and has an established set of situational cues, it also needs to have the following: established rewards, opportunities for future teachers to practice the actions that will eventually solidify into habits over and over again, and finally, obstacles in the path of enacting those habits should be removed as much as possible so that enacting the desired habits at the beginning is easy. However, all of the above are not enough to develop and sustain productive habits. The same way as it is easier for all of us to keep exercising with a group of friends than alone, it is important to have a community to develop and sustain good habits. The role of the teacher community is to support its members in the moments of doubts and struggles, to provide help when a specific difficulty arises (how to solve a specific physics problem, how to conduct a specific physics experiment, how to explain to a parent why you don't 'lecture' and so forth), but most importantly, to provide and receive honest feedback. Therefore, if you wish teachers to develop productive habits, you need to also think of the development of a supporting community.

While some habits are productive for all teachers, some are specific for the subject matter and some are unique for the teaching/learning approach that we wish our pre-service teachers will implement when they become teachers. Etkina, Gregorcic, and Vokos (Etkina *et al* 2017) grouped productive physics teaching habits into three categories: habits of mind, habits of practice, and habits of maintenance and improvement. Inside this grouping, they separated the habits of mind into habits of mind of a physicist and habits of mind of a physics teacher. In the following sections of this chapter, we will describe these types of habits as they apply to all physics teachers and to the teachers who are using the ISLE approach. In the subsequent chapters, we will focus on the dispositions, knowledge, and skills necessary to form these habits in pre-service teachers (PSTs) and discuss how to help physics teachers form specific habits during their pre-service teacher preparation programs and in-service professional development activities.

3.2 Habits of mind

According to Etkina, Gregorcic, and Vokos (Etkina *et al* 2017) habits of mind 'are the examples of habits of thinking like a physicist and thinking like a physics teacher in the new environment' (p 010107-6). There are many habits of mind characteristic of physicists. One of them is the spontaneous noticing of physics applications in the world around them and thinking about those applications. Think of the situational cues that trigger this habit—walking on ice, shivering in wet clothes in the wind, seeing round droplets of water on leaves—all of those trigger our thinking about frictional forces, latent heat of evaporation, surface tension, and so forth. Once we mentally explain (or even note) to ourselves what is going on, we experience this good feeling of secret knowing (all of the world is physics). As the natural world is all around us, and every phenomenon can be a cue, we practice this habit multiple times every day. But most importantly, there is no specific place or condition for practicing it—we can do it anytime and anywhere as long as we are not thinking about something else. It is a great habit and we hope you enjoy it every day too.

But noticing and thinking about the physics principles in the phenomena that surround us is not enough for a physics teacher. In addition to having this habit, a physics teacher needs to develop a habit of capturing these phenomena for future analysis by their students.

For example, we all drop small rocks in lakes or ponds and observe the resultant circles in the water that they form. When dropping a rock, a physicist notices the spherical shape of the circles and thinks about wave fronts, the speed of the waves, the wavelength, etc. A physics teacher takes out their phone and makes a movie of the circles to serve as an illustration of wave motion for their students. An ISLE teacher makes the same movie but is thinking of using it as an observational experiment for their students to start thinking about wave motion with the goal of developing multiple explanations. How does the disturbance propagate through the water? Does the water rush out in circles around the rock? Or, does the disturbance move outward (without moving the layers of water outward) and the water only moves up and down? Thinking about these two explanations, the ISLE teacher quickly drops a small piece of wood or paper and makes another movie. They put the piece of paper on the surface of the water, drop another rock, and video the outcome. This is the video of a testing experiment. Before students conduct or watch the video of the testing experiment, they will make a prediction based on each explanation separately. If the water rushes outward along with the disturbance, then the piece of paper will move outward. Instead, if the water only moves up and down, then the piece of paper will bob up and down. By having this habit, an accidental observation of dropping a rock in a pond becomes the beginning of unintentionally planning the way in which students will eventually investigate waves (the following movie was taken by the authors on an ocean beach: https://mediaplayer.pearsoncmg.com/assets/_frames.true/sci-phys-egv2e-alg-11-1-4).

Therefore, the main difference between these two habits of mind is that, while both a physicist and a physics teacher notice the physics in the world around them, the physics teacher immediately connects this physics to their students' learning.

What is needed for a person to develop these habits? First, they need to believe that this is the right thing to do (dispositions), then they need to know physics well enough to notice the instances and think of explanations, then they need to be skilled to record and edit the videos to clearly show the important details (e.g. finding the proper angle to make the circular waves most visible), and they need to be skilled into connecting those instances to student learning. However, dispositions, knowledge, and skills are not enough to develop productive teacher habits. The four magic conditions of existence of contextual cues, rewards, repetition, and obstacle removal need to be fulfilled and a supporting community needs to be present if we wish for these habits to persist. In this chapter, we focus on some of these habits without delving into how we can help future teachers develop them. The development is the subject of the rest of the chapters.

Etkina and colleagues (Etkina *et al* 2017) gave several other examples of physics teacher habits of mind (pp 010107-7):

- Treating all students as capable of learning physics and contributing to the generation of physics knowledge (as opposed to treating learning physics as a weed-out competition).
- Approaching problem solving as a physicist (napkin calculations, drawing a sketch before solving any problem, being able to do an order of magnitude estimation, being able to do a long calculation without a calculator just using powers of 10, etc).
- Using mathematics in a physics-specific way. Specifically, mathematics plays a different role in physics compared to other sciences. Physics is much less oriented towards statistics than biology and more oriented towards mathematical modeling and internal consistency of multiple representations than chemistry.
- Being aware of the 'surroundings' (nature, current events, breakthroughs in science or socio-scientific issues such as climate change, energy, health, etc) as a source of teaching and learning physics (e.g. visiting a garbage plant to take a video of an eddy current waste separator to use it in a lesson on electromagnetic induction) by building on the inherent ease of experimentation that physics affords; habitually thinking of how to use everyday objects and widely available modern technology (such as mobile phones and the Internet) to help students notice, wonder about, and learn something (e.g. stumbling upon a video on YouTube and immediately incorporating it into the lesson on the following day; dumpster diving—not passing by something in a dumpster that can be used for helping students learn physics through a conviction that cheap, readily-available materials can serve as the basis of a good physics lesson).

There are several ISLE-specific physics habits of mind of a physics teacher who teaches physics through the ISLE approach in addition to all of the habits listed above. The most important habit is thinking of physics as a process, not as a set of rules or a collection of facts. To develop this habit, a teacher needs to seek to understand how physics ideas have been constructed historically, which ideas were rejected, and which survived multiple testing. Basically, an ISLE-focused teacher needs to habitually see the development of physics through the ISLE lens.

In general, examples of such 'physics epistemological habits of mind include inductive (experiment-based) and 'spherical cow' reasoning, analogical reasoning, establishing causality, questioning claims, quickly assessing coherence of suggested ideas with the rest of the physics body of knowledge, and being able to spontaneously think of an experiment to test an idea when it is proposed (hypothetico-deductive reasoning)' (Etkina *et al* 2017). Similar ISLE-focused epistemological habits of mind include inquiring whether a particular historical experiment could be considered an observational, a testing, or an application experiment, whether there were any competing hypotheses explaining certain evidence, how those hypotheses were tested and so forth. We will focus on the analysis of the historical development of physics ideas through the ISLE lens in the interlude for chapter 4.

Another ISLE-specific habit of mind connected to the example above is habitually asking yourself when reading scientific papers or popular announcements of scientific discoveries on social media: What were the observational experiments? What were the competing hypotheses? What were the testing experiments? Did they do any application experiments? Did they identify and validate assumptions? It is interesting to read the articles about the work of recent Nobel laureates to find all those elements in the reports (see the article 'Pioneering quantum physicists win Nobel Prize in physics' by Wood (2022, p 1004)). In chapter 4, we will return to the physics habits of mind as they relate to physics teacher preparation and professional development.

Additional ISLE-specific physics teacher habits of mind include (although this list is not exhaustive):
- When planning a unit or a lesson, always start by choosing an appropriate 'need to know' (Knowles 1980). The 'need to know' is the code for something *that motivates students to learn*—a cool thing, an unanswered question, or whatever you think is going to motivate them. The 'need to know' is for motivating the students but NOT to elicit anything or to have any discussion (e.g. a person in a car struck by lightning—one of the coolest videos to create the need to know for the concept of electric field!).
- Making sure that among the goals of the unit/lesson, there are procedural (or scientific abilities-based goals)[3] as well as conceptual goals. Are the students going to learn how to use hypothetico-deductive reasoning? Are the students

---
[3] See the chapter on scientific abilities in the first ISLE-based book (Etkina *et al* 2019) and the original paper by Etkina *et al* (2006).

going to learn how to evaluate assumptions? Are the students going to learn to represent their ideas in multiple ways?
- Habitually asking yourself 'How do I know this?' or 'How did physicists figure this out?' before deciding how your students are going to learn the material.
- Habitually looking for the observational experiments and testing experiments for your students for any concept you plan for them to construct.
- Habitually choosing a system and the objects in the environment when using force, momentum, or energy frameworks to analyse physical phenomena (see more about systems in chapter 5).
- Habitually thinking of physics phenomena, physical quantities, and their relations using the language and symbolism adopted by the ISLE approach. Examples are: double subscript force notation, using the terminology of 'force exerted by object A on object B', differentiating between the terms 'conserved' and 'constant' when dealing with such quantities as momentum and energy, making a distinction between the term electric field and the physical quantity, $\vec{E}$. For more examples of ISLE-specific symbols and language, see the textbook CP: E&A and the IG that describe the usage and explain the reasons for it. See chapter 5 for more on the use of language.
- Habitually looking for natural phenomena that can be used to help students generate multiple explanations. One such example is a streak of alcohol put on a piece of paper. The streak dries and disappears slowly. How can this slow disappearance be explained? (See chapter 12 in CP: E&A.) Or you pour ice-cold water in a glass. The glass becomes wet on the outside. How can the appearance of the water be explained? (See chapter 1 in the ALG.) These two simple phenomena are familiar to the students and they probably know the scientific terms of 'evaporation' and 'condensation', but what they do not commonly do is ask themselves what those terms mean and *how* they know what is happening. For example, how do they know that the water comes from the air in the second example? We found that students can generate multiple plausible explanations for the phenomena described above and they are also capable of designing and conducting testing experiments for those explanations. It is the habit of ours to examine phenomena around us to find more of those simple ones that the students can hypothesize about without having any prior physics experience.

In chapter 5, we will return to the habits of mind in connection to physics teacher preparation and professional development.

3.3 Habits of practice

According to Etkina, Gregorcic, and Vokos (Etkina *et al* 2017) the habits of practice 'include (a) the habits that involve spontaneous decisions during lesson planning and (b) the habits, which enacted in the classroom, lead to student learning. The habits of practice are therefore intertwined with the habits of mind and cannot be separated definitively' (pp 010107-6). Below we list selected productive habits of practice for all physics teachers and after that, we will focus on the examples of habits of practice of the teachers who implement the ISLE approach.

- *Positioning students and yourself.* This habit involves the set-up of your classroom and your movements during the lesson. Traditional teaching involves students positioned individually in their desks facing the teacher and the teacher standing in the front of the classroom by a big board. Reformed teaching or student-centered teaching involves tables organized for group work. Equipment is set up on those tables or at the special locations around the room with small whiteboards for student groups to record their work. The teacher is usually invisible as they roam about the classroom listening to student conversations. When they talk to a group of students, they kneel to level with the students so that they do not project authority by towering above the group. The photos below (see figure 3.1) show such classroom set-ups and teacher behaviors, with a yellow arrow pointing to the teacher.

- *Listening (or not) to students.* This is one of the most difficult habits of any teacher. When students talk, do we focus on what they are saying and try to understand even when a student does not use correct language or do we plan what we are going to say/do next? Habitual focusing on what the students are saying is one of the important features of student-centered teaching (that is why it is called student-centered).

- *Helping students connect new ideas to their existing ideas and applying them to real-world phenomena.* This habit requires teachers to be familiar with student ideas. Such knowledge comes from the research literature (see, e.g., Driver and Warrington (1985), Wittmann *et al* (1999)). However, in physics and science education, there are two fundamentally different approaches to student ideas. One approach sees them as robust incorrect theories (called misconceptions) which need to be weeded out or removed from students' brains through careful instruction (the elicit–confront–resolve approach (McDermott 1990) or predict–observe–explain approach (White and

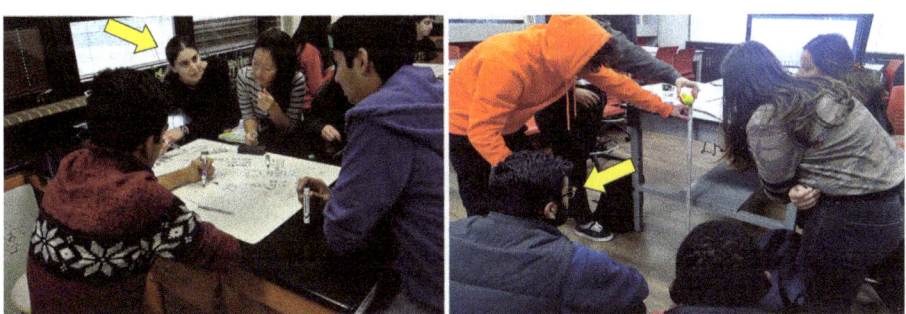

Figure 3.1. Students working in groups. Notice how the teachers are positioned.

Gunstone 1992)). A fundamentally different approach (to which the authors of this book subscribe) takes the view that students have pieces of knowledge, disconnected ideas, that they infer from everyday life experiences and the language that we use. Depending on the question and its context, they put these ideas together. While these ideas can be correct in one context, they would not be applicable to another context. However, we can always find productive ideas (or resources, see Hammer 2000, Hammer and Elby 2003, Hammer *et al* 2005) on which we can build new understanding. As we will discuss in chapter 5, it is impossible to 'remove' any incorrect ideas from a human brain. It is only possible to rewire connections between those ideas and new ones. In chapter 5, we will discuss this issue in more depth. Here, we only wish to make the reader aware that the habit of connecting new ideas to existing ones cannot be developed without a clear disposition concerning student ideas and the knowledge of those from the literature.

- *Approaching problem solving as a physicist.* When physicists try to solve a new problem, they do not start by searching for the right formula to plug in the givens. They draw a sketch of the situation and think of it conceptually first without searching for equations. They focus on the fundamental principles and not on the surface features. They solve equations symbolically before plugging in any numbers, and finally evaluate the reasonableness of the solution. As a teacher, one needs to cultivate this habit in oneself as a learner and then systematically develop it in the PSTs during their training.
- *Preparing each lesson while systematically focusing on three elements: goals, formative assessments, and sets of activities in which to engage students so that they achieve the goals.* As Etkina, Gregorcic, and Vokos wrote in their 2017 paper:

To develop this habit the future physics teachers need generic knowledge of instructional planning (units and lessons), but they also need to know what this means for specific lessons. Therefore, they need to be familiar with the documents that discuss the goals and assessment of physics instruction (e.g. NGSS Lead States 2013 and German National Education Standards 2004) and with the resources for physics activities and assessments, such as:
- PhET Interactive Simulations, https://phet.colorado.edu/.
- Physics Union Mathematics (PUM), http://pum.rutgers.edu/.
- American Modeling Teachers Association, http://modelinginstruction.org/.
- Diagnoser, http://www.diagnoser.com/.
- Assessment resources (FCI, CSEM, BEMA etc) https://www.physport.org/Assessment.cfm.
- Physics Instructional Resource Association (PIRA), https://pira.wildapricot.org/.
- European Physical Society National educational resources https://www.eps.org/page/edu_resources_nat.

- AAPT Resources https://www.aapt.org/resources/.
- PhysPort, https://www.physport.org.

In addition, the future physics teachers should have a strong knowledge of the curriculum as a map that allows students to build new ideas on what that they already know. For example, PSTs need to be able to explain why one needs to be familiar with the kinetic molecular theory to understand sound waves and why one needs to understand the concept of a system to learn energy. This is especially relevant given research results that show that the quality of new teachers is determined by the amount of experience they have with day-by-day implementation of instructional units in their preparation programs.

- *Changing (or not) the path of the lesson based on what the student said.* Assume that you listened to a student making a comment or asking a question. You can say in response that the comment/question will be addressed later or you can change the path of the planned lesson immediately to address the comment or the question. Treating your lesson plan flexibly and being ready to change it in response to a student is a sign of a truly constructivist approach to student learning.
- *Answering (or not) students' questions.* When a student asks a question, our first impulse is to answer it (when we know the answer). It might seem like a good habit, but providing an answer deprives the student and the rest of the class of an opportunity to construct the answer themselves. A good habit here would be to toss the question back to the class and ask them to work in groups to come up with possible answers before helping.
- *Responding (or not) to student ideas (correct and incorrect).* When a student gives a correct answer to a question, it is very tempting to validate it and confirm the answer. However, a different habit would be to toss the answer back to the class and ask if they agree. Another habit would be to ask the student who provided the answer to elaborate and argue why the answer is correct. When the answer is incorrect, again, it is a good habit to ask the student to explain their reasoning to check whether it is the language or the context that need to be tweaked. Another possibility is to accept the answer as correct for the time being and ask the student or the class how they could test it experimentally.
- *Helping students when they are stuck (leading or falling back).* Many teachers have a habit of asking leading questions to help students overcome a difficult situation. It is important to remember that this type of questioning follows the neuronal connections in the brain of the teacher. Their students might not have the same connections and, while they are able to follow the line of questioning, the big picture evades them. In other words, even if the students follow leading questions and eventually come to the correct answer, they will not construct the big picture/coherent knowledge about the topic. The leading questions often result in students trying to guess what the teacher has in mind rather than revising their own thinking. A better habit is 'falling back'

questions, which is when the teacher helps the students to go back to something that they know, from which they can make the conceptual leap themselves. We will come back to these types of questions in chapter 4.
- *Organizing (or not) group work (whiteboards for sharing, grouping students).* We recommend habitually placing small whiteboards on student tables and giving them dry erase markers of different colors (one color per student) to help start group work and establish accountability (see the whiteboards in figure 3.1).
- That being said, it is important to remember that the one-color-per student idea will not work when the colors are used for physics representations, such as blue for negative electric charges and red for positive. Additionally, it is important for the teacher to develop a habit of using different colors when drawing on the board. For grouping, it is important to remember to avoid situations when one female is placed with several males to ensure that all voices are heard. Having groups of mixed ability is also beneficial (Heller and Hollabaugh 1992).
- *Making sure that all equipment functions.* This is the habit that one of the authors (EE) did not have for many years. For a lab, she would check one set-up (in DC circuits, e.g., she would check the power supply, ammeters, voltmeters, light bulbs, connecting wires, etc), but she would not check *every single piece* of equipment. It led to the need to troubleshoot during the lab if a perfectly built circuit would not function properly. Was it a dead battery? A broken wire? A burnt ammeter? Since developing the habit of checking every single piece of equipment for every group, this problem went away. For whole class experiments, when there is only one set-up and the students come to the teacher's desk to observe it, a good habit is to prepare the set-up a few days (or better, a week) before, so that if some parts are missing or broken, there is enough time to order or build new ones.

The habits described above are productive habits of practice for all physics teachers who engage in student-centered learning. However, there are specific additional habits of practice that are common to teachers who implement the ISLE approach. All of the above habits apply to ISLE too. In the list below, we added relevant literature/resources where you can see examples of such habits in action.
- Habitually asking students who just watched an observational experiment to describe what they saw in simple words that a five-year-old can understand (see activities in chapter 12 in the OALG/ALG).
- Habitually asking the groups of students to list 'wild' ideas to explain what they observed or to discuss how to represent the data to find patterns (see activities in chapters 1 and 12 in the ALG).
- Habitually asking students to choose a system in order to analyse a process or a phenomenon of interest (see chapters 3, 6, 7, 9, 10, 12–20, 26–30 in the textbook CP: E&A).
- Habitually asking students to represent their knowledge in multiple ways and check for consistency of representations (see any chapter in the textbook CP: E&A).

- Habitually labeling forces with two subscripts (see chapters 3 and 4 in CP: E&A and any later chapter in which forces are discussed).
- Habitually asking groups of students to design experiments to test their ideas, explanations, patterns, relations, etc (see any chapter in the textbook).
- Habitually asking students to evaluate their answers, solutions, and/or experimental results themselves (see the problem solving strategy steps used in worked examples throughout CP: E&A textbook).
- Habitually asking students to represent numerical experimental findings or results of calculations as intervals—not as discrete numbers—with units (see Rubric G 'Data analysis rubric' in the rubrics for the development of scientific abilities at https://sites.google.com/site/scientificabilities/rubrics).
- Habitually asking students 'what should we do next' in the lesson (expecting them to follow the ISLE progression of elements).
- Habitually asking students 'How do you know this?' (see the Observational and Testing Experiment tables in every chapter of the textbook **CP: E&A**).
- Habitually asking students to reflect at the end of the lesson or unit about what they learned and how they learned it (e.g. see Etkina 2000, Harper *et al* 2003, May and Etkina 2002). This can be done through simple questioning of every student at the end of the lesson: 'Name one thing that you learned today and describe how you learned it.' Give students a minute to collect their thoughts and then go around and listen to each student with the rule that the next person cannot repeat what has already been stated previously. Another approach is to ask student groups to list on their whiteboards what they learned and how they learned it. Then, each group presents one thing until all lists are exhausted. Alternatively, such reflections can be a part of their homework. Research shows that productive reflection needs to be taught through continuous feedback (May 2002). As James Zull said, 'We need reflection to develop complexity. The art of directing and supporting reflection is part of the art of changing a brain.' (Zull 2002)

Note again, that to form any of these habits, the teacher needs to feel that it is the right thing to do, have knowledge of why it is an important thing to do, and the skill to implement this knowledge in practice.

Finally, to develop any of these habits, a teacher must also have one habit of practice that is vital for a teacher of any subject and of any age. It is the habit of proper time management. If you cannot manage your time outside of school, then you will not be able to prepare lessons well in advance. You will not be able to check the equipment and order or fix missing or broken parts. You will not be able to provide feedback to the students the day after they submit their work, when they still care about it. During the lesson, you will not notice that the time is close to the end and will miss out on having students reflect on what they learned. These are just a few examples of issues that might come up if a teacher does not possess good time management habits.

3.4 Habits of maintenance and improvement

According to Etkina, Gregorcic, and Vokos (Etkina *et al* 2017):

> habits of maintenance and improvement are the habits that involve continuous learning on the part of the teacher as an individual and as a member of the community, as she organizes her professional life to give priority to maintaining the community, actively sharing new findings, and using the findings of other teachers (p 010107-6).

There are two conditions for the development of these habits: availability of a professional community and time management habits that allow the teacher who has a community to actively participate in it. Such communities can be formal or informal.

Examples of formal communities are the American Association of Physics Teachers (AAPT), local chapters of the AAPT in different states and regions, the American Modeling Teaching Association (AMTA), the International Research Group in Physics Teaching (GIREP), the Latin American Physics Education Network (LAPEN), national societies of physics teachers around the world, and many others.

An example of an informal community is the Facebook group 'Exploring and Applying Physics' for those who implement the ISLE approach (see https://www.facebook.com/groups/320431092109343343). Another example is a community of the graduates of the Rutgers physics teacher preparation program (Etkina 2015).

From the above list of different communities, it is clear that without good time management habits it is very difficult for a teacher to find time for these activities. Therefore, reading physics teaching journals (such as *The Physics Teacher*, *Physics Education*, *American Journal of Physics*, *European Journal of Physics*, etc) and participating in professional communities needs to become a habit developed during teacher preparation. If the program itself creates a community of pre-service teachers and program graduates, then developing this habit becomes much easier. In addition, such a community ensures that it consists of like-minded people who share the same dispositions and foundational knowledge. This shared vision helps overcome initial fears that sharing one's difficulties will be seen as a weakness. For a teacher preparation program, having such a community is a huge advantage. It helps the leaders of the program to stay in touch with practicing teachers. It also helps them place their current PSTs for student teaching internships with like-minded teachers who are trained in the same method of instruction. It ensures continuous professional development for those who graduated and continuous feedback from the field to the leaders of the program. See an example of creating a community of graduates and the benefits of such a community in chapter 7.

While it is important for a teacher to be a member of a community, we hope that ISLE-minded teachers take it a step further—that they become leaders in their own local communities and in the larger formal communities. Examples can be serving as members on GIREP or AAPT committees, running local and national workshops, submitting articles to the *Physics Teacher* and *Physics Education*, serving as members in national exam committees and committees for curriculum development, participating in international projects, and many others. Recently, one of the ISLE physics teachers, Debbie Andres, was elected vice-resident of the American Association of Physics Teachers (AAPT) to become president a year later. This is a great example of leadership.

3.5 *Interlude*: Eugenia and Gorazd on habits and routines

Eugenia: While habits are the things that a teacher has, the existence of these habits leads to the creation of the routines for the students. A routine can be seen as a sequence of steps or procedures that your class follows on a regular basis. Here, I want to talk about how routines follow from habits, which in turn follow from skills, built on the knowledge and based on the dispositions. Specifically, I want to share a story that in retrospect is the logical progression of the elements above.

When I was a student in the Soviet educational system, one of the routines that I experienced was the following. At the beginning of a lesson, the teacher (any teacher, in fact), sitting at their front desk, scrolls through the list of names of the students in the class. Once the finger of the teacher stops at someone's last name, she says: 'Today to the board goes Etkina!' And Etkina, with her head down, gets up and slowly moves to the chalkboard, expecting the mental torture that she would have to endure in a minute. The rest of the class exhales: 'Whew, it's not me today! It is Etkina who will be questioned at the board, and in the meantime, I can talk to my friends or just relax in peace.' The teacher then asks Etkina to explain what an electric field is and how to find the quantity of the E field. Etkina, who is a good student, read the textbook at home and she happily draws electric field lines of a single charge on the board and states the definition of the E field. The rest of the class is supposed to be listening to her, but of course, the rest of the class are busy discussing their own lives. But Etkina knows that it is not enough to describe what she knows. After she finishes reporting, the teacher will ask her a question. It will be something new. And she might not be able to answer it. She knows it and she is scared in advance. And the question comes: 'What is the E field at the location right in the middle between two point-like charges with q_1 and q_2 that are separated by a distance d?' It seems like a simple question for those who are fluent with superposition principle, but Etkina only read about it in the textbook, and she does not know how to apply it. She does not even know that it is the superposition principle that she needs to use to answer the question. She stares at the board while thinking about her friend, the boy named Pete, who she wants to impress but she knows she is failing miserably. After about 4 minutes of struggling, she admits: 'I have no idea how to do it.' 'How disappointing,' says the teacher. She gives Etkina a four, which is not a bad mark, just one point below the highest—5, but Etkina knows that the teacher thinks that she is stupid and so does the rest of the class (at least those few who were listening). She slowly walks to her seat, hating the teacher and herself.

This is a true story. It happened to me over 45 years ago and I still remember the humiliation and anger I felt while walking to my seat. But that routine was common for all subjects and all teachers. In addition, the teachers would randomly announce a quiz at the end of the class and we had to write it, receive a grade, and this grade would affect the final grade in the course. I hated those quizzes. I wish I knew in advance that a quiz would be happening as I would have studied for it. And more, once I received the grade, there was nothing I could do to improve it. Even if I learned the material and could solve the problem that I could not solve yesterday, it would make no difference. Nobody cared.

Therefore, when I decided to become a teacher, I thought intensely about what kind of a teacher I wished to be. I did not know at that time that I was contemplating my dispositions, but now I know that those were exactly the dispositions that I struggled with. Based on my experience as a student, I made several disposition-like decisions (which I implemented every day of my 40 year long teaching life).

First, I was thinking about this 'going to the board' practice. Every lesson, a teacher would ask 3–4 students to the board, similar to the example I described above. The goal was clear—to assess what we knew and to give us grades. But what else did this practice achieve? First, it instilled fear in all of the students. For some teachers, we would 'calculate' who is going to be called tomorrow and those people would study a lot before that day. But some teachers called on random people. There was no pattern and everyone was scared to be called. The issue with being called was in those additional questions that would trip you even if you did the homework and studied, as if they were designed on purpose to humiliate you in front of the rest of the class. What about the rest of the class? This practice made them feel happy and relieved when somebody else was called to the board. It made people happy when somebody else was suffering. As you can see, it was not a good practice at all. So, I decided that I would *never* call anyone to the board unless a student expressed that they wanted to do so. I also thought that I would try to avoid all instances that make students fear physics or myself. And as fear comes from not knowing what awaits you, I thought that I would make physics lessons predictable.

I knew that a teacher could not help the students learn if they did not know where the students were at the moment. How would I know what my students knew and what they struggled with? I decided to give them daily short quizzes at the beginning of each lesson. Three–five minutes max. This way, students would know that the quiz is every day, but there are no trick questions. If you participated in class yesterday and you did your homework, then a '5' is guaranteed. But if you did not do well, there is still a chance to improve. Come to the physics classroom early in the morning before school starts and you will get a similar question to answer. The grade you get goes into my gradebook; the previous grade is crossed out.

To implement this system, I needed to develop certain skills. One was a part of lesson planning. In every lesson plan, I had to write a short quiz that would not take more than 3–5 min and would provide enough information for me to address students' difficulties. When the students came to class, I needed to develop a skill of actually having them spend only 5 min on the quiz. I abandoned taking attendance and used the submitted quizzes as a record of attendance. While the students were taking the quiz, they had their notebook open along with their homework so that I could walk along their desks and quickly check who did some work at home and who did not. Another skill I needed was the development of a grading system that would not eat up all my time. I had about 100 students every day—100 quizzes. I developed a skill of spending about 30–40 min per day grading them. Having high reading speed and focusing only on the important aspect of their answers helped a lot. Finally, the hardest skill was time management in the morning. Having one small child first and then two, it was difficult to juggle the morning routine to get them ready for daycare before I had to rush to work. It took a lot of planning and practicing to remove the morning stress and be at my desk by 7:45 am every morning (school started at 8:30 am, so I had about 40 min to spend with those who came to improve their previous work).

With time, I developed the following habits and routines related to formative assessments and the improvement of student work:
- Putting the text of the quiz into my lesson plans (habit).
- Printing out the quizzes for the students (routine).
- Grading the quizzes every day (habit).
- Coming to school at 7:45 am, 45 min before the school day started (routine).
- Meeting with students at 7:45 am so that they can improve their work (habit).
- Having the papers with the text of the quizzes on students' desks before the start of each lesson (routine).
- Keeping an accurate record of everyone's resubmissions (habit).

These habits led to the following habits and routines of my students:
- Doing homework after every lesson as they knew that there would be a quiz the next morning (habit).
- Starting each lesson with doing the quiz (routine).
- Trying to get to the physics classroom as quickly as possible because there would be more time for the quiz then (habit).
- Coming to school at 7:45 am if they decided to improve their work (routine).

As you can see, many of my habits and routines led to a few habits and routines of my students. These were not the only habits and routines that I developed. More are coming up in the following chapters.

Gorazd: I took over the physics education program after almost 15 years of work in experimental condensed matter physics. Like any traditional teacher, I viewed experiments as demonstration experiments (demos). I enjoyed designing new demo experiments and used them extensively in my lessons. During this time, experiments had two roles for me: a motivational role and an explanatory role. My view at that time was also reflected in the way I wrote articles in which I described the experiments I was developing.

I was mostly describing how I observed the phenomenon, how I arrived at the explanation, and how I 'proved' or verified the explanation. I almost never described several different explanations in the articles, although I remember having them while working on the experiment. I never described in the article how I disproved an explanation. I thought I only needed to show the correct explanation—the final, best product. I rarely predicted results based on explanations and I rarely designed another, independent experiment to measure the same physical quantity. I described how I experienced the experiments and their development as a creator and as a teacher. In these articles, I wrote about students mainly as my audience and used sentences such as: 'I will explain how I presented ... to the students', 'if you do this ... the students can see ....', 'you can show to the students ...', '... and here is a puzzle for the students', 'To give a correct explanation, students must understand ...'.

However, even then I was looking for ways to engage students more in the lessons, especially when conducting experiments. That is how I came across the predict–observe–explain (POE) method. At first, I was excited, but I soon realized that there was something wrong with the POE. The method only worked for some topics, but it did not allow me to extend it to others, especially in the introductory lessons, when the students were supposed to construct new knowledge. I also noticed that the method only worked at the beginning when it was something new for the students, and less and less later on as they got used to the method.

In my search for new ideas, I came across the ISLE approach, which immediately attracted me with a completely different view of the role of experiments in student learning compared to the ones I had known up until then. Although the division into three roles (observational, testing, and application experiments) looked completely natural and logical, I could not imagine how these roles would really come to life in the classroom. It was only later, when I observed Eugenia and her students do ISLE and when I started using the ISLE approach myself, that I gradually developed the habit of looking for and seeing these roles of experiments everywhere.

Of course, not everything went smoothly at the beginning. The most common mistake I made was that after the students proposed testing experiments, I proceeded to conduct those experiments and make judgments about the hypotheses. I skipped a very important step—asking students to make predictions about the outcomes of the experiments before conducting the testing experiment. Once I developed this habit, I realized that predictions (based on the hypotheses under test and not based on intuition) are the most exciting and motivating step for the students.

But the roles of the experiments, as set out in ISLE, only achieve their purpose when students work in small groups using whiteboards. And that was my next problem. Although I understood and saw the benefits of working in small groups, I initially had many concerns about having the whole class run this way. I was worried if all the students would see the board, would they even listen to me, how would they take notes, etc. Again, I was helped by Eugenia's example and encouragement, as well as by the students, who took the small group work very well and showed me that my worries were unnecessary. Now having students work in groups during my lessons is a habit. Organizing the tables with whiteboards for group work before each lesson is my regular routine.

There is another element of ISLE that I had considerable difficulty with at the beginning—student reflections about what and how they learned at the end of the lessons (see more about reflections in chapter 6). Reflections were something completely new to me. The only experience I had were my observations of reflections at the end of Eugenia's lessons. I usually attended her classes in January or February when the students were already used to reflections. These were in-depth reflections from which even I learned a lot. But when I tried to get my students in the physics education program to reflect on the lesson, it was completely different. They might praise some experiment or some new tool they learned about, but nothing deeper and nothing about teaching or learning—a far cry from what I saw with Eugenia's students. First, I blamed 'different cultures' but I also suspected that maybe I was doing something wrong, so I decided to observe Eugenia's classes again. I realized that I needed to perfect my questions (with which I encourage students) and my responses to their reflections. I also realized that students needed a certain amount of time to get used to reflecting. Eugenia's students were so good also because I observed them in the middle of the year. But I also realized that there are differences between the cohorts. Some students are more relaxed and don't find it difficult to speak in public, while some are more reserved. In such cases, the following trick (which I also took from Eugenia) helped: I ask students to first reflect in groups and write their ideas on whiteboards and then the whole class discussion starts. Today, I enjoy my students' reflections, which are as good as those in Eugenia's classes. And I realized that the 'different cultures' argument was empty. Encouraging students to reflect on their learning (and be successful in doing it) is yet another productive habit that I have developed.

References

Dewey J 1922 *Human Nature and Conduct* (New York: Henry Holt)
Driver R and Warrington L 1985 Students' use of the principle of energy conservation in problem situations *Phys. Educ.* **20** 171–6
Etkina E 2000 Weekly reports: a two-way feedback tool *Sci. Educ.* **84** 594–605
Etkina E 2015 Using early teaching experiences and a professional community to prepare pre-service teachers for every-day classroom challenges, to create habits of student-centered instruction and to prevent attrition ed C Sandifer and E Brewe *Recruiting and Educating Future Physics Teachers: Case Studies and Effective Practices* (College Park, MD: American Physical Society) pp 249–66
Etkina E, Brookes D and Planinsic G 2019 *Investigative Science Learning Environment: When Learning Physics Mirrors Doing Physics* (San Rafael, CA: Morgan and Claypool)
Etkina E, Gregorcic B and Vokos S 2017 Organizing physics teacher professional education around productive habit development: a way to meet reform challenges *Phys. Rev. Spec. Top. Phys. Educ. Res.* **13** 010107
Etkina E, Van Heuvelen A, White-Brahmia S, Brookes D T, Gentile M, Murthy S, Rosengrant D and Warren A 2006 Developing and assessing student scientific abilities *Phys. Rev. Spec. Top. Phys. Educ. Res.* **2** 020103
German National Education Standards Concerning Physics for Middle School Graduation 2004 *Bildungsstandards im Fach Physik für den Mittleren Schulabschluss Beschluss vom 16.12. 2004* (München: Luchterhand) https://kmk.org/fileadmin/veroeffentlichungen_beschluesse/2004/2004_12_16-Bildungsstandards-Physik-Mittleren-SA.pdf
Hammer D 2000 Student resources for learning introductory physics *Am. J. Phys.* **68** S52–9
Hammer D and Elby A 2003 Tapping epistemological resources for learning physics *J. Learn. Sci.* **12** 53–90

Hammer D, Elby A, Scherr R E and Redish E F 2005 Resources, framing, and transfer *Transfer of Learning From a Modern Multidisciplinary Perspective* ed J Mestre (Greenwich, CT: Information Age) pp 89–119

Harper K, Etkina E and Lin Y 2003 Encouraging and analyzing student questions in a large physics course: meaningful patterns for instructors *J. Res. Sci. Teach.* **40** 776–91

Heller P and Hollabaugh M 1992 Teaching problem solving through cooperative grouping. Part 2: designing problems and structuring groups *Am. J. Phys.* **60** 637–44

Knowles M S 1980 *The Modern Practice of Adult Education* (Englewood Cliffs, NJ: Cambridge Adult Education)

Lally P, Van Jaarsveld C H, Potts H W and Wardle J 2010 How are habits formed: modelling habit formation in the real world *Eur. J. Soc. Psychol.* **40** 998–1009

May D B 2002 *How Are Learning Physics and Student Beliefs about Learning Physics Connected? Measuring Epistemological Self-reflection in an Introductory Course and Investigating its Relationship to Conceptual Learning* (Columbus, OH: The Ohio State University)

May D B and Etkina E 2002 College physics students' epistemological self-reflection and its relationship to conceptual learning *Am. J. Phys.* **70** 1249–58

McDermott L C 1990 Millikan Lecture 1990: what we teach and what is learned—closing the gap *Am. J. Phys.* **59** 301–15

NGSS Lead States 2013 *Next Generation Science Standards: For States* (Washington, DC: National Academies Press)

Schoenfeld A H 1998 Toward a theory of teaching-in-context *Issues Educ.* **4** 1

Schoenfeld A H 2010 *How We Think: A Theory of Goal-Oriented Decision Making and Its Educational Applications* (New York: Routledge)

Schön D 1983 *The Reflective Practitioner: How Professionals Think in Action* (New York: Basic) 1st edn

White R and Gunstone R 1992 *Probing Understanding* (London: Falmer)

Wittmann M C, Steinberg R N and Redish E F 1999 Making sense of how students make sense of mechanical waves *Phys. Teach.* **37** 15

Wood C 2022 Pioneering quantum physicists win Nobel Prize in physics *Quanta Magazine* 4 October 2022 https://quantamagazine.org/pioneering-quantum-physicists-win-nobel-prize-in-physics-20221004/ (Accessed 7 May 2023)

Wood W 2019 *Good Habits, Bad Habits: The Science of Making Positive Changes that Stick* (London: Pan Macmillan)

Zull J E 2002 *The Art of Changing the Brain: Enriching Teaching by Exploring the Biology of Learning* 1st edn (Sterling, VA: Stylus)

IOP Publishing

The Investigative Science Learning Environment: A Guide for Teacher Preparation and Professional Development

Eugenia Etkina and Gorazd Planinsic

Chapter 4
ISLE and the development of physics habits of mind

4.1 Treating physics as a process and not as a set of rules

In order to implement the ISLE approach, a teacher needs to habitually think like a physicist[1]. One of the most important habits of such thinking is treating physics as a process—not as a set of rules. This habit relates to the physics epistemology. If we think about epistemology as the field of knowledge that investigates the elements of knowledge and how knowledge is constructed, then the habit of treating physics as a process is an epistemological habit. What does it mean to have physics epistemology?

First, let's consider the elements of physics knowledge. We could group the normative knowledge (the knowledge that our students need to develop independent of the level of a physics course that they are taking) into the following categories:
- physical phenomena and physical objects,
- models of phenomena, objects, systems, and interactions,
- physical quantities and their relationships,
- measuring instruments,
- physics devices,
- testing experiments,
- predictions of the outcomes of testing experiments,
- application experiments, and
- assumptions.

---

[1] We remind the reader that the abbreviation CP: E&A stands for the textbook *College Physics: Explore and Apply*, ALG stands for the *Active Learning Guide*, OALG stands for the *Online Active Learning Guide*, and IG stands for the *Instructor Guide*. Proper citations for these materials are in chapter 1.

If you think of *anything* that you know in physics, your knowledge would fall into one of those categories. We will analyse each of them separately and give examples of how to help teachers and students think about each as a physicist.

4.1.1 Physical phenomena and physical objects

Physical phenomena and physical objects are things that happen (exist) and can be observed directly or indirectly. Examples of phenomena include the mechanical motion of objects, waving of a string, water flow in a river, light shining on a surface, and clothes sticking to each other after being pulled from a dryer. When we observe physical phenomena, we conduct *observational experiments* with no expectations of the outcome. The goal of doing such experiments is to identify some pattern, which we will later model or explain. These observational experiments can be qualitative or quantitative, involve some measuring instruments or be done with our direct senses. When the students observe phenomena, it is important to ask the following questions:
- What did you observe?
- What data can you collect?
- How can you represent the data to find patterns?
- What are the important patterns?
- What patterns can we explain?

Here is an example of a physical phenomenon (see the video https://youtu.be/gT5_KYmOgKs and figure 4.1):
- The students observe the disturbance that they created and how it travels along the Slinky™ (a helical spring toy). They also observe each coil moving up and down while the disturbance propagates to the right.
- There are many patterns that they can find by varying different parameters. They can collect data concerning the speed of the propagation of the disturbance in various situations. They can change how taut the spring is, they can attach lead beads or small magnets to a spring, and they can also change the amplitude of the disturbance and measure the speeds.
- They can explain the propagation qualitatively by the interactions of the coils and use this explanation to account for different speeds for different conditions.

To help students conduct observational experiments with the ISLE approach, we have developed relevant self-assessment rubrics. They can be found at https://sites.google.com/site/scientificabilities/rubrics (Rubric B).

4.1.2 Models of phenomena, objects, systems, processes, and interactions

While the word 'model' is ubiquitous in educational vernacular, the definitions of models vary in textbooks and science education literature (https://plato.stanford.edu/entries/models-science/). In the ISLE approach, we adopt the definition of a model as a simplified version of a phenomenon, object, system, or process

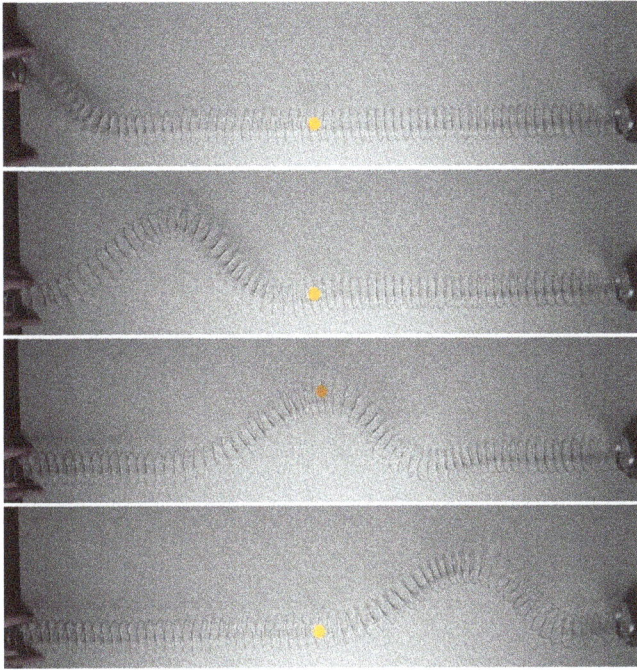

Figure 4.1. Four successive photos of a traveling transversal pulse on a Slinky™.

(Etkina *et al* 2006). Examples of models are free fall (the model of a phenomenon when you consider objects falling in the absence of air—a simplification) and point-like objects (a model of an object when you disregard the object's size—a simplification). Etkina and colleagues discuss different types of models including models of systems (ideal gas), models of processes (constant motion model, isobaric process, isothermal process), or models of interactions (electric field model and magnetic field model). It is possible to think of what we call hypotheses or explanations of phenomena as models because these are simplifications in some ways too.

Explanations can be causal or mechanistic. A causal explanation (a causal model) shows how one physical quantity depends on another quantity, but it is not concerned with a mechanism (for example, $F = G\frac{m_1 m_2}{r^2}$, see cause–effect relationships on the next page). A mechanistic explanation involves a mechanism explaining the relationship (for example, that liquids cool during evaporation as the fastest molecules leave and the average kinetic energy of the remaining molecules decreases). When students develop models, it is important to ask the following questions:
- What phenomenon, object, or system are you trying to simplify?
- What are your simplifications?
- How can you justify these simplifying assumptions? When is the model applicable?
- Is the model qualitative (if yes, is it causal or mechanistic?) or quantitative?
- What experiments can you conduct to test (and possibly reject) the model?

- What predictions can you make about the outcomes of these experiments using the model?
- What are the limitations of the model?

Here is an example of analysis of an ideal gas model using the above questions:
- We are trying to simplify real gases.
- The simplifications are that (1) the particles are identical point-like objects that have mass but negligible size, (2) they interact with each other and with the walls of the container only during collisions but not at a distance, and (3) their motion obeys Newton's laws.
- As the interactions of microscopic particles decrease dramatically when the distance between them is larger than their sizes, we can estimate that, for example, the average distance between the particles in air at normal conditions is about 30 times larger than their diameter. Therefore, their size is negligible and there is no interaction at a distance (see figure 4.2).

We know that at normal conditions one mole of gas occupies 22.4×10^3 cm^3. If we assume that this volume is equally divided between the gas particles (the cubes in the figure), then each particle is contained in a cube whose volume is D^3. We calculate this volume as follows:

$$D^3 = V_{\text{per particle}} = \frac{V_{\text{one mole}}}{N_{\text{particles in a mole}}} = \frac{22.4 \times 10^3 \text{ cm}^3}{6.02 \times 10^{23}} = 37.2 \times 10^{-21} \text{ cm}^3,$$

therefore

$$D = \sqrt[3]{37.2 \times 10^{-21} \text{ cm}^3} = 3.3 \times 10^{-7} \text{ cm}.$$

Recall that the typical diameter of an atom is about $d = 0.1$ nm $= 10^{-8}$ cm. Thus, the average distance between the particles in air at normal conditions is about 30 times larger then the diameter of the particles.

Figure 4.2. How we can justify simplifying assumptions when modeling air as an ideal gas.

- The model is both qualitative (in describing the mechanism) and quantitative because it allows for using Newton's laws to derive the expression for the pressure that an ideal gas exerts on the walls of the container ($P = \frac{1}{3}nm\overline{v^2}$).
- We can test this model to predict how the pressure of the same gas should depend on the volume or temperature. See the relevant experiments in chapter 12 of CP: E&A.
- The model is limited to rarefied gases. For example, a teacher should be able to realize that the ideal gas law cannot be used to solve the following problem (CP: E&A, p 382): 'A 5000-l cylinder is filled with nitrogen gas at 300 K and is closed with a movable piston. The gas is slowly compressed at constant temperature to a final volume of 5 liters. Determine the final pressure of the gas.'

There is another important point that we need to consider. It is the relationship between the models and multiple representations. We can think of a force diagram or an energy bar chart, for example, as an abstract model of a phenomenon of interaction of an object or a system with other objects. If the students view a force diagram or a bar chart in this way, they can answer the same questions about the diagram as listed above. Let's say the students need to model the phenomenon of dropping a tennis ball using a bar chart (to learn how to draw bar charts, use chapter 7 in CP: E&A) based on position-versus-time and velocity-versus-time graphs.

- First, we need to observe the phenomenon of dropping a tennis ball (figure 4.3(a)) and study the graphs (figures 4.3(b) and (c)). From the observations and from the $y(t)$ graph, we see that the ball rises to a height that is lower than the height from which it was dropped. From the $v_y(t)$ graph, we learn that the speed of the ball after the rebound is smaller than before it. From the $v_y(t)$ graph, we also see that the magnitude of the acceleration of the falling ball is smaller, and the magnitude of the rising ball is larger than $g = 9.81$ ms^{-2}.
- We can explain the different accelerations by drawing a force diagram for the ball (figure 4.3(d)). In order to do this, we need to realize that the air drag is not negligible and that the direction of the drag force is always in the opposite direction of the velocity.

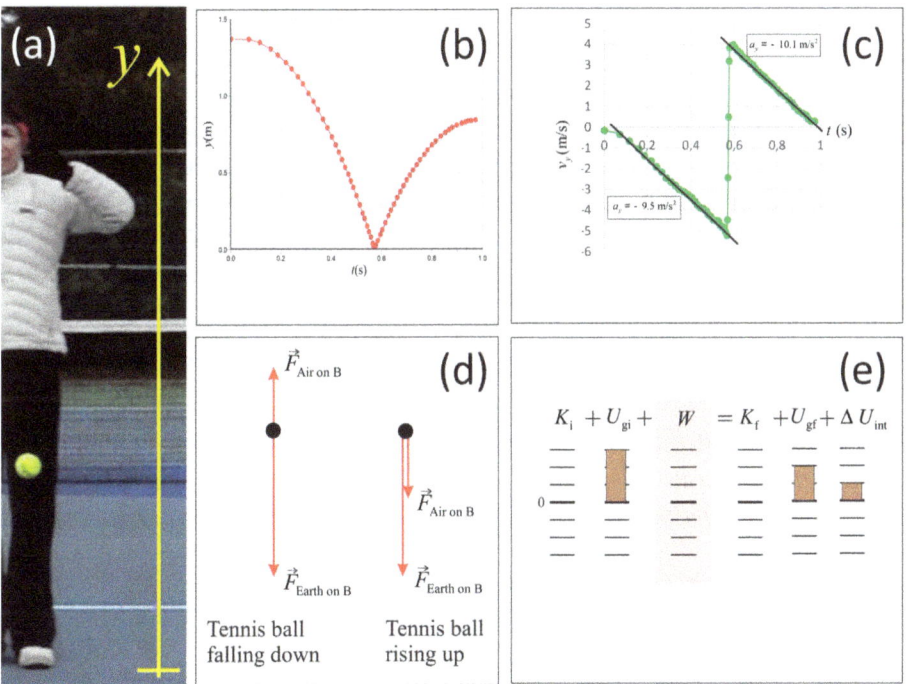

Figure 4.3. Multiple representations of a dropped and bouncing tennis ball.

- Now we are ready for the energy analysis. We can choose the ball, Earth, air, and the ground as the system. The initial state is when the person releases the ball and the final state is when the ball comes up to the highest point after bouncing from the ground.
- We simplify the situation by ignoring the motion of Earth. The simplification of Earth as stationary is easily validated as its mass is infinitely larger than that of the ball.
- The model is quantitative. According to it, some of the ball's initial gravitational potential energy is converted to the internal energy of the ball, air, and the floor during their interactions ($U_{gi} = U_{gf} + \Delta U_{int}$). The ball that was originally released from a height of 1.37 m rises to a new height of 0.84 m after rebounding against the floor. Knowing the mass of the ball (58 g), we can calculate the change of the internal energy of the system:

$$\Delta U_{int} = U_{gi} - U_{gf} = 0.058 \text{ kg} \times 1.37 \text{ m} - 0.84 \text{ m} \times 9.81 \text{ ms}^{-1} = 0.65 \text{ J}.$$

- This analysis also leads to predictions for the outcome of testing experiments. For example, what will happen if a person pushes on the ball at the beginning or what will happen if we repeat the same experiment in vacuum? We can see how an abstract model of an energy bar chart allows us to explain phenomena and predict new phenomena.
- The model is limited to the case when the distance between the ball and the ground is much less than Earth's radius.

4.1.2.1 Physical quantities

A physical quantity is a feature or characteristic of a physical phenomenon or a model that can be compared to some unit using a measuring instrument (see below). Examples of physical quantities are your height, your body temperature, the speed of your car, the force that Earth exerts on you, or the temperature of air or water.

Physical quantities that contain information about the direction of some quantity are called *vector quantities* and are written using symbols with an arrow on top ($\vec{F}, \vec{v}$). Force and velocity are vector quantities. Physical quantities that do not contain information about direction are called *scalar quantities* and are written using italic symbols (m, T, and K). Mass is a scalar quantity as are temperature and kinetic energy. Scalar quantities can be positive or negative.

It is important to recognize the difference between the operational definition of a physical quantity ($a_x = \frac{\Delta v_x}{\Delta t}$) and the cause–effect relationship ($a_x = \frac{\Sigma F_{on\,A x}}{m_A}$)—see the causal explanations above. An operational definition tells us how to determine a specific quantity, but does not tell us why it has a specific value as the variables in the definition cannot be changed independently (when we double the time interval, the change in velocity doubles). A cause–effect relationship tells us what the quantity depends on as we can change the variables independently (the mass of an object can be double without changing the net force exerted on it).

One of the important features of physical quantities is that they are not exact numbers with units (or points on a number line)—they are intervals. The value of the interval within which we know the quantity is predicated on the method that we used to determine it. An excellent document describing how to help students learn about experimental uncertainties in the ISLE approach was created by M Gentile and A Karelina and can be found at https://drive.google.com/drive/folders/1bYPf4GzCTtETFT7C9tz4g791MQpTQbdb. When the students are developing a new physical quantity, it is important to ask them the following questions:
- What phenomenon or model does this quantity describe?
- How do you define the quantity operationally?
- What are the units of the quantity?
- How do you know *if* this quantity depends on other known quantities? How *does* this quantity depend on other known quantities?
- How can we test this relation experimentally?
- How do we determine the quantity: can we measure it directly or do we need to calculate it using other measurable quantities?

Here is an example of analysing the physical quantity of acceleration:
- The quantity describes the changes in motion of an object. Operationally, acceleration is defined as $\vec{a} = \frac{\Delta \vec{v}}{\Delta t}$ where $\Delta \vec{v}$ is the change of velocity vector and Δt is the time interval during which this change occurred. Depending on the Δt, we can talk about instantaneous acceleration or average acceleration.
- The units of acceleration are s^{-2}.
- For an inertial reference frame observer, the acceleration of an object (or system) depends on its mass and the sum of forces exerted on it, $a_x = \frac{\Sigma F_{\text{on A}x}}{m_A}$ and similar for other components. We can test this cause–effect relationship using an Atwood machine set-up (see the video https://youtu.be/sUQPlAGbyMo) and predict how far the object on the left side will move in the first second after letting go of the object on the table.
- We can 'measure' the quantity using a simple accelerometer (see figure 4.4).

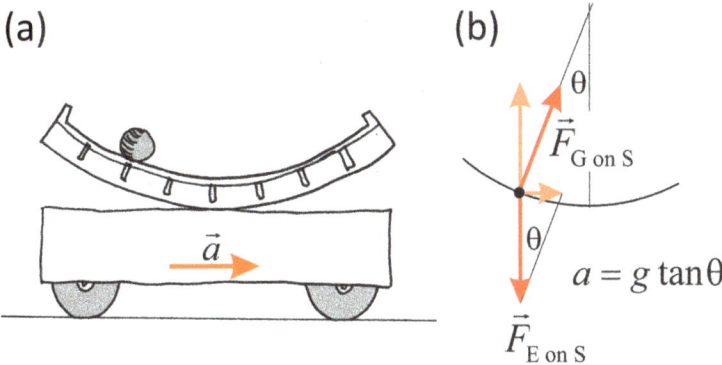

Figure 4.4. (a) A small sphere on a circular gutter can be used as a simple accelerometer; (b) using a force diagram for the sphere, students can derive the equation for its acceleration.

To help student answer the above questions, we have developed self-assessment rubrics: https://sites.google.com/site/scientificabilities/rubrics (Rubric G).

4.1.3 Measuring instruments

We use measuring instruments to measure physical quantities (such as a meterstick, a clock, an ammeter, etc). Some quantities can be measured directly using an appropriate device (time, length, electric current) and some can only be calculated using directly measurable physical quantities (kinetic energy, entropy). Here is the list of questions that the students should be able to answer about any measuring instrument:
- What physical quantity does this instrument allow us to measure?
- What are the physics principles behind the operation of the instrument?
- What are the rules of operation of the instrument?
- What are the safety procedures (if applicable)?

Below is an example of the analysis of a measuring instrument such as a spring scale to determine unknown forces (see figure 4.5):
- A spring scale allows us to determine an unknown force.
- The principles of operation of a spring scale are Newton's second law and superposition of forces. The scale shows the stretch of a spring which is calibrated in newtons (the spring needs to obey Hooke's law). The scale is used to balance an unknown force. For example, when we use the scale to hang an object at rest (zero acceleration), the forces that the spring scale and Earth exert on the object add to zero (superposition), or balance each other (Newton's second law, as acceleration is zero). Therefore, the magnitudes of these forces are the same and the reading of the scale is equal in magnitude to the force that Earth exerts on the object.

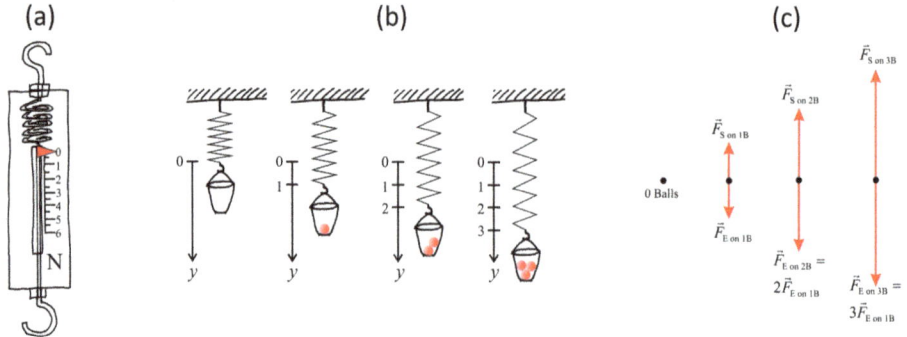

Figure 4.5. Using a spring scale to measure the force that Earth exerts on an object.

- To have an accurate measurement of an unknown force (in our example, the unknown force is the force that Earth exerts on the object), the scale and the object must have zero acceleration. To have an accurate reading of the scale, the person reading the scale needs to have their eyes at the level of the dial.
- Do not use the scale for any force larger than the largest force that can be seen on the instrument.

4.1.4 Physics devices

Physics devices are systems that people create as they develop an understanding of physics. Examples are a motor, a generator, a mirror, a lens, a battery, and so forth. When the students encounter a new physics device, it is important to ask the following questions:
- What is the name of the device?
- What are the main parts of the device?
- What are physics principles/relations that govern the operation of the device?
- What are the safety procedures?

As an example, we can consider a Van de Graaff generator (see figure 4.6). Here are possible answers:
- The name of this device is a Van de Graaff generator.
- The main parts are a metal dome, a rubber belt, two rollers made of different material, and two metal combs.
- The principle of operation is charging by separating two objects made of different materials (triboelectric effect): as the rubber belt runs between the two rollers that are made of different materials, electrons move from one

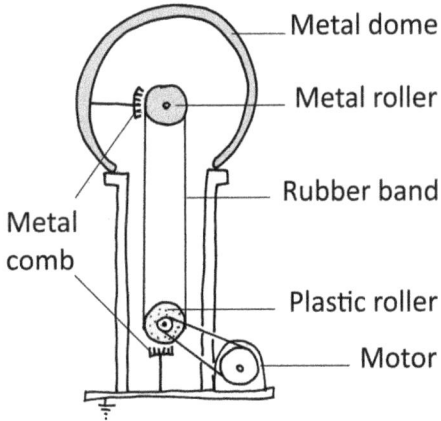

Figure 4.6. A Van de Graaff generator.

roller to the rubber, and then from the rubber onto the other roller, which is inside the metal dome.
- Only use the instrument in dry rooms in which there is no risk of explosion.

4.1.5 Testing experiments

These are experiments specifically designed to test a model, hypothesis, and/or explanation. Testing experiments are different from observational experiments. However, testing experiments also involve physical phenomena as we predict the experiment's outcome using the explanation under test before conducting the experiment. An example of such a testing experiment can be the famous Galileo's inclined plane experiment when he was testing his hypothesis that all objects fall at constant acceleration without air resistance.

Here are the questions that will help our students to design and conduct testing experiments:

- What is the model, hypothesis, and/or explanation to be tested?
- Brainstorm possible experiments whose outcome you can predict using the model, hypothesis, and/or explanation under test. What quantities can you measure and what quantities can you calculate?
- Make predictions about the outcomes of the experiments that you designed (before conducting them) as if the hypothesis that you invented is correct. How does the prediction follow from the hypothesis? Explain your thought process.
- What additional simplifying assumptions are you making in your predictions?
- How might these assumptions affect the predicted outcome—will they make the result smaller or larger than expected?
- After you conducted the testing experiment, how do you know whether the prediction and outcome match or do not match?
- What is your judgment about the model, hypothesis, and/or explanation under test?

We will analyse Galileo's inclined plane experiment as an example.

Galileo was testing his hypothesis that when objects fall freely, their speed increases proportionally to time, in other words $v = at$. As he could not measure the speed directly and the time of fall was always rather short for the times when people did not have handheld watches, Galileo decided to measure the distance instead. He designed an experiment where a small ball rolled down an inclined plane and he made a prediction of what he should observe if his hypothesis was correct. While he used graphical methods to make the prediction, we can use our modern mathematical methods to do it: if $v = at$ (where a is the coefficient of proportionality) then the average speed is $at/2$. Therefore, the distance traveled during time t is $\frac{at^2}{2}$. From this reasoning follows that the distance increases as the square of the time for an object that starts from rest. Therefore, if the ball covers a distance of d during the first

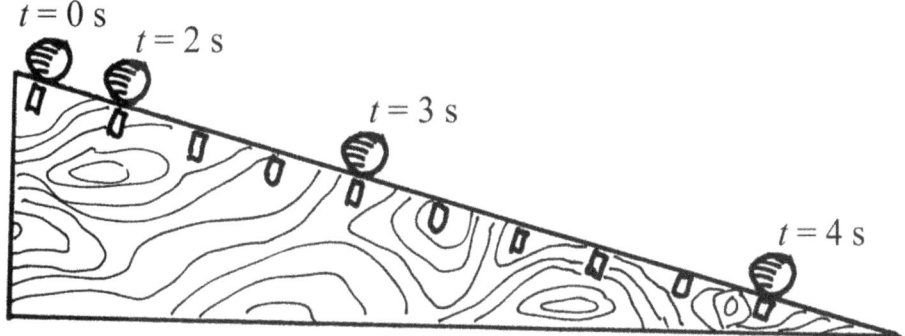

Figure 4.7. Galileo's testing experiment.

second, in 2 s it would cover a distance of 4d, in 3 s it would cover a distance of 9d and so forth (see figure 4.7).

Galileo could measure these distances and compare them to those that he predicted before running this testing experiment. He also assumed that the motion of a rolling ball down an inclined plane is similar to the motion of a falling object. It is not clear how Galileo validated his assumption, but today we can do multiple testing experiments for his hypothesis that for a falling object without air resistance, $v = at$. We can use a motion detector to obtain a position versus time graph or design an indirect experiment with a string and lead beads (fishing weights) as described in figure 4.8. Before running the experiment, students can predict that the sound pulses produced by the falling beads that hit the ground will be separated by equal time intervals. You can see the outcome of the experiment at https://youtu.be/FY0pmUDHXKU.

To help students learn to design testing experiments, we have developed a set of rubrics that can be found at https://sites.google.com/site/scientificabilities/rubrics (Rubric C).

4.1.6 Predictions

What we have in mind are predictions of the outcomes of the testing experiments (not to be confused with hypotheses or models). A prediction is a statement of the outcome of a particular experiment (before you conduct it) based on the hypothesis being tested. It says what should happen in a particular experiment if the hypothesis under test is correct. A prediction is not a guess. Without knowing what the experiment is, one cannot make a prediction. A prediction is *not* equivalent to a hypothesis, but should be based on the hypothesis being tested. In the above example of Galileo, the hypothesis being tested was that the velocity of the objects is proportional to the duration of the fall. For the same experiment, the prediction that was based on this hypothesis was that when the objects roll down the ramp, the distance that they cover is proportional to the time squared. Rubrics for the

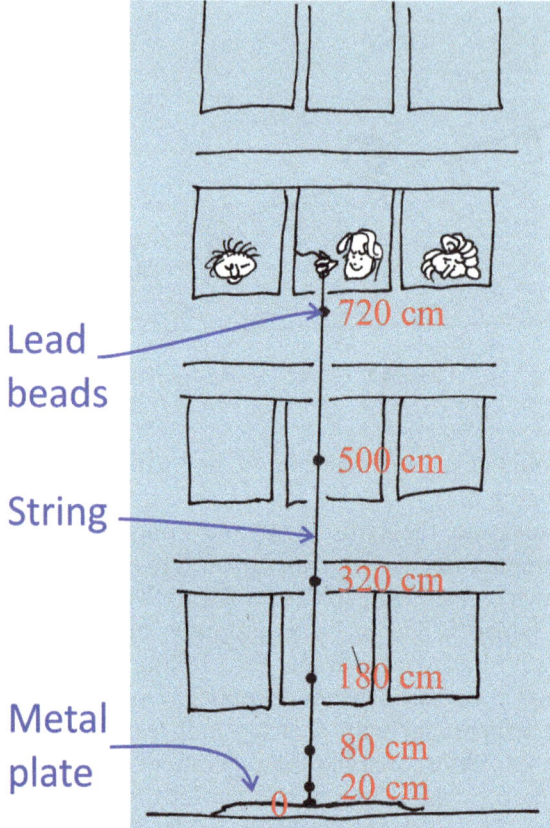

Figure 4.8. Fix lead beads on an 8 m string at the following distances from one end of the string: 0, 20, 80, 180, 320, 500, and 720 cm. Hold the string from a window so that the bead at 0 cm is touching the ground. Release the string and listen to (and record) the sound produced by the lead beads hitting the ground. You can make the sound louder by letting the beads fall on a metal plate.

development of the ability to make predictions are at https://sites.google.com/site/scientificabilities/rubrics (Rubric C).

4.1.7 Application experiments

These experiments are different from observational and testing experiments because, to design them, one needs to put several tested models together to achieve a specific goal. Application experiments make a bridge between physics and engineering. The goal can be the measurement of a physical quantity (How many significant figures of g do we know? How much energy does a square meter of Earth receive from the Sun? What is the coefficient of static friction between your shoe and the carpet?) or to build a specific device (to power an LED without burning it, or to build an electric motor, a telescope, etc). Here are the questions that help students design and analyse application experiments:

- What is the problem that you are trying to solve?
- What experiment can you design to solve the problem?
- What equipment will you use? What physical quantities will you measure and how will you measure them?
- What is the mathematical procedure that will allow you to use the measured quantities to solve the problem?
- What additional assumptions are you making? How can you validate them?
- How will you evaluate the results of your experiment?
- What independent experiment can you design to solve the same problem?
- After you compared the results of the two experiments, are the results consistent with each or not? How do you know?

The following example shows what students do when they conduct application experiments (full text of the lab and relevant rubrics are at https://drive.google.com/drive/folders/1wSTN5mYDzBFGN-VNVRCqrMw1eGTP_Ve2). The lab is called Newton's laws and circular motion.

Title: Application Experiment: Coefficient of friction between the shoe and the floor tile

Brainstorm two independent experiments to determine the maximum coefficient of static friction between your shoe and the sample of floor tile provided. Once you have done this, call your TA over and discuss your experiments with them.

Available equipment: Shoe of your choice, spring scale, ruler, protractor, floor tile, tape, string, digital balance, and an assortment of objects of different masses.

Include the following in your report (a–g are for each experiment; h–i are for after you have performed both experiments):

  a. Describe your experimental procedure. Include a sketch of your experimental design. Explain what steps you will take to minimize experimental uncertainty.
  b. Decide what assumptions about the objects, interactions, and processes you need to make to solve the problem. How might these assumptions affect the result? Be specific.
      Considering each of the relevant assumptions, separately evaluate the effect that making each assumption will have on the results. For example, evaluate how the coefficient of static friction will change if you pull the shoe at an angle of 50° above the horizontal, rather than exactly horizontally.
  c. Draw a force diagram for the shoe (recall your assumptions). Include an appropriate set of coordinate axes. Use the force diagram to devise a mathematical procedure to determine the coefficient of static friction.
  d. What are the sources of experimental uncertainty? Which measurement is the most uncertain? How did you decide?
  e. One of the assumptions you have likely made is that the coefficient of friction does not depend on the normal force that the surface exerts on the shoe (it's okay if you didn't come up with that on your own). In order to determine if this assumption is reasonable, perform a quick experiment on the side to evaluate the assumption. Decide if the assumption is or is not reasonable. Explain how you made your decision.

f. Give two different methods for measuring an angle with the available equipment. Which method is likely to provide a result with less uncertainty? Explain your reasoning.
g. Perform the experiment and record your observations in an appropriate format. What is the outcome of the experiment?
h. When finished with both experiments, compare the two values you obtained for the coefficient of static friction. Taking into account experimental uncertainties and the assumptions you made, decide if these two values are consistent or not. If they are not consistent, explain possible reasons for how this could have happened.
i. Describe the shortcomings you noticed in the experiments. Suggest specific improvements.

The self-assessment rubrics to help students develop abilities necessary to design, conduct, and interpret application experiments can be found at https://sites.google.com/site/scientificabilities/rubrics (Rubric D).

4.1.8 Assumptions

Assumptions, as they are understood by physicists, are the things that we choose to ignore (the air resistance, roughness of surfaces, mass of pulleys, rotation and velocity of Earth, etc). The ISLE approach activities encourage students to articulate assumptions and evaluate their effect on the predicted outcome of a testing experiment. The prediction is made based on the hypothesis under test, but additional assumptions that the prediction does not take into account might make the prediction larger or smaller that the experimental outcome. Evaluating the effects of additional assumptions is crucial for testing models and hypotheses. To learn more about assumptions, read the following document written by D T Brookes at https://docs.google.com/document/d/0By53x8SYAF1lbVN5Sk9PdS14Nmc/edit?resourcekey=0-4lqoBJMhnvAk6kivtSoSLg#heading=h.gjdgxs

Above, we cited scientific abilities rubrics to help students navigate the elements of physics. The rubrics can be seen as specific tools for helping students operate with most of those elements effectively. In our first book on the ISLE approach (Etkina *et al* 2019), all of chapter 5 was dedicated to the abilities and rubrics. The work on rubrics is described in several papers (Etkina *et al* 2006, 2008, 2009, 2010, Buggé and Etkina 2016, 2020). The rubrics are available at https://sites.google.com/site/scientificabilities/rubrics. The rubrics help students design observational, testing, and application experiments, to develop explanations/hypotheses, to make predictions, to collect data and determine physical quantities, to evaluate assumptions and uncertainties, and to communicate.

4.1.9 How are the above elements connected to the physicist habits of mind?

When a physicist encounters a new phenomenon or reads a research paper, they habitually ask themselves the questions listed above for each of the elements. Perhaps they do not ask all of the questions at once, but all of them eventually. It would be great if physics teachers not only ask themselves these questions habitually, but engage the students in answering these questions when they encounter new ideas in the course. This practice might prevent students from seeing physics as collection of facts and instead allow them to see the above elements as the building blocks of physics, similar to the bricks, windows, banisters, toilets, sinks, etc, as building blocks of a house. One way to do it is to make posters in the classrooms with the questions to be answered about each element and when the students encounter a specific element, focus on the questions and the similarity of all representatives of such element. For example, when the students learn a new physical quantity, they should habitually ask themselves its operational definition, units, the instrument to measure it, and so forth.

We can think of the elements discussed above as the building blocks of physics. How do physicists put them together or use them to develop the huge body of knowledge that our students need to learn? In their paper introducing the habits framework, Etkina, Gregorcic, and Vokos (Etkina *et al* 2017) give the following examples of reasoning used by physicists:

> inductive (experiment-based) and 'spherical cow' reasoning, analogical reasoning, establishing causality, questioning claims, quickly assessing coherence of suggested ideas with the rest of the physics body of knowledge, and being able to spontaneously think of an experiment to test an idea when it is proposed (hypothetico-deductive reasoning). (p 010107-7)

Below we will elaborate on some of these examples and connect them to physics teacher preparation and professional development.

4.1.10 Inductive reasoning

In physics, inductive reasoning means finding patterns in the experimental data, constructing models and/or explanations of collected data, and making generalizations and extrapolations based on data (Copi *et al* 2006). This type of reasoning is crucial for the implementation of the ISLE approach as observational experiments, their analysis, and inference of patterns is the first step in the student construction of any concept.

An example of inductive reasoning could be moving a magnet with respect to a coil connected to a galvanometer and deducing the pattern inductively.

Activity for the students

Observational experiment. Your group has the following equipment: a whiteboard, markers, a coil with several turns, a bar magnet, and an analogue galvanometer.

    a. Examine the equipment that you have on your desk (see figure 4.9). The galvanometer registers current through the coil. It needs to be connected directly to the coil (note, there is no battery). Now that you have connected the galvanometer to the coil, work with your group members to find out what you can do to make the galvanometer register a current

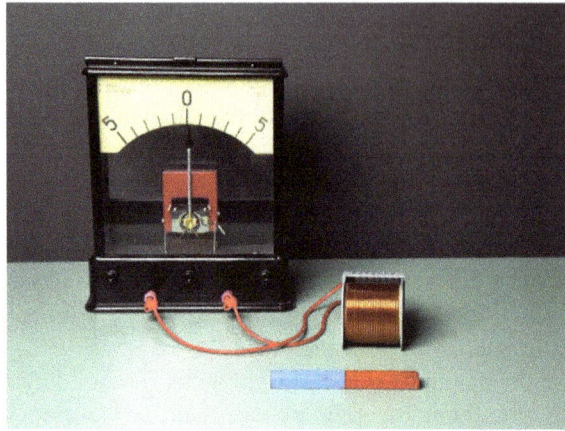

Figure 4.9. A galvanometer, coil, and magnet for observational experiments.

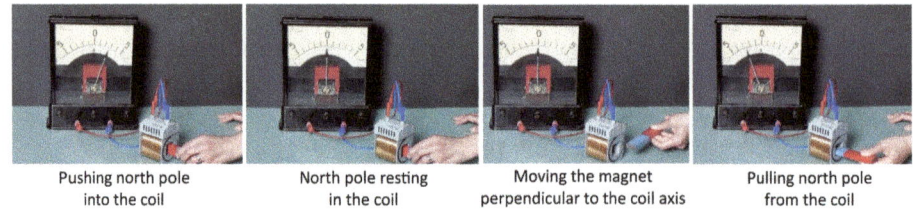

Figure 4.10. Situations which lead or do not lead to an induced current.

through the coil. Once you figure out one way, look for other ways so that at the end you can formulate a pattern for the cases in which the current is induced. Describe your experiments and findings with words and sketches.
b. Develop a rule: Devise a preliminary rule that summarizes the condition(s) needed to induce a current in a coil.

The students conduct a series of experiments and find that for certain motions the current is induced through the coil (see the series of photos in figure 4.10), but for other motions and when the magnet is at rest with respect to the coil, there is no current through the coil.

This finding is a pattern, but it can be generalized to form a hypothesis. The generalization (hypothesis) would be that one needs to have relative motion between a coil and a magnet to induce a current in the coil. We can think of this hypothesis as a causal explanation. It connects the cause (motion of a magnet) to the effect (the appearance of the induced current).

This was a qualitative example. To see how we can help students develop quantitative inductive reasoning, we use an example from electrostatics. Imagine that your students are investigating interactions of electrically charged objects. You do not have equipment for them to collect quantitative data, so you offer them the following activity, which helps them to construct Coulomb's law.

Activity for the students

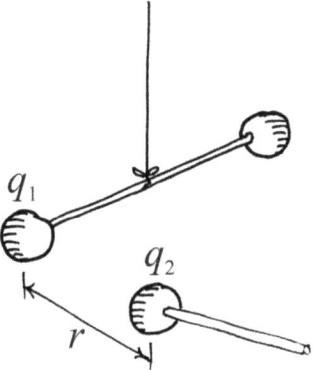

The figure above shows a schematic of the experimental set-up that Charles Coulomb used to determine the force that one charged ball exerts on another charged ball. He wished to find how the force that two electrically charged objects exert on each other depends on the magnitudes of the charges and on their separation. Coulomb did not have tools to measure the absolute magnitude of the electric charge on the metal balls. He used an ingenious method that helped him estimate the relative charges. He divided charges on the balls in half by touching a charged metal ball to an identical uncharged ball. The table that follows (table 4.1) provides data that resembles what Coulomb might have collected. Represent the data graphically by collaborating with your group members on a whiteboard. Discuss with your group: which are the independent variables and what is the dependent variable in Coulomb's experiment? Then, analyse the changes in the dependent variable as you change only *one* independent variable at a time. Use this analysis technique (controlling variables) to find patterns in the data and devise a mathematical relationship based on these observations. Put your final mathematical representation on a whiteboard and share it with another group.

Notice that the students need to graph the data separately for each variable (see figure 4.11) to infer the pattern of the force being directly proportional to the product of the electric charges and inversely proportional to the square of the distance between their centers. If the students do this activity, they would determine exactly from *where* Coulomb's law originates.

As you can see from the above examples, inductive reasoning is used for the analysis of the observational experiments. The elements of physics used in inductive reasoning are the phenomena (observational experiments), physical quantities, and models/explanations/hypotheses. Sometimes two other elements are used too—measuring instruments and physics devices, but those are not always necessary.

Table 4.1. Inductive reasoning to develop Coulomb's law (adapted from CP: E&A, p 152).

Charges (q_1, q_2)	Distance r	Force F
1, 1 (unit)	1 (unit)	1 (unit)
1/2, 1	1	1/2
1/4, 1	1	1/4
1, ½	1	1/2
1, ¼	1	1/4
1/2, ½	1	1/4
1/4, ¼	1	1/16
1, 1	2	1/4
1, 1	3	1/9
1, 1	4	1/16

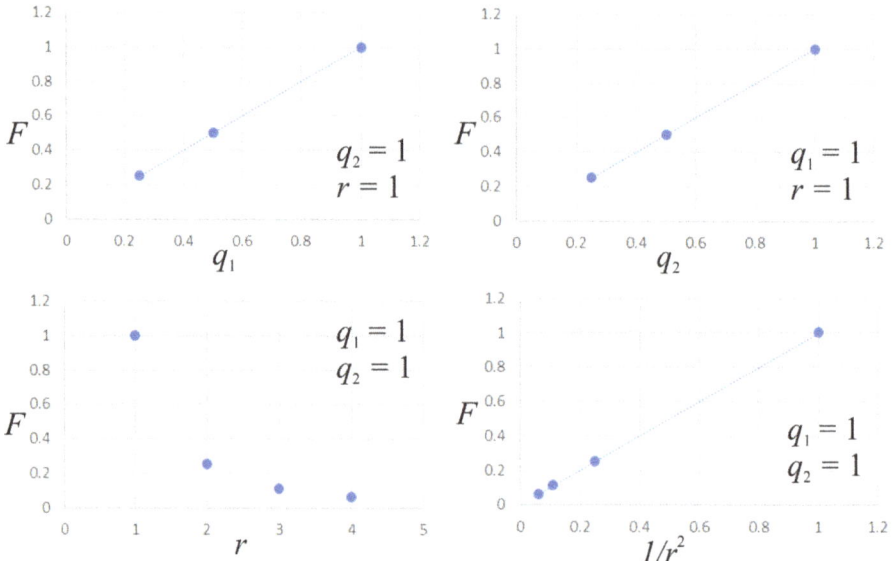

Figure 4.11. Graphical representations of Coulomb's data to help students find a quantitative pattern using inductive reasoning.

We have developed specific language for the prompts that would elicit students' qualitative and quantitative descriptions of phenomena. Although we listed some of them when discussing phenomena/observational experiments, some prompts are new:
- Describe what you observe using simple language that a five-year-old will understand.
- Describe what you observed without trying to explain.
- Describe the experimental set-up with words and with a sketch.

- Decide what is/are the independent variable(s) and what is the dependent variable. Represent the data graphically.
- Find a pattern in the graph.
- Represent the pattern mathematically.
- Devise an explanation for the pattern (or devise a hypothesis explaining the pattern).

When students collect data, there is an issue of experimental uncertainty. We will return to this issue later in the chapter.

4.1.11 Spherical cow reasoning

Wikipedia defines spherical cow reasoning as:

> a humorous metaphor for highly simplified scientific models of complex phenomena. Originating in theoretical physics, the metaphor refers to physicists' tendency to reduce a problem to the simplest form imaginable in order to make calculations more feasible, even if the simplification hinders the model's application to reality. (Wikipedia n.d.a)

Although Wikipedia considers the spherical cow metaphor humorous, it lies at the heart of physics—simplifying an incredibly complicated world to be able to explain and predict it. While the language of spherical cows is not used in education, we engage our students in 'spherical cow' reasoning every time we ask them to ignore friction, air resistance, the curvature of Earth's surface, the dependence of g on the distance from the center of Earth and so forth. We teach them that Newton's laws are only applicable in inertial reference frames while being on a rotating and revolving Earth means that we, as observers, are clearly in a non-inertial reference frame. We do it as the effects of the above factors are small compared to the main factors affecting the phenomena and, if ignored, we still arrive at predictions supported by the experiments. The problem is that our students (and even teachers) will very often not be aware that a particular approach ignores some factor. A great example is the formulation of Newton's first law in most American introductory textbooks. Here are two examples:
- If the net force on an object is zero, the object does not accelerate and the velocity of the object remains constant. If the object is at rest, it remains at rest; if it is in motion, it continues in motion in a straight line at constant speed. (Stewart *et al* 2019).
- Every object continues in its state of rest, or of uniform motion in a straight line, unless it is compelled to change that state by forces impressed upon it (Hewitt 2014).

This wording leads the students to believe that both statements are always true. But if an observer is accelerating themselves, then this law does not work.

We teach our students Newton's laws while often forgetting their biggest limitation—the observer! The second and third laws do not work for most observers on Earth and our students experience this every day when they are on a school bus or in a car going to school. They feel being pushed forward and backward without any extra objects interacting with them when the bus starts or stops. They see other objects accelerating without any extra forces exerted on them. How often do we bring these experiences in class? The reason for them is that when you are an observer in an accelerating reference frame, Newton's second law does not work. And therefore, the role of the first law is to limit the observers to only those who do not accelerate themselves. This nuanced meaning of the law is often lost in a traditional statement about *every object continues, blah, blah, blah*. No, every object does not continue. It only does if the person observing this object is not accelerating themselves. We can see Newton's first law as the statement of the existence of the inertial reference frame observers. Sounds wild, right?

The above example of Newton's first law shows the importance of discussing with students what simplifications they are making when developing explanations/models or solving problems. Helping physics teachers develop the awareness of the ever-present spherical cow reasoning is infinitely more important.

4.1.12 Analogical reasoning

Analogical reasoning is often used when we wish to explain a new phenomenon using something that we know by using the word 'like'. An analogy is a cognitive device and should not be confused with a metaphor, which is a figure of speech. When the word 'like' is dropped, an analogy becomes a metaphor. For example, we can say that time is precious like money is. In this case, money is an analogy for time. But when we say 'don't waste my time', we do not refer to money as an analogy, but we speak about time as we would speak about money.

When we make analogies, we are trying to understand or explain some new phenomenon using our knowledge of a familiar phenomenon. The new phenomenon is called a 'target' and the known phenomenon is called a 'base'. In a way, making an analogy is mapping the objects and their relationships in a target to the known objects and their relationships between objects in the base (Glynn *et al* 2012).

For example, a common analogy is to say that electric charge flows in a circuit (target) like water flows through a closed pipe system (base) (Gentner and Gentner 2014) with the battery (object) being analogous to the pump (object). While a battery does not look like a pump, its relationship to other parts of the electric circuit is like the relationship of a pump to the parts of a closed water system. There are two important habits of mind when using analogical reasoning:

    1. We should not forget that every analogy has limitations or situations when the analogy is no longer valid. For example, a battery provides a constant voltage for a wide range of resistances including the case when the resistance is infinite (an open switch). The pressure difference provided by the pump on the other hand strongly depends on the resistance of the system to flow and it increases significantly when the flow is stopped (a closed valve). The

electrical resistance of the wires is inversely proportional to the cross-sectional area of the wire while the resistance of the tube to liquid flow is inversely proportional to the square of the cross-sectional area of the tube (Poiseuille law).
2. Research shows that it is much more effective when the students come up with analogies to help them understand something instead of the instructor providing them with analogies. This is because sometimes what is a base for the teacher is not familiar to the students (Gentner and Gentner 2014).

Research also shows that in physics, analogies with time become metaphors (Brookes and Etkina 2007). For example, when physicists were first figuring out what electric current was, they had a model of a weightless fluid of positive electric charge moving through the circuit. Then, as water is also a fluid, it became an analogy for the movement of the electric charge. But now we simply say 'current flows', which is a metaphor based on the original analogy.

Analogical reasoning plays an important role in the ISLE process. When the students are developing a mechanistic explanation of a phenomenon, they almost always use analogical reasoning. Such reasoning allows them to relate the new explanation or mechanism to something that they already know.

For example, in the case of explaining how a disturbance can propagate along a string, they come up with an analogy of multiple springs strung together. When one spring is pulled, it pulls on the next one, and so forth. This is how the disturbance spreads along the string without carrying any material.

When preparing future physics teachers, it is important to encourage them to come up with their own analogies and when they do so, ask for the mapping of objects, relationships, and most importantly, the limitations. Habitual searching for analogies is a common habit of mind of physicists.

4.1.13 Hypothetico-deductive reasoning

We already touched upon this reasoning when we discussed testing experiments and predictions in the ISLE process. According to Encyclopedia Britannica, the hypothetico-deductive (H-D) is a

> procedure for the construction of a scientific theory that will account for results obtained through direct observation and experimentation and that will, through inference, predict further effects that can then be verified or disproved by empirical evidence derived from other experiments. (Encyclopedia Britannica n.d.)

Galileo used the H-D method when he made predictions of the outcome of his inclined plane experiment and then made a judgement about the motion of falling objects based on the outcomes of the experiment. Arthur Eddington used H-D reasoning when he predicted the deflection of the position of stars using Albert Einstein's newly proposed general relativity. H-D method used in science is crucial as is separates science from religion. When a scientist proposes a hypothesis, even if it explains all known phenomena, the hypothesis needs to be potentially falsifiable through some experiment. If there is no experiment that can potentially reject the hypothesis, the hypothesis is not scientific.

The H-D method is used in the ISLE process when we make predictions of the outcomes of the testing experiment using a hypothesis under test and then make a judgement about the hypothesis based on the outcomes of the experiment. This is how the logical progression works: *If such and such* [hypothesis] is true **and** I do *such and such* [testing experiment], **then** *such and such* should happen [prediction] **because** [explicit connection between the hypothesis and the prediction]. **But**, *such and such* [prediction] did not happen [outcome of the testing experiment], **therefore** *such and such* [hypothesis] is not true [judgement]. Or: **And** *such and such* [prediction] happened [outcome of the testing experiment], **therefore** *such and such* [hypothesis] has not been rejected yet.

The following prompts help students develop H-D reasoning:
- What is the phenomenon(a) that your hypothesis explains?
- What is the hypothesis? Can you come up with a different one?
- Can you design an experiment whose outcome you can predict using this hypothesis?
- What should be the outcome of the experiment if your hypothesis is true? How do you know? Explain.
- After you conduct the experiment, how would you know if the outcome matched or did not match the prediction, especially when the experiment is quantitative?
- What are experimental uncertainties in your testing experiment? How do they affect your prediction?
- What are additional assumptions (additional to the hypothesis under test) that you made in your prediction? Will the assumptions, if not validated, lead to the outcome being larger or smaller than predicted?
- How can you validate your assumptions?
- After you conducted the experiment, what is your judgement about the hypothesis?

In chapter 1, we used H-D reasoning when we showed how the students test a hypothesis that each point of an extended source emits one ray of light. In this example, the reasoning was as follows (see table 1.1 in chapter 1):

If such and such [our hypothesis that each point of an extended light source sends one light ray] is true **and** I do *such and such* [cover the light bulb with aluminum foil with a tiny hole poked in it], **then** *such and such* should happen [we should see a tiny light spot on the wall] **because** [when that one light ray hits the wall, some of this light will reflect to our eyes]. **But**, *such and such* [the tiny light spot] did not happen [in fact the whole wall was brightly lit], **therefore** *such and such* [the hypothesis that each point of an extended source of light emits only one ray] is [not true and rejected].

How can we help pre-and in-service physics teachers develop H-D reasoning? As is the case with every other scientific ability, they need to first learn the steps of the logical chain when testing simple hypotheses. After they design a first testing experiment and make a prediction about its outcome, they need to reflect on the steps and record those steps as shown in the example above. Then, when they test every new idea that they devise, they need to carefully word the prediction using the language shown above. Over the years, we observed several difficulties with this process:

- Teachers confuse explanations/hypotheses with predictions.
- They wish to run testing experiments to 'see what happens' without making predictions.
- They do not use hypotheses to make predictions.
- They do not follow the H-D progression in their reasoning by putting the description of the experiment after *if*: '*if* I do such and such, *then* such and such will happen'. This logical chain does not follow the logical flow of hypothetico-deductive reasoning as there is no hypothesis and no deduction based on the hypothesis.

In chapter 1 of the CP: E&A and in the ALG/OALG, we have three activities that will help future teachers practice such reasoning (see their brief description in the section below). In addition, we suggest reading activities that help students develop H-D reasoning skills. Such activities consist of the students reading a scientific text that describes a process of the development of a specific idea in physics and annotating it to note the elements of physics.

Below is an example of such an activity (the solutions are italicized):

Spiders can fly (adapted from Yong 2018, original article Morley and Robert 2018)

Match the sentences with the elements of scientific reasoning (see the table below) by writing the number that you find at the beginning of the sentence into the corresponding row in the table.

1. On 31 October 1832, Charles Darwin walked onto the deck of his ship and realized that thousands of tiny red spiders had boarded the ship. 2. The ship was 60 miles offshore, so the creatures must have floated over from the Argentinian mainland. Spiders have no wings, so how can they be taken into the air? 3. For a while, it was commonly believed that the spider silk catches on the wind, dragging the spider with it. 4. But then scientists noticed that spiders only 'fly' during light winds. 5. An alternative idea was first proposed in the early 1800s: spiders fly by electrostatic repulsion. Recently, Erica Morley and Daniel Robert from University of Bristol tested the idea with actual spiders. 6. They decided to put the spiders on vertical strips of cardboard in the center of a plastic box in which they could generate electric fields similar in strengths to what the spiders experience outdoors. 7. If the electrostatic repulsion idea was correct, they expected to see some spiders taking off when switching the electric field on. 8. When they conducted the experiment, many of the spiders actually managed to take off, despite being in closed boxes with no airflow within them. 9. Based on this result, scientists agreed that electrostatic

repulsion plays an important role in spider flying. 10. However, researchers from the Technical University of Berlin recently showed that spiders prepare for flight by raising their front legs into the wind. 11. Therefore, it seems that the complete explanation for the phenomenon might include a combination of electrostatic repulsion and the motion of air.

Elements of scientific reasoning	Sentence numbers
Observations, collecting data, identifying patterns	1, 4, 10
Proposing a hypothesis, explanations, models	3, 5, 11
Making assumptions	2
Designing/proposing testing experiments	6
Prediction for the outcome of the testing experiment	7
Outcome of the testing experiment	8
Making judgments	9

4.1.14 Theory

You have probably been wondering why we never mentioned the word 'theory' in our discussions of the elements of physics or physics models of reasoning. We provide an explanation below.

The word 'theory' is probably the most misunderstood and misused word in physics and in everyday language. When looking up the word 'theory' in search engines, we find the following definitions: 'a supposition or a system of ideas intended to explain something, especially one based on general principles independent of the thing to be explained', 'an idea used to account for a situation or justify a course of action' and the synonyms that are listed are hypothesis, conjecture, speculation, etc. From these definitions and synonyms, it looks like we are speculating when we say 'theory' and that there is no 'proof'. However, if we look up the word 'scientific theory', the definition comes back completely different. From Wikipedia:

> A scientific theory is an explanation of an aspect of the natural world and Universe that has been repeatedly tested and corroborated in accordance with the scientific method, using accepted protocols of observation (read observational experiments), measurement, and evaluation of results. Where possible, theories are tested under controlled conditions in an experiment (read testing experiments). In circumstances not amenable to experimental testing, theories are evaluated through principles of abductive reasoning (astronomy for example) that seeks the simplest and most likely conclusion from a set of observations. Established scientific theories have withstood rigorous scrutiny and embody scientific knowledge. (Wikipedia n.d.b)

This is quite different from speculation, right?

If we think of physics, then we have for example:
- *Newtonian theory*, which explains the wide range of mechanical phenomena involving macroscopic objects at speeds much smaller that the speed of light; has its own set of physical quantities, conceptual and mathematical models which were tested in numerous experiments, measuring instruments and physics devices, and has its own limitations.
- *Kinetic molecular theory*, which explains a wide range of mechanical and thermal phenomena involving microscopic objects, has its own set of physical quantities, conceptual and mathematical models which were tested in numerous experiments, measuring instruments and physics devices and has its own limitations.
- *Special relativity theory*, which explains a wide range of mechanical phenomena involving macroscopic and microscopic objects moving at the speeds close to the speed of light, has its own set of physical quantities, conceptual and mathematical models which were tested in numerous experiments, measuring instruments and physics devices and has its own limitations.

From the examples below, we see that the word theory in physics is much more specific than even the definition of a scientific theory in Wikipedia. To be called a 'theory' in physics, a set of knowledge has to account for a wide variety of observational data, has its own set of physical quantities, mechanistic and causal explanations, and mathematical models relating those quantities, has been tested in numerous testing experiments, and applied for practice. All kinds of reasoning—inductive, hypothetico-deductive, analogical, and abductive—are used to create the conceptual and mathematical structure of the 'theory'. The development and testing of these models is impossible without special measuring instruments and physics devices. Therefore, all of the elements of physics knowledge discussed above and all types of reasoning come together to form a 'theory' in physics. Finally, each theory has its own very carefully defined set of limitations—statements when it can be applied to produce predictions that match the outcomes of the testing experiments. Thus, the word 'theory' in physics and in science in general is as far from a 'hypothesis' or a 'speculation' as a finished house is far from individual bricks, doors, window frames, etc. Only a proper understanding of the word 'theory' would allow teachers and students to argue against those who think that science education should only involve 'facts'.

4.2 Noticing physics everywhere

The physics habit of mind of noticing physics everywhere means thinking of the laws of physics when engaged in regular activities. For example, when walking, one can spontaneously think of the static friction force that allows us to start and stop every step. Another example is watching warming water in a transparent glass tea pot. At the same time, spontaneously thinking about gas and vapor pressure inside the bubble along with atmospheric pressure and pressure in the water when the bubbles first decrease in size when going up in the pot. Lastly, as the water gets hotter and hotter, they start growing bigger and bigger until they explode on the top.

To develop such a habit, one needs to connect physics principles to the surrounding phenomena. To help teachers develop such a habit, a teacher educator needs to first bring up their real-life experiences when their students are learning new material, and then gradually start asking them which of their experiences are relevant for a specific physical phenomenon or relationship. For example, when the students are learning about series and parallel electric circuits, an example of a former question (teacher generated) could be: 'We can turn off a light in our house without affecting other lights. What does it tell us about how the house is wired?' And an example of the latter could be: 'We learned that when more appliances are connected in parallel to a battery, the electric current through the battery increases. What real-life experiences related to this phenomenon do you have?' The best example of a habitual noticing of physics in everyday life was given to me by one of my pre-service physics teachers. He said: 'I was running trying to catch the bus that was about to pull off the bus stop and I was thinking: here my chemical energy is being converted to my kinetic energy.'

The ISLE approach helps teachers develop this habit naturally as most of the observational experiments are simple and involve phenomena with which their students are already familiar. For example, one of the first introductory activities in the ALG asks students to observe ice cold water being poured in a glass and then devise several different mechanisms to explain how the glass gets wet on the outside. Another introductory example is poking an inflated air balloon and observing the loud sound. What makes this sound? Experiments with scales, bowling balls, etc, help students to 'see' how physics principles explain motion of different objects. One of the best examples of helping teachers apply physics principles to real-life phenomena is using rollerblades or skate boards when teaching mechanics (Etkina 1998). In the Rutgers program, Eugenia not only teaches her classes while on rollerblades to help future physics teachers learn how to help their students learn mechanics, she also trains them on the safety and technical aspects that are important when helping high school students construct such concepts as relative motion, projectile motion, Newton's laws, collisions, etc, using rollerblades. They rent roller skating rinks and conduct those trainings there. See some photos of such experiences below in figure 4.12.

There are many rollerblading experiments in the CP: E&A, ALG/OALG, in the paper 'Physics on rollerblades' (Etkina 1998), in the **OALG** activities, and on the website http://islevideos.net/.

4.3 Approaching problems as an expert

We are all familiar with the approach to problem solving that our students often use. They first write down the known and unknown variables using the problem statement. Next, they search for an equation which has the same variables as the givens in the problem and the quantity that needs to be determined. This is called the back inference technique (Larkin *et al* 1980). Experts use a forward inference technique—they read the problem statement and think about the physics concepts

Figure 4.12. Physics on rollerblades. (a) Eugenia pushes Danielle and both of them start moving, although Danielle did not visibly push Eugenia. (b) A group is holding metersticks and will start moving in a circle. People closer to the stationary person at the center move much slower than the ones on the outside. The metersticks make sure that everyone has the same period of revolution. (c) Mike is throwing a heavy medicine ball to Eugenia and Eugenia accelerates back when catching it. The ball exerts a force on Eugenia.

involved that can be used to solve the problem and then they choose strategies based on concepts that are relevant to the problem (Chi *et al* 1981). Experts also utilize a larger number of heuristics or experimentally derived cognitive 'rules of thumb' (Abel 2003). Specifically, they start solving a problem with visualizing the situation and drawing sketches of the process described in the problem. When experts use heuristics, they 'chunk' the information together while novices look at the problem in pieces (Simon 1974). When stuck, experts return on qualitative analysis. When finished, they evaluate the final result, often using an alternative method of solving the same problem (Gerace *et al* 2001).

Poklinek Čančula, and Etkina (Čančula *et al* 2015) studied how experts and novices solve experimental problems. They found that experts spontaneously adopt the ISLE approach to solving the problem (observing, developing hypotheses, testing them, revising, and going through the same process again and again) while the students take a while before utilizing the same process. Jones continued this line of work and found that while solving experimental problems, experts not only utilize the logical progression of the ISLE approach, but add 'contrasting cases' to every step of their reasoning (Jones *et al* 2014). A contrasting case is when we compare two things to learn what one is and what one is not. We often ask students to solve examples when they continuously operate with the same idea. However, it turns out that knowing when this idea does not work is important as well. For example, contrasting velocity and acceleration helps students to learn that a rocket can have a very high velocity and zero acceleration. Or vice versa, when we start moving, we have zero velocity, but non-zero acceleration.

Based on the research findings described above, there are several aspects of physicists' approach to problem solving that can be thought of as habits of mind:
- When reading a problem, visualize it and draw a sketch or some other physics representation.
- Think of what physics models and relations are relevant for the situation described in the problem.
- After using these models/relations to solve the problem, evaluate the solution by either using a different approach to solve it or one of the evaluation methods (extreme case analysis, unit analysis, reasonability of the answer).

We can also think of problem solving as the ISLE process. This way, the approach works for traditional paper-and-pencil problems and for experimental problems. The situation described in the problem can be considered an observational experiment. Representations (sketches and physics representations) can be seen as helping to find patterns. The mathematical relations can be seen as hypotheses. And the final answer (if one is needed) can be seen as a prediction. From the above, it follows that each problem needs to end with a testing experiment where the solver compares the prediction to the outcome. If an experiment is possible, then it is important to set it up for the students. But if the experiment is not possible, then the students need to learn other means of evaluating their solutions. The following questions will help students to learn the evaluation techniques:
- Is the answer reasonable?
- Did you conduct a unit analysis? Did you use the unit analysis to correct the equation that is not self-consistent?
- Did you conduct a relevant special case analysis? Did you use the special case analysis to correct the model, the equation, or a claim?
- Did you identify assumptions in the mathematical model that you used to solve the problem?

Rubric I of the scientific abilities' rubrics will help students self-assess their work and improve it: https://drive.google.com/file/d/0By53x8SYAF1lTUpfYTZGbE90T2c/view?resourcekey=0-DezS6AJ_K63jt54stc1RYA.

Combining two approaches of solving a problem and evaluating the solution, we devised a problem solving strategy that helps students develop an expert-like problem solving approach (we already shared an example of applying such an approach in the interlude for chapter 1). It has four steps that one needs to follow:
- Sketch and translate.
- Simplify and diagram.
- Represent mathematically.
- Solve and evaluate.

To show how this strategy works, we present an example (see figure 4.13). While in chapter 1 we focused on the use of different representations, here our goal is to discuss the mental process that the students need to follow in order to develop expert-like problem solving techniques.

In this example, we see that in order to 'sketch and translate', a student needs to imagine the problem situation and draw a sketch. They need to decide what system to analyse and figure out the known and unknown variables. This translation from words to physical quantities is sometimes difficult for the students as physicists often use specific words to identify specific conditions. For example, when we say an object is *dropped* from a certain height, it means that the initial speed of the object is zero. When we say that a sled moves *on smooth snow*, it means that the kinetic friction force exerted on the sled is zero. When students are translating the language into the symbols of physical quantities, they need to reflect on the physics meaning of some words. It is also important to decide the system that is being analysed. In this case, the system is Cheryl as the problem asks to find a force exerted on her by the car seat.

The term 'simplify' means that the student needs to decide whether certain simplifications need to be made. In the solution below, Cheryl was simplified as a point-like object as her size was much smaller than the radius of the circle.

Physical representations in this case were the motion diagram (to determine the direction of acceleration) and the force diagram. Those need to be consistent with each other. Here, we see that because Cheryl's acceleration points down, the sum of the forces exerted on her should point down too. We identify objects interacting with Cheryl—those that are either touching her or, in the case of Earth, the force is exerted at a distance. We find two objects—the car seat and Earth. Thus, there are two forces being exerted on Cheryl. We also add the direction of a positive axis—it will help us later to write necessary equations. Notice that we still did not ask the students what concept is applicable to solving the problem. While the experts do this at the beginning of solving the problem, sometimes it is difficult for the students to identify such a concept. That is why representing the problem situation in different ways might help them figure out this concept. In the case of this problem, it is the force diagram that helps.

We can represent the force diagram mathematically to find the force exerted on Cheryl. As Newton's second law is written in a vector form, we use force components and here the positive direction of the axis comes in handy. From the sum of the force components and the knowledge of the relationship between speed, radius, and acceleration, we can find the component of the force exerted by the car seat on Cheryl.

The last step is to plug in the values of the physical quantities while keeping track of the units. We find the value of the force to be 451 N. How can we evaluate this answer? In this case, we use the first method—is the answer reasonable (as the units are correct)? We compare the force exerted by the seat to the force exerted by Earth and remind the students of the feeling they have when they go over a bump in a car on a bicycle. We could also use a special case analysis. When the road is flat ($r \to \infty$ and $a = 0$), $N = mg$.

	Cheryl drives her car at a constant 15 m/s speed over the top of circular hump. What is the direction and the magnitude of the force exerted by the car seat on Cheryl as she passes the top of the 50-m-radius arc? Her mass is 85 kg.
Sketch and translate • Sketch the situation described in the problem statement. Label it with all relevant information. • Choose the system and a specific position to analyze its motion. • Identify the unknown that you need to find and label it with a question mark.	Put all of the relevant information on the sketch: Cheryl's mass, her speed, and the radius of the circle arc along which the car moves. Cheryl is the system. $v = 16$ m/s $m = 85$ kg $R = 50$ m
Simplify and diagram • Decide if the system can be modeled as a point-like object. • Determine if the constant speed circular motion approach is appropriate. • Indicate with an arrow the direction of the object's acceleration as it passes the chosen position. • Draw a force diagram for the system at the instant it passes that position. • On the force diagram, draw an axis in the radial direction toward the center of the circle.	Assume Cheryl to be a point-like object and analyze her as she passes the highest part of the vertical circle along which she travels. We use a velocity change diagram as she passes the top of the circular path to find that she has a downward radial acceleration toward the center of the circle. The force diagram must have the sum of the forces pointing down. Thus, the upward normal force that the car seat exerts on Cheryl must be smaller in magnitude than the downward gravitational force that Earth exerts on her. A radial r-axis points down toward the center of the circle. $\vec{N}_{\text{S on C}}$ $\vec{F}_{\text{E on C}}$ r, radial
Represent mathematically • Convert the force diagram into the radial r-component form of Newton's 2nd law. • For objects moving in a horizontal circle (unlike this example), you may also need to apply a vertical y-component form of Newton's 2nd law.	We use the force diagram to write the radial form of Newton's second law $ma_r = \Sigma F_r$ In the radial direction the components of the two forces exerted on Cheryl are $F_{\text{E on C } r} = +mg$ and $N_{\text{S on C } r} = -N_{\text{S on C}}$. The magnitude of the radial acceleration is $a_r = v^2/r$. Therefore, the radial form of Newton's 2nd law is $m\dfrac{v^2}{r} = +mg + (-N_{\text{S on C}})$. Solving for $N_{\text{S on C}}$, we have $N_{\text{S on C}} = mg - m\dfrac{v^2}{r}$
Solve and evaluate • Solve the equations formulated in the previous two steps. • Evaluate the results to see if they are reasonable (the magnitude of the answer, its units, limiting cases).	We substitute the known information into the previous equation $N_{\text{S on C}} = (85 \text{ kg})(9.81 \text{ m/s}^2) - (85 \text{ kg})\dfrac{(15 \text{ m/s})^2}{50 \text{ m}} =$ $= 834 \text{ N} - 383 \text{ N} = 451 \text{ N}$ We find that the seat exerts a smaller upward force on Cheryl than Earth pulls down on her. You have probably noticed this effect when going over a smooth hump in a roller coaster or while in a car or on a bicycle when crossing a hump in the road—it almost feels like you are leaving the seat or starting to float briefly above it. This feeling is caused by the reduced upward normal force that the seat exerts on you.

Figure 4.13. An example showing a multiple-representation based problem solving strategy. A similar problem can be found in CP: E&A, chapter 5.

4.4 Treating mathematics in a physics way

A long time ago, Galileo said that a field of study becomes science when it starts using the language of mathematics. But while physics uses mathematics as a language through which it communicates its laws, the use of mathematics in physics is very different from its use for pure mathematical purposes. In their paper, 'Obstacles to mathematization in introductory physics', S Brahmia, A Boudreaux, and S Kanim wrote: 'University students taking introductory physics are generally successful executing mathematical procedures in context, but often struggle with the use of mathematical concepts for sense making' (Brahmia *et al* 2016, p 1). In this section, we will explore some of the physics habits of mind related to using mathematics.

As Brahmia and colleagues put it:

> Mathematizing in physics involves translating between the physical world and the symbolic world in an effort to understand how things work. Specific skills include representing concepts symbolically, defining problems quantitatively, and verifying that solutions make sense. Physicists develop and communicate ideas through the shared meanings they have built around these strong connections between mathematics and physics. (Brahmia *et al* 2016, p 1).

We can argue that mathematics, being the language of physics, has its own vocabulary and grammar. We can see physical quantities and other symbols as a part of physics vocabulary and their arrangement and rules using which we combine them as the grammar of mathematics in physics. It is in the vocabulary and grammar that the important differences and details of using mathematics in physics lie.

Let's start with the first part of the vocabulary—physical quantities. Similar to mathematics, in introductory physics we have numbers and vectors. An important difference is that in physics, all quantities have units and therefore we cannot have certain operations with the quantities that have different units. Physical quantities that have different units cannot be added or subtracted from each other. They can only be divided or multiplied.

However, multiplication and division of quantities in physics are very different from multiplication and division in mathematics. In mathematics, multiplication means addition. When we multiply the number 5 by the number 3, it means that we will either have the number that is made of three fives ($5 + 5 + 5 = 15$) or of five threes ($3 + 3 + 3 + 3 + 3 = 15$). In physics, when we multiply two physical quantities, the new quantity cannot be made by summing anything. For example, momentum of an object is a product of mass and velocity of an object. To obtain it, we multiply the magnitude of velocity (speed in m s^{-1}) by the mass of the object (in kg). The new quantity has the units of kg $\times$ ms^{-1} and we assign the direction of this quantity to be the same as the direction of the velocity vector. Nowhere in this progression do we find any addition.

The same is true for the division procedure or ratio quantities. In mathematics, division means a sequence of subtractions. For example, to divide 15 by 5 we take away 5 from 15, then another 5, and then one final 5. It took a total of three times to be left with nothing (15 : 5 = 3). We know how many fives is contained in 15—three fives. However, when we do a similar operation in physics, the result is different. For example, if we divide the distance that a person traveled (15 km) by the time they took (5 h), we will not find 15 km to be made of three 5 h intervals. Instead, we will find the average speed of the travel of 15 km in 5 h, 15 km : 5 h = 3 km h^{-1}. The result is a new quantity—average speed—and it has new units, km h^{-1}.

Physicists execute these operations habitually while remembering the importance of units. However, we very rarely discuss the differences in these mathematical procedures in physics and mathematics. It is difficult to separate the vocabulary and grammar when we talk about physical quantities. We can think of basic quantities (those that have SI units) as vocabulary and the compound quantities as sentences—combining both vocabulary and grammar.

In addition, we can also think of mathematical signs as the grammar of mathematics. Let's examine a few meanings of signs in physics. The plus or a positive sign in an equation can mean many different things: adding quantities that have the same units ($m_{total} = m_1 + m_2$), the positive sign in front of a force or velocity vectors scalar component (often omitted) means that the vector itself points in the direction chosen as positive (when the y-axis points down, the component of the force that Earth exerts on the system of interest is positive and equal to $+m_{system}g$), the positive sign in front of work done by a force (also often omitted) means that this work adds to the total energy of a system ($W = +10$ J), and a plus sign in front of an electric charge means that it is a positive charge ($q = +10^{-9}$ C). It is interesting to notice that when physicists use the plus or positive sign, they only use it when it signifies an operation of addition whereas in all other cases mentioned below, the sign is dropped and is assumed to be positive.

The situation with the negative sign or a minus sign is even more obscure. In their paper 'Framework for the natures of negativity in introductory physics' (Brahmia et al 2020), Brahmia and colleagues discuss the following use of the negative sign in physics (p 010120–1):

- 'In the equation, $x(t) = -40$ m $+ (-5$ m s$^{-1})t + 1/2(9.8$ m s$^{-2})t^2$ the minus sign shows that initial position, velocity, and acceleration of the object all pointed in the negative direction of the chosen positive axis. Thus, the sign signifies direction in relation to the *a priori* chosen positive direction.
- In the equation, $F_{1 \text{ on } 2} = -F_{2 \text{ on } 1}$, the negative sign signals that the force exerted by object 2 on object 1 is in the exact opposite direction as the force exerted by 1 on 2.
- In the expression, $0 - (-5 \, \mu C)$, the first negative sign indicates that a quantity of electric charge is being removed from an electrically neutral object. However, the second negative sign indicates which of the two different types of electric charge is being removed.

- In Faraday's law, $\varepsilon = -\frac{d\Phi_B}{dt}$, the negative sign reminds the expert that the EMF induced by a changing magnetic flux opposes (rather than reinforces) the change that created it.'

Additionally, the negative sign in front of work, $W = -5\,\text{J}$, means that the work done by the force in question decreased the energy of a chosen system. Also, the negative sign in front of gravitational potential energy, $U_g = -5\text{J}$, means that it is the energy of a bound system, and one needs to do positive work to separate two interacting objects. Conducting a deep theoretical analysis of the nature of negativity, Brahmia and colleagues developed a categorization scheme that encompasses all possible uses of the negative sign in physics (Brahmia *et al* 2020, p 010120-6).

From the above discussion, it is clear that while the sign conventions might be 'transparent' for physicists, the students need additional discussions every time they meet a positive or a negative sign in front of a quantity or in an equation.

Brahmia and colleagues recommend focusing on details to help those students who otherwise might not have grasped the meaning of the negative sign. Specifically, they suggest keeping the positive and negative signs in front of the quantities in addition to operation signs. When the teachers are writing equations or solving problems in their courses or PD programs, it is crucial that they practice this technique of keeping all the signs in front of the physical quantities—positive or negative. This approach will help them and, consequently, their students develop the habit of mind that signs carry specific meaning.

This step (of keeping the sign in front of the quantity) is especially important when the addition sign is dropped and the negative sign in front of a quantity becomes an operational sign (for example when you add two force components, one of which is in the negative direction). In this case, the plus sign is dropped and the operation looks like a subtraction. See examples of this careful keeping of the plus sign when adding a negative component in the textbook CP: E&A (p 68). They propose using the term 'minus' for the operation of subtraction and the term 'negative sign' to describe the symbol.

Brahmia and colleagues also recommend switching what has become the standardized positive and negative directions of chosen axes by not necessarily pointing the vertical axis up or the horizontal axis to the right. Doing this helps students pay attention to the chosen positive direction and do not assume that it is the same as in mathematics (see examples in CP: E&A, p 31, the y-axis points downward).

Another sign, the equal sign, deserves our attention too. In their paper, 'How physics textbooks embed meaning in the equals sign', Dina Zohrabi Alaee and colleagues (Alaee *et al* 2022) discuss the differences between the meaning of the equals sign in mathematics and in physics. They analyse physics textbooks of different levels and found the use of the equal sign for the following purposes:

- To communicate the operational definition of a quantity ($v = \lim_{\Delta t \to 0} \frac{\Delta x}{\Delta t}$) and the equation reads left to right (they call it 'definitional').

- To communicate the causality of a relationship of physical quantities (see the cause and effect relationships of physical quantities discussed in section 4.1 on the elements of physics knowledge). An example would be $a_x = \frac{\sum F_{\text{on sys } x}}{m}$. These equations are usually read right to left—the sum of the forces and the mass of the system (cause) affect the acceleration of the system (they call this role of the equal sign 'causality').
- While operational definitions and cause–effect relationships are fundamental to physics, Alaee and colleagues posit that sometimes there is a need to associate different quantities with each other temporarily (they call it 'assignment'). For example, the sum of $F = F_{\text{Earth on sys}} + (-F_{\text{Rope on sys}})$. They discuss how sometimes 'assignment' might be turned into 'causality'.
- Another way of using the sign is 'balancing', which means dynamic equilibrium. Dynamic equilibrium happens when two quantities are the same in magnitude and opposite in direction (e.g. for a book sitting on the table, $|mg| = |N|$, which means that the magnitude of the force that Earth exerts on the book is equal to the magnitude of the force that the table exerts on the book and these two forces balance each other to cause zero acceleration of the book).
- The final category of using the equal sign is 'calculate', which basically says what operation needs to be completed to calculate the desired quantity. In physics, the 'calculate' equal sign is usually used at the very end of solving a problem when we plug the known values into the derived equation.

As we can see from the above classification by Alaee and colleagues, the equals sign, similar to the plus and minus signs, can have multiple meanings and, while physicists habitually and often subconsciously reflect on those meanings, the teachers and students might not. Therefore, to develop the habit of differentiating between the meanings of the equal sign, we need to explicitly focus our teachers' and students' attention on that meaning.

Finally, when thinking about any kind of a mathematical equation in physics, physicists habitually turn it into a story. For example, seeing $x = (-10 \text{m s}^{-1})t / + (2 \text{m s}^{-2})t^2$ a physicist would immediately visualize an object (very likely a car based on the high speed of 10 m s^{-1}) that is passing the zero point of a chosen coordinate axis while moving in the negative direction and slowing down at a rate of 4 m s^{-1} s^{-1} (a relatively large acceleration). This ability to visualize the situation and tell a story about it is one of the habits constantly used by physicists when interpreting mathematical representations. One of the goals of learning physics is to help our students see the 'story' behind physical quantities and mathematical symbols. Therefore, the teachers need to practice telling such stories during their training. An example of activities that can help address this goal are Jeopardy questions and problems (Van Heuvelen and Maloney 1999). Jeopardy problems present a mathematical representation of a solution to a problem and the students are asked to describe the problem situation for which the representation shown is a solution (see the example in figure 4.14).

Equation Jeopardy The solution to a problem is represented by the following equation:

$$(9.0 \times 10^9 \, \text{N} \cdot \text{m}^2/\text{C}^2)(+24.0 \times 10^{-9} \, \text{C})(+36.0 \times 10^{-9} \, \text{C})/(0.08 \, \text{m}) =$$
$$= \frac{1}{2}(0.008 \, \text{kg})v_x^2 + (9.0 \times 10^9 \, \text{N} \cdot \text{m}^2/\text{C}^2)(+24.0 \times 10^{-9} \, \text{C})(+36.0 \times 10^{-9} \, \text{C})/(0.18 \, \text{m})$$

Sketch a situation that the equation might represent and formulate a problem for which it is a solution.

Figure 4.14. Example of equation Jeopardy.

As you can see from the above discussion, the topic of mathematization in physics is bottomless. Below, we recommend a few more references in addition to the ones we used above that summarize the work in this area of PER:
- White Brahmia S, Olsho A, Smith T, Boudreaux A, Eaton P and Zimmerman C 2021 Physics inventory of quantitative literacy: a tool for assessing mathematical reasoning in introductory physics *Phys. Rev. PER* **17** 020129 10.1103/PhysRevPhysEducRes.17.020129
- Sherin B L 2001 How students understand physics equations *Cogn. Instr.* **19** 479 10.1207/S1532690XCI1904_3
- White Brahmia S 2019 Quantification and its importance to modeling in introductory physics *Eur. J. Phys.* **40** 04400 10.1088/1361-6404/ab1a5a
- Domert D, Airey J, Linder C and Kung R L 2007 An exploration of university physics students' epistemological mindsets towards the understanding of physics equations *Nordic Stud. Sci. Educ.* **3** 15–28 10.5617/nordina.389

4.5 *Interlude by Eugenia*: History of physics, physics habits of mind, physics teacher preparation, and professional development (why should physics teachers know where physics rules come from?)

In this chapter, we described some of the habits of mind of physicists that are relevant for future and practicing physics teachers. How and when will physics teachers develop these habits if they did not have an opportunity to conduct authentic research? One might think that this development occurs during undergraduate education, but this is not the case. First, only about 40% of those who teach physics in the US (for example) have a major or a minor in physics, and second, traditional teaching in most universities around the world utilizes the transmission mode of education and does not help students develop many of the above habits. Therefore, it is the physics teacher preparation programs and professional development activities that need to address this issue. We have many examples in this chapter on how to structure activities to help future and in-service teachers develop some of these habits. In chapter 7, we provide details of the physics teacher preparation programs that are focused on the development of productive habits. In this interlude, we describe one other approach to help teachers develop some of these habits. This approach involves the history of physics. Knowing the history of physics from the epistemological perspective will help teachers learn *where* the models and relations that their students need to learn came from. However, we are not proposing a course in the history of physics. We are proposing to *use* the history of physics to help teachers develop physics habits of mind.

If we study the history of physics, we will find that physicists used inductive and hypothetico-deductive reasoning to come up with new ideas and later rule them out or support them through experimentation. Many of the observational experiments were accidental as some were done when the scientists were testing some other ideas. New mathematical tools made many complicated derivations very simple (such as the invention of logarithmic scale) and new graphical representations made new ideas accessible for many more scientists (such as the Cartesian coordinate system and Feynman diagrams). Many physicists struggled with the same issues that our students struggle with today. In this interlude, we will not go through the history of physics systematically, but we will only highlight several examples connected to the development of the habits of mind. We then share the summary of the syllabus for a course that can help develop such habits and describe student work in the course.

4.5.1 Historical examples of the development and use of the physics habits of mind

1. *Galileo Galilei*

    Galileo, in 'Dialogues concerning two new sciences' (Magie 1963) discusses how we can deduce—from observations—that falling objects speed up as they fall down. He wondered how the speed changes as the object falls. This is what he wrote:

    When, therefore, I observe a stone initially at rest falling from an elevated position and continually acquiring new increments of speed, why should I not believe that such increases take place in a manner which is exceedingly simple and rather obvious to everybody? If now we examine the matter carefully, we find no addition or increment more simple than that which repeats itself always in the same manner. Thus we readily understand when we consider the intimate relationship between time and motion; for just as uniformity of motion is defined by and conceived through equal times and equal spaces (thus we call a motion uniform when equal distances are traversed during equal time intervals), so also we may, in a similar manner, through equal time intervals, conceive additions of speed as taking place without complication; thus we may picture to our mind a motion as uniformly and continuously accelerated when, during any equal intervals of time whatever, equal increments of speed are given to it. Thus, if any equal intervals of time whatever have elapsed, counting from the time at which the moving body left its position of rest and began to descend, the amount of speed acquired during the first two time intervals will be double compared to that acquired during the first time interval alone; so the amount added during three of these time intervals will be treble; and that in four, quadruple that of the first time interval.

And thus, it seems, we shall not be far wrong if we put the increment of speed as proportional to the increment of time; hence the definition of motion which we are about to discuss may be stated as follows: A motion is said to be uniformly accelerated, when starting from rest, it acquires, during equal time intervals, equal increments of speed.

From here you can see how Galileo uses the patterns in simple observational experiments of falling objects to propose a hypothesis of how the speed of a falling object changes using analogical reasoning and looking for a simplest possible description of the change in speed. However, Galileo does not stop hypothesizing. He continues with the derivation of a consequence of this reasoning that 'the distances traversed, during equal intervals of time, by a body falling from rest, stand to one another in the same ratio as the odd numbers beginning with unity.' Then he proceeds with the design of a testing experiment with an inclined plane that we described above and makes a prediction about the outcome using the above consequence. Therefore, in his quest to understand how falling objects move, Galileo engages in inductive, hypothetico-deductive, and analogical reasoning and invents the concept of a testing experiment in physics.

2. *Benjamin Thompson, Count Rumford*
In the paper, 'An inquiry concerning the source of the heat which is excited by friction' (Thompson 1798), Thompson describes puzzling observational experiments:

Being engaged, lately, in superintending the boring of cannon, in the workshops of the military arsenal at Munich, I was struck with the very considerable degree of Heat which a brass gun acquires, in a short time, in being bored; and with the still more intense Heat (much greater than that of boiling water, as I found by experiment) of the metallic chips separated from it by the borer.

Here we see that Benjamin Thompson used the habit of mind of noticing physics everywhere, even when he was occupied superintending the boring of canons. Interestingly, he proceeded to more quantitative observations and then testing experiments. These led him to believe that it is the motion of the borer that leads to the heating of the cannon—not the transfer of caloric fluid as it was believed before him.

3. *James Joule*
In his paper, 'On the changes of temperature produced by the rarefaction and condensation of air' (Joule 1845), Joule describes an experiment of gas expansion in a vacuum and records no change in temperature of this gas. He had two containers connected to each other and connected by a 'coupling nut, which had a piece attached in the center of which there was a bore of 1/8 of an inch in diameter, which could be closed perfectly by means of a proper stopcock'.

One of the containers had dry air in it and the other one was completely evacuated using a pump. When he removed the stopcock and allowed the air to expand into an evacuated chamber, he did not notice any change in temperature. While he does not say why he conducted this experiment, it is clear that he was testing his idea of energy constancy of an isolated system. When the gas expands in a vacuum, it contains the same number of particles and does not transfer any of its energy to the surroundings (the surroundings do not do any work on the expanding gas). Therefore, the gas temperature should remain the same. This is exactly what he observed.

We can go on with additional examples of physicists using the habits of mind and ways of reasoning that we discussed in this chapter. Below, we show the syllabus of a course called 'Development if Ideas in Physical Science' that helps future physics teachers learn this material and simultaneously practice using it with their future high school students.

The course uses the ISLE framework to analyse historical progression of the development of several fundamental ideas. This means that the students learn about initial observational experiments, different hypotheses explaining them, testing experiments that rejected some of the hypotheses, and ones that supported the hypotheses. Future teachers practice hypothetico-deductive reasoning and learn how to engage their future students in similar activities. We invite the reader to first examine the syllabus in the appendix and then come back to read how the course meetings and student homework are organized.

4.5.2 Overview of the 'Development if Ideas in Physical Science'

The course meetings are structured as high school physics lessons (for 2.5 h) and 0.5 h of reflections and discussion. During the first nine weeks, the instructor of the course acts as a teacher and the students (pre-service physics teachers) act as high school students. They work in groups, use whiteboards to share their solutions to the problems, and participate in class-wide discussions. The course instructor prepares all of the equipment and designs all of the activities for these lessons. Activities include doing observational experiments that resemble the historical experiments, discussing possible explanations, designing and conducting testing experiments, reading original writings of the scientists, and comparing their experiments and thoughts to the ones that the students have just conducted. Examples of activities are in chapter 7.

After the 'lesson' ends, the 'students' become reflective teachers—they share what they learned about physics and what they learned about teaching physics. What were the methods used? Why were they successful? What was the specific vernacular? How was participation encouraged? How were all opinions and experiences respected? After the first month of the course passes and the pre-service teachers get used to the format of the lessons, they sign up for a specific project (see the table below). They first start working on understanding the physics behind the familiar words (static electricity, electric charge, conductors, insulators, for example) and passing a content interview. Then they start designing a 2.5 h lesson which will help their peers learn how specific ideas were developed. The 'teachers' need to decide what goals the lesson will have, what activities the 'students' will work on, how they will be motivated to do those activities, what the 'ideal answers' are and what other responses the teachers might receive, and so forth. They submit the first draft to the instructor of the course via Google docs, the instructor provides feedback, and the students work on revisions. Once the flow of the activities is finalized, the students start preparing equipment and rehearsing parts of the lesson with each other. The instructor continues to provide feedback and gives help with equipment and experiments. Once the date of the lesson comes, they completely take over the class for 2.5 h. The last 30 min of the reflection are led by the course instructor. The 'teachers' first reflect on their experiences and then each student in the course shares their reflections. During the lesson led by the pre-service teachers, the course instructor implements the 'flight simulator method', which is described in detail in chapter 7. In chapter 7, we also show examples of class and homework activities.

Appendix
'Development of Ideas in Physical Science' course

Course description

Learning outcomes
At the end of the course the students will be able to:
1. Explain how scientists devised the most important ideas and relations that constitute the content of a general physics course.
2. Use similar processes in a classroom to help students construct physics concepts and relations and to connect this process to the Next Generation Science Standards.
3. Provide examples of how student learning of physics/physical science relates to the scientists' learning.
4. Design and implement a coherent unit of instruction during which high school students construct one of the fundamental ideas following the historical progression.

The main goal of the course is to help you understand the epistemology of physical science and its implications to science instruction. Epistemology is the study of the construction of knowledge. Basically, in this course you will learn how scientists know what they know, how they approach problems, and how they decide what to keep and what to discard. We will focus on the process that leads to the laws of physics and chemistry that we teach our students and how learning of our students sometimes resembles that of real scientists. You will learn how to use the knowledge of epistemology and history of science to design physics/chemistry lessons.

Class materials
Textbooks required:
    Holton G and Brush S 2001 *Physics, The Human Adventure* (New Brunswick, NJ: Rutgers University Press) ISBN 0-8135-2908

Not required but will be helpful:
Shamos M 1987 *Great Experiments in Physics* (New York: Dover) ISBN 0-486-25346-5

Magie W F (ed) 1963 *A Source Book in Physics* (Cambridge, MA: Harvard University Press) ISBN 9780674823655

Required and provided by the instructor:
Etkina E, Planinsic G and Van Heuvelen A 2019 *College Physics: Explore and Apply* 2nd edn (London: Pearson) ISBN 0134601823

Description of activities
Participation in class discussions. Class work will be primarily group work. You will explore contemporary versions of classical experiments, read and interpret original papers of scientists, explore how scientists chose one theory over another, and discuss how to adapt some of the historical materials for high school physics instruction. At the same time, you will learn how students construct similar concepts. We will also discuss the readings that you will do at home. Each week you will read several of the chapters of the text and additional articles. We will discuss these chapters in class.

Homework (individual assignment). (a) Each week you will combine the material from class, from the *Human Adventure* and *College Physics* books, and from any other convenient sources (I encourage you to use the Shamos book, and *A Source Book in Physics*, and resources on the Web) to write a report reconstructing the inductive, analogical and hypothetico-deductive reasoning and experimental evidence used by scientists to construct a particular idea. In your report try to make clear distinctions between *initial observational experiments,* reasoning (*hypotheses*), *experiments* conducted *to test* hypotheses, *predictions of the outcomes* of these experiments, based on the hypotheses, and the outcomes of the testing experiments. Try not to confuse experimental evidence with hypotheses/explanations. Also, do not confuse hypotheses/explanations with predictions. The glossary of terms is at the end of the syllabus.

At the end of *each* report you need to reflect on how your personal understanding of the concept changed because of the learning of the history. After you submit your homework, you will receive feedback and you will improve your work based on this feedback as many times as necessary.

Interview (individual). As one of the major skills of a teacher is to be able to listen to a student, you need to practice listening. To do this, you will choose one concept whose historical development we will trace in the course and interview two people—an expert in the field of physics and a person who is not familiar with physics. The goal of the interview is to find out what the person understands about the concept and how she/he can apply it. You need to: 1. Design the interview questions. 2. Submit them to the course instructor a week prior to the interview. 3. Revise the questions and resubmit them to the course instructor. 4. Conduct the interview and record it. 5. Write a report. In the report you need to show that you can connect what you heard during the interview to the history of the development of the concept and to your lesson. Make sure you have enough time for steps 1 though 4. Note, that step 4 might require improvements based on the feedback of the instructor.

The Physics Teacher (individual). As a future physics teacher you need to be a member of the American Association of Physics Teachers (aapt.org) and receive and read *The Physics Teacher*—the main journal for physics teachers in the US. Join ASAP and choose one article (or more!) from any issue that is relevant to your project. Read and write a brief summary of the article. Can you incorporate any of the material from it into your lesson? Or can you build a piece of equipment based on the article? Based on your choice of the article and the material it is your decision how you will incorporate it into your own library of activities. Once you submit the summary of the article, be prepared to revise it after feedback.

History project (group assignment). You and two of your teammates will choose one fundamental idea in physical science (from the list in the table below) and trace its historical development following the ISLE cycle. Together you will write *a paper* describing the development of the idea and *prepare a lesson and submit the lesson plan* to teach in class in which *parts* of the historical cycle will be recreated. Preparation for the project takes more than a month, thus you need to start asap. The first step is to learn the physics content. Use both textbooks to make sure you understand the tiniest details of the material. After that you will need to pass the content interview. Once you pass the content interview you will start working on the history and on the lesson. You will submit both (the paper and the lesson) for feedback to the course instructor, revise (be prepared to do several revisions) and then teach a lesson in class. In your lesson you should use at least one experiment that is analogous to a historical experiment important for the development of the idea (or present data from a historical experiment). The lesson plan should reflect what you learned during the interview and the material from *The Physics Teacher* article.

Story-telling (individual assignment, part of the history project above). You will choose a physicist who contributed to the development of your history project idea (see the history project assignment) and research personal information about them and their scientific achievements (you need to find a book dedicated to this person; Internet materials are not sufficient). You will write a story about them that you will tell in class or record with a screen cast and narration. In the story your character should become alive.

Instructional materials (group assignment). The story-telling project and the lesson curriculum materials *after final revisions* will be shared with all class participants.

Tentative list of topics for discussions and homework assignments (by week, chapters are from Holton and Brush (2001)).

Week	Topic	Assignment (chapter in *Physics, the Human Adventure*)
1	Epistemology of physics. The study of motion. Galileo's work.	1, 3, 12, 13, 14
2	The study of motion. Galileo's story.	6, 7, 8
3, 4	Newtonian world. Newton's work and story.	9, 10, 11
5, 6, 7	The laws of conservation, caloric theory. Lavoisier and Madam Lavoisier, Black, Thompson (Count Rumford), Leibnitz, Huygens, Mayer, Joule, Helmholtz—contributions and stories.	15, 16, 17, 18
8, 9	Atomic theory, gases, pressure Dalton, Torricelli.	19, 20, 21, 22
10, 11	Electromagnetism (students microteaching; see table below)	24, 25
12, 13, 14, 15	Modern physics (students microteaching; see table below)	23, 26

Ideas and scientists for microteaching projects (groups of two/three).

Ideas	Scientists
Class 10: Static electricity, electric charge, conductors and insulators; November 12	Franklin, Coulomb, DuFay, Gray
Class 11: Battery, Ohm's law; November 19	Galvani, Volta, Ohm
Class 12: Cathode rays, electron; November 26	J J Thomson; Millikan
Class 13 Quantum model of light; December 3	Einstein, Lennard, Stoletov
Class 14 Radioactivity; December 10	Mari Curie
Class 15 Fission;	Lise Meitner, Ida Noddack

(End of syllabus)

References

Abel C F 2003 Heuristics and problem solving *New Dir. Teach. Learn.* **2003** 53–8

Alaee D Z, Sayre E C, Kornick K and Franklin S V 2022 How physics textbooks embed meaning in the equals sign *Am. J. Phys.* **90** 273–8

Brahmia S, Boudreaux A and Kanim S E 2016 Obstacles to mathematization in introductory physics arXiv:1601.01235

Brahmia S W, Olsho A, Smith T I and Boudreaux A 2020 Framework for the natures of negativity in introductory physics *Phys. Rev. Phys. Edu. Res.* **16** 010120

Brookes D T and Etkina E 2007 Using conceptual metaphor and functional grammar to explore how language used in physics affects student learning *Phys. Rev. Spec. Top. Phys. Edu. Res.* **3** 010105

Buggé D and Etkina E 2016 Reading between the lines: lab reports help high school students develop scientific abilities *Physics Education Research Conf. Proc.* D Jones, L Ding and A Traxler pp 52–5

Buggé D and Etkina E 2020 The long term effects of learning in an ISLE approach classroom *Proc. Physics Education Research Conf.* ed S F Wolf, M B Bennett and B W Frank

Čančula M P, Planinšič G and Etkina E 2015 Analyzing patterns in experts' approaches to solving experimental problems *Am. J. Phys.* **83** 366–74

Chi M T, Feltovich P J and Glaser R 1981 Categorization and representation of physics problems by experts and novices *Cogn. Sci.* **5** 121–52

Copi I M, Cohen C and Flage D E 2006 *Essentials of Logic* 2nd edn (Upper Saddle River, NJ: Pearson Education)

Ecyclopedia Britannica Editors n.d. hypothetico-deductive method *Ecyclopedia Britannica Editors* https://www.britannica.com/science/hypothetico-deductive-method

Etkina E 1998 Physics on rollerblades *Phys. Teach.* **36** 32–5

Etkina E, Brookes D and Planinsic G 2019 *Investigative Science Learning Environment: When Learning Physics Mirrors Doing Physics* (IOP Concise Physics) (San Rafael, CA: Morgan and Claypool Publishers)

Etkina E, Gregorcic B and Vokos S 2017 Organizing physics teacher professional education around productive habit development: a way to meet reform challenges *Phys. Rev. Phys. Edu. Res.* **13** 010107

Etkina E, Karelina A, Murthy S and Ruibal-Villasenor M 2009 Using action research to improve learning and formative assessment to conduct research *Phys. Rev. Spec. Top. Phys. Edu. Res.* **5** 010109

Etkina E, Karelina A and Ruibal-Villasenor M 2008 How long does it take? A study of student acquisition of scientific abilities *Phys. Rev. Spec. Top. Phy. Edu. Res.* **4** 020108

Etkina E, Karelina A, Ruibal-Villasenor M, Jordan R, Rosengrant D and Hmelo-Silver C 2010 Design and reflection help students develop scientific abilities: learning in introductory physics laboratories *J. Learn. Sci.* **19** 54–98

Etkina E, Van Heuvelen A, White-Brahmia S, Brookes D T, Gentile M, Murthy S, Rosengrant D and Warren A 2006 Developing and assessing student scientific abilities *Phys. Rev. Spec. Top. Phys. Edu. Res.* **2** 020103

Etkina E, Warren A and Gentile M 2006 The role of models in physics instruction *Phys. Teach.* **43** 15–20

Gentner D and Gentner D R 2014 Flowing waters or teeming crowds: mental models of electricity *Mental Models* (New York: Psychology Press) pp 107–38

Gerace W J, Dufresne R J, Leonard W J and Mestre J P 2001 Problem solving and conceptual understanding *Proc. Physics Education Research Conf.* p 33

Glynn S M, Duit R and Thiele R B 2012 Teaching science with analogies: a strategy for constructing knowledge *Learning Science in the Schools* (Abingdon: Routledge) pp 259–86

Hewitt P 2014 *Conceptual Physics* (New York: Pearson)

Holton G and Brush S 2001 *Physics, The Human Adventure* (New Brunswick, NJ: Rutgers University Press)

Jones D C, Malysheva M, Richards A, Planinšič G and Etkina E 2014 Resource activation patterns *2013 PERC Proc. Expert Problem Solving* ed P V Engelhardt, A D Churukian and D L Jones (Portland, OR: PER-Central) pp 201–4

Joule J 1845 On the changes of temperature produced by the rarefaction and condensation of air *Philos. Mag.* **26** 369

Larkin J, McDermott J, Simon D P and Simon H A 1980 Expert and novice performance in solving physics problems *Science* **208** 1335–42

Magie F (ed) 1963 *A Sourcebook in Physics* (Cambridge, MA: Harvard University Press)

Morley E L and Robert D 2018 Electric fields elicit ballooning in spiders *Curr. Biol.* **28** 2324–30

Simon H A 1974 How big is a chunk? By combining data from several experiments, a basic human memory unit can be identified and measured *Science* **183** 482–8

Stewart G, Freedman R A, Ruskell T and Kesten P R 2019 *College Physics for the AP Physics 1 Course* 2nd edn (New York: W H Freeman)

Thompson B 1798 An inquiry concerning the source of the heat which is excited by friction *Philos. Trans. R. Soc. Lond.* **88** 80–102

Van Heuvelen A and Maloney D P 1999 Playing physics jeopardy *Am. J. Phys.* **67** 252–6

Wikipedia n.d.a Spherical cow *Wikipedia* https://en.wikipedia.org/wiki/Spherical_cow#:~:text=Originating%20in%20theoretical%20physics%2C%20the,the%20model's%20application%20to%20reality

Wikipedia n.d.b Scientific theory *Wikipedia* https://en.wikipedia.org/wiki/Scientific_theory

Yong E 2018 Spiders can fly hundreds of miles using electricity *The Atlantic* 5 July https://www.theatlantic.com/science/archive/2018/07/the-electric-flight-of-spiders/564437/

… IOP Publishing

The Investigative Science Learning Environment: A Guide for Teacher Preparation and Professional Development

Eugenia Etkina and Gorazd Planinsic

Chapter 5

ISLE and the development of physics teacher habits of mind and practice

In chapter 3 we identified several important habits of mind of physics teachers, including specific habits of mind of those who teach physics through the ISLE approach. In chapter 4 we discussed physics epistemology and related habits of mind. Here, we focus on a selected list of physics teacher habits of mind and practice that are particularly important for those who implement the ISLE approach and discuss how to develop them in a physics teacher preparation program or a professional development program for in-service teachers[1].

5.1 Development of inductive and hypothetico-deductive reasoning

We discussed the elements of both types of reasoning and the questions that help develop these types in chapter 4. Here we focus on what activities help develop both types of reasoning, for both pre-service teachers in their teacher preparation programs and in-service teachers in their professional development (PD) programs. As with every scientific ability, we develop different types of reasoning through multiple learning and teaching opportunities in different contexts and receiving feedback. These opportunities occur during class work in physics teacher preparation programs, in workshop activities in PD programs, through scaffolded practice, and then through independent teaching practice.

We can use the ISLE approach that we use in physics classes in teacher preparation programs and PD activities: students learn through collaborative activities (Brookes *et al* 2020) using carefully designed materials (such as the ALG and OALG). Future teachers and PD participants should work in groups

---
[1] We remind the reader that the abbreviation CP: E&A stands for the textbook *College Physics: Explore and Apply*, ALG stands for the *Active Learning Guide*, OALG stands for the *Online Active Learning Guide* and IG stands for the *Instructor Guide*. Proper citations for these materials are in chapter 1.

on the activities that develop these types of reasoning, present their findings on whiteboards, share the whiteboards with the rest of the class for the whole class discussion, and then reflect on the process in which they were engaged. The key here is to choose activities that allow the participants to practice both kids of reasoning. We have several introductory activities that engage pre- and in-service physics teachers in the development of inductive and hypothetico-deductive reasoning; if you are teaching in a university context, note that these are also helpful to train teaching assistants (TAs). These activities are presented in the ALG, chapter 1. Here are two examples.

5.1.1 Cool glass

Equipment per group: dry glass, ice-cold water, paper, oil, access to the refrigerator, high precision scale (to 1 mg), whiteboards, markers. Note: the higher the humidity in the room, the more clearly visible will be the outcomes of the experiments (in our experience the humidity of the air should be 60% or higher). Alternatively, the teacher may show slides with photos.

   a. A teacher will put a dry and empty glass on your group's desk and pour ice-cold water into the glass. Carefully observe the glass for few minutes. Describe in simple words what you observe. Share with the rest of the class—did everyone else observe the same things?
   b. Work with the members of your group to propose different explanations for the observed patterns. Try to devise as many explanations as possible. Put them on the whiteboard.
   c. How can you find out which explanation is correct? In science we conduct testing experiments. A testing experiment is an experiment whose outcome you predict before conducting it using the idea under test. You do not need to agree with the idea but the prediction of the outcome must be based on it. After you design the experiment and make predictions based on all explanations that you devised, you will conduct the experiment and compare the outcome to the prediction. Work with your group members to propose testing experiments that you can run to test the proposed explanations. Try to propose as many as you can.
   d. For *each* testing experiment, make prediction for its outcome based on *each* explanation that you proposed in step b. Indicate any assumptions that you made when making predictions. (Note: The best testing experiments are those that give different predictions for different explanations).
   e. Perform the testing experiments that you proposed in step b (if necessary, ask teacher for additional equipment).
   f. Compare the outcomes of the testing experiments with the predictions that you made in step c. What can you say about the explanations under test now? Can you reject some explanations? Do not forget to include the assumptions when making any judgements. Can you verify some assumptions?

Below we describe the approximate progression of student thought and discussions (in this case the students are either pre-service teachers or in-service teachers):

The students observe the glass after cold water was poured into it (figure 5.1). They do not need to predict anything, but observe.

Figure 5.1. Observational experiment (from left to right).

Teacher: What did you observe? Please answer using the words that a five-year-old can understand. Please do not use any scientific terms.

Students: After cold water was poured into the glass, the glass got wet on the outside.

We can see water droplets on the outside. (Note: there is no fancy scientific language here. Anyone can be successful.)

Teacher: Now come up with several possible explanations of where this water came from. Try to think of how you could test each explanation experimentally.

The students work in groups coming up with explanations. The teacher looks at the whiteboards and decides with which explanation to start first. We recommend starting (if the students come up with it) with the explanation that 'Water goes out from the top and settles on the outside'.

Teacher: Group A came up with the explanation that water goes out from the top and settles on the outside. How can we test it?

Students (after working in groups for a minute): Take a glass filled with cold water and cover it.

Teacher: What should happen if your explanation is correct?

Students: It should not get wet. (Note that the explanation does not need to be correct, it just needs to be experimentally testable).

Here either the students perform the experiment or the teacher shows the outcome (see figure 5.2).

Figure 5.2. Outcome of the testing experiment 1.

Teacher: What is the outcome of this testing experiment? What does it tell us about the explanation under test?

Students: The glass is still wet! So, we can reject this explanation.
Teacher: Group B has a different explanation. Can you please describe it?
Student from Group B: The water seeps through the glass. And we have the testing experiment for it! We will take an empty dry glass, put it into a cold place and then take it out.
Teacher: What should happen *if* the water seeps through the glass?
Student from Group B: *If* the water seeps through the glass *and* we do not have any water in the cold glass, *then* it should not get wet *because* there is no water to go through! (Notice the *if-and-then-because* statement which shows the hypothetico-deductive reasoning chain).

Teacher (who already had an empty glass in the fridge), takes the glass out: What do we observe (see figure 5.3)?

Figure 5.3. Outcome of the texting experiment 2.

Students: The glass is still wet! We can reject the second explanation!
Teacher: I see that Group C has yet another explanation. Please share!
Students from group C: The material of the glass 'sweats' and the water comes from the material of the glass, similar to us sweating when we run.
Teacher: How can we test this explanation?
Students get together in groups and think of a testing experiment.
Student from group C: Pour another cold liquid into the glass. If the glass sweats water and we pour another cold liquid (let's say cold oil) into the glass, then the glass should still have water on the outside, not oil because in our explanation it sweats water.

The teacher already has prepared cold liquid (can be oil, or vinegar). The students observe the experiment (see figure 5.4).

Figure 5.4. Outcome of the testing experiment 3.

The glass has water on the outside, not oil!

Teacher: We have the outcome that matches the prediction. Does it mean that we just proved the explanation that water on the outside is the result of 'sweating glass'?

Here the teacher has a short discussion about the impossibility of one testing experiment whose outcome matches the prediction to prove the hypothesis that was used to make that prediction. How many experiments does it take to prove something? The students should come up with the answer that this number should be infinite.

Students from group D: We have another explanation! The water comes from outside! We could test it and possibly rule it out or rule out the sweating glass explanation!

Teacher: Great! Think of an experiment for which you can make two different predictions using these two different explanations!

Student from group D: We thought of using a scale. Weigh the glass right after we poured cold water into it. If 'glass sweats', then the mass should stay the same or decrease (in case some water evaporates). If the water comes from the outside air, then the mass should increase.

The teacher has already the scale ready (or the video). The students observe the experiment (see figure 5.5; you can watch the video of the same experiment at https://youtu.be/x6bNApNaeQA).

Figure 5.5. Outcome of the testing experiment 4.

The students observe increase of the mass. They agree to reject the sweating glass explanation and continue testing the explanation that water comes from the air.

Here is another example of the activity for ALG that help the students start developing inductive and hypothetico-deductive reasoning. We will not go through the whole process, just share the activity (ALG, chapter 1, pp 1–2):

5.1.2 Popping the balloon

Equipment per class: rubber balloons, needle, plastic bags (thin plastic bags for vegetables and fruit work best), access to water (a tap), bucket for catching water. Optional: small embroidery hoop (about 12 cm in diameter).

A teacher is holding a fully inflated balloon. The teacher asks the students to observe carefully while she pops the balloon using a needle.

a. Describe what happened when the teacher popped the balloon.
b. What makes the sound so loud? Work in your groups to propose several explanations using only simple words.
c. Propose testing experiments that you can run to test the proposed explanations. Try to propose as many as you can.
d. For *each* testing experiment, make prediction for its outcome based on *each* explanation that you proposed in step b. Indicate any assumptions that you made when making predictions. (Note: The best testing experiments are those that give different predictions for different explanations).
e. Perform testing experiments that you proposed in step b (if necessary, ask teacher for additional equipment).
f. Compare the outcomes of the testing experiments with the predictions that you made in step c. What can you say about the explanations under test now? Can you reject or revise some explanations? Do not forget to include the assumptions when making any judgements. Can you verify some assumptions?

While the above activities are excellent for helping develop inductive and hypothetico-deductive reasoning, they are not enough to develop the habits. Therefore, we have observational experiments and pattern recognition activities for every physics concept in the ALG and OALG to help students develop inductive reasoning habits. Similarly, we have testing experiments with predictions activities to help develop hypothetico-deductive reasoning habits. The repetition and consistency in the requirements for these types of reasoning help students develop the habits.

5.2 Documenting physical phenomena in the outside world

Without connections to everyday phenomena, students think that physics laws only work in the physics classroom. How do we help them see physics working everywhere around them? This will only happen if the teacher themself notices this physics and brings it back to the classroom. While taking photos of exciting physical events is one way to do this, there are many mundane events occurring every day that we can document and use to develop physics problems and laboratory exercises. Below we show two examples of physics problems that were developed by documenting physical phenomena that everyone observes but few people capture.

The first phenomenon happened in the house when the children were playing with stuffed animals and one of them got shoved across the floor (see figure 5.6).

Problem: Natasha and Gregor played with a squishy toy. They pushed it along a smooth wooden floor and noticed that the toy slows down in a repeatable way. What if the toy slows down at constant acceleration that is not dependent on its initial velocity? They set up an experiment to measure distances that the toy traveled after being pushed and released. They recorded the initial speeds for different tries and the time of travel (you can see their data in the table below). What do the data say about their hypothesis? If you think that the hypothesis is supported, find the toy's average acceleration and its maximum speed.

Experiment	Distance (m)	Time (s)
1	0.68	0.58
2	1.06	0.72
3	1.89	0.96
4	2.78	1.16
5	3.12	1.23

Figure 5.6. A problem involving a toy.

The second occurred when we were hanging one of Gorazd's pictures (yes, he draws!). I (Eugenia) accidentally bumped the picture that he just put in place and to my surprise it remained in its tilted position (figure 5.7). I remembered that it happens often with the pictures on walls. What a great problem for static equilibrium!

Problem: You have a framed picture hanging on the wall (see left photo). You noticed that when you rotate the picture a little and let it go, it stands still in its new position (see the photo on the right). Explain how the picture can stand still in its new position. Support your explanation with a force diagram. Indicate any assumptions that you made.

Figure 5.7. A problem involving a tilted framed picture. (The framed drawing is by G Planinsic. Similar problems can be found in CP: E&A).

Note that both phenomena do not represent anything spectacular or out of the ordinary. The most interesting thing about them that they are 'unnoticed' by us every day and that is why they were not previously captured and used for learning. How can we help develop this habit of documenting physical phenomena in our future physics teachers? One way can be creating specific assignments in our physics teacher preparation programs that teach our students how to notice, document, and analyse simple physical phenomena. These assignments can follow a progression from being heavily scaffolded (the phenomenon is identified, and the students need to capture and analyse it) to giving a list of possible phenomena and providing requirements for documentation, to finally having an open-ended assignment to find phenomena themselves and decide what to document about it and how to use it in the classroom.

Table 5.1 shows some examples of phenomena to notice, what to document or which data to record, and suggestions for the activities to use in the classroom.

An important thing to add here is that often the phenomena that we encounter in the outside world require knowledge from multiple physics lessons and therefore can be used at different times. For example, to explain and analyse the phenomenon number 5 in table 5.1, students need to combine the ideal gas law with knowledge about hydrostatic pressure.

Table 5.1. Everyday phenomena and related activities.

Phenomenon/ experiment	Requirements to the documentation/data collection	Types of the activities	References
1. Automatic sliding doors during operation.	Make a video of the doors opening and then closing. Measure the dimensions of the door.	Ask the students to draw $x(t)$ and $v(t)$ graphs for door motion using video analysis software. Let them discuss how the graphs relate to each other and how door operation is related to typical speed of walking.	CP: E&A, ch 2, p 50 (problem using real data)
2. A water strider moving on the water's surface.	Make a video of water strider moving approximately along a straight line. Estimate the length of the water strider.	Ask the students to draw $x(t)$ and $v(t)$ graphs for the water strider's motion using video analysis software. Let them discuss how the graphs relate to each other. What is the maximum acceleration of the water strider?	CP: E&A, ch 2, p 49, (problem using real data)
3. Suspend a hairdryer by holding a cord and then switch on the hairdryer	Determine the angle between the cord when the hairdryer is on and the	Ask the students to draw a force diagram for the hairdryer in both cases and	CP: E&A, ch 6, p 174, (problem

(first turn the disconnected hairdryer on and then plug the connection into a socket).	vertical. You can take two photos of the hairdryer from the same position (one when it is off and one when it is on) and determine the angle from the photos. Measure the mass of the hairdryer.	explain the outcome of the experiment using momentum arguments.	using real data)
4. Seesaw (such as shown in the photograph below) or some other piece of playground equipment.	Take a photo of the seesaw. Make sure the dimensions of the seesaw on the photo are not affected by the perspective (take photo from far away, facing perpendicular to the seesaw; see the photo on the left).	Ask the students to invent a physics problem using the photograph. The problem should involve forces and torques. Provide the students with the masses of the people and ask them to try to occupy seats on the seesaw so that the seesaw is in equilibrium.	CP: E&A, ch 8, p 246, (problem)
5. Close an empty soda bottle while flying in an airplane or hiking in the mountains. Observe the bottle after landing/descending the mountain. When coming home, immerse the closed bottle in water, open the cap, and let water fill the bottle, so it obtains its original shape.	Take a photograph of the bottle during flight and when landing. Measure the volume of water that went into the bottle and measure the volume of the empty bottle.	Ask the students to determine the air pressure in the airplane/at the top of the mountain indicating the assumptions that they made. Ask them to compare the values with those obtained with an independent experiment or found in the literature.	CP: E&A, ch 12, pp 371, 375, (solved example)
6. Dive with an upside down empty soda bottle to a certain depth in the sea or in a pool and then close the bottle.	Measure the volume of water that went into the bottle and measure the volume of the empty bottle. If possible, measure also the depth to which you dived (for example, using a long rope).	Ask the students to determine the depth to which a person dived indicating the assumptions that they made. Ask them to compare the values with those obtained from an independent experiment.	CP: E&A, ch 13, p 413, (problem)
7. (Tibetan tea) Compare how hot water in a cup cools if there is a thin layer of oil on the surface and how it cools if there is no oil.	Use two identical cups filled with the same amount of hot water. Record the temperature of the water in both cups at regular time intervals, until the water cools down. Plot the data	Ask the students to explain the difference in the cooling curves. Ask the students to propose one or more testing experiments for their explanations, predict the outcome of	CP: E&A, ch 15, p 474, (problem)

(Continued)

Table 5.1. (*Continued*)

Phenomenon/ experiment	Requirements to the documentation/data collection	Types of the activities	References
	for both cups in a temperature-versus-time graph.	those experiments based on the explanations and then run the experiments.	
8. Obtain a shallow tray, put it on a table that is illuminated by sunlight or by a small distant lamp. Observe the shadow of the edge of the tray that forms at the bottom of the tray when the tray is empty and when the tray is filled with water.	Draw a scale on the bottom of the tray (see photo below) to measure the size of the shadow. Measure also the dimensions of the tray.	Ask the students to determine the index of refraction of water (including uncertainties) using the data that they collected. Ask them to compare the value with the value from the literature.	OALG 22.3.6, (activity)
9. Observe a partly immersed sign on the side wall of a swimming pool (see an example below).	Take a photograph.	Ask the students to explain (using ray diagrams) why the letters look different although they are all painted the same size and why we see multiple letters.	ALG 23.6.5, (activity)
10. Place a compass (or a mobile phone with an application for measuring magnetic field) on the floor of an underground train. Observe the compass needle or phone readings while riding the train. Do the same in a car (check different places in the car). Observe the needle or phone readings while starting and stopping the car engine.	Take a video of the compass or record the measurements on the mobile phone.	Ask students to explain the movements of the compass needle or changes in magnetic field components recorded by the mobile phone. Ask them to estimate any physical quantity that they can using the measured data and any assumptions that they made.	

5.3 Treating students' ideas as resources for the development of normative concepts, not as misconceptions that need to be weeded out

5.3.1 How do we treat student ideas?

The disagreement of how to treat student ideas is as old as the idea that the students are not blank slates when they come to us. The recognition that students come to class with existing knowledge is the first step in helping them learn (Zull 2002, Dehaene 2020). Investigating those ideas is the next important step. Often, these ideas look 'wrong' at a first glance as they do not match the normative physics understanding that we wish our students to develop. However, after that the viewpoints of educators diverge.

Some educators label the ideas that students bring into the classroom as 'misconceptions', sometimes labeled as preconceptions, alternative conceptions, etc (Clement 1982, McCloskey 1983, Halloun and Hestenes 1985, Kaltakci-Gurel *et al* 2016), while some label them as 'resources' (Smith *et al* 1994, and more below).

When researchers or teachers talk about misconceptions (or whatever label they give to them) they usually mean that students strongly hold firm cognitive structures (conceptions) that are different from experts, affect how students understand and explain natural phenomena, and must be overcome or replaced. In other words, if a student has a misconception about something, first, it means that they have some robust conception, and second, it means that this conception is wrong and needs to be cleared out from their mind and hopefully replaced by the correct conception. As D Hammer put it a long time ago (Hammer 1996), 'This view frames research designed to identify misconceptions and instruction designed to reveal, confront, and replace them.'

As we said above, this view appears consistent with the constructivist idea that students are not blank slates. Plus, it agrees with our experience: students do come up with incorrect ideas, don't they? We all have experiences when our students express views completely inconsistent with the laws of physics. For example, many researchers have documented the student 'misconception' that 'motion implies a force, and when there is no force, motion ceases' (impetus theory as described by Hestenes *et al* (1992)). It looked like in many instances (including standardized assessment instruments, such the Force concept inventory described in the paper cited above) students have this robust wrong idea.

Yet this 'misconception' viewpoint imposes certain tasks on the teacher, namely to identify such misconceptions in students' minds and help them get rid of those ideas. That last step breaks the connection with constructivism. If we remove those prior 'wrong' ideas from students' minds, then how do we help our students build new ideas which (as brain research tells us, see Dehaene (2020)) can only be developed if they connect to previously existing ideas? In fact, when Brookes and Etkina (2009) conducted a linguistic analysis of student responses about the 'misconception' that motion implies a force, they found that the students do not see the force as a cause of motion but treat it as a property of motion (similar to momentum or kinetic energy). Thus their responses are completely correct and only require some language correction, not the replacement of the conception. Based on similar analysis, other researchers (diSessa 1982, 1993, Smith *et al* 1994) have challenged the idea of a discontinuity between student and expert knowledge, arguing that it conflicts with the constructivist account of how we develop new understanding.

5.3.2 Knowledge in pieces

> In focusing only on how student ideas conflict with the expert concepts, misconceptions approach offers no account of productive ideas that might serve as resources for learning. Because they are fundamentally flawed, misconceptions themselves must be replaced.... An account of useful resources that are marshalled by learners is an essential component of a constructivist theory, but the misconception perspective fails to provide one. (Smith *et al* 1994, p 124, as cited in Hammer 1996).

At the same time, several researchers started investigating student ideas in detail and found that they are not robust ideas but depend on the context and the wording of a question (Schuster 1993 and many others). For example, the students have small experience-based ideas that they put together when asked a scientific question.

Is it possible that students (and all people) construct cognitive structures based on their everyday experience and then apply these structures to answer physics questions? diSessa answered this question positively when he developed the concept of 'phenomenological primitives' or 'p-prims', which refer to simple ideas that grew out of generalizations of everyday phenomena (diSessa 1993).

As Hammer writes: 'In diSessa's model, intuitive physics is made up of smaller, more fragmentary structures diSessa called phenomenological primitives, or p-prims for short. The misconceptions perspective, diSessa argued, confuses emergent knowledge, acts of conceiving in particular situations, for stable cognitive structures' (Hammer 1996, p 98).

diSessa identified several p-prims in student reasoning. Some of them are:

- Maintaining agency (for example food is needed on a hike).
- Actuating agency—the consequence of something lasts longer than this something (for example, if you burn your tongue eating hot food, the pain lasts longer than the contact with the hot beverage).
- Closer means stronger (the closer you are to the stove, the hotter it is).
- Ohm's p-prim—the stronger the cause, the stronger the effect, the stronger the resistance or impediment, the stronger its effect on the electric current, and several others.

Now, you can see that p-prims can be used to explain many physical and social phenomena, but they are not connected to normative physics knowledge. When we ask a student a physics question, they activate one of the available p-prims and sometimes the answer is correct (current though a resistor is directly proportional to the potential difference across it and inversely proportional to its resistance) and sometimes it is not (velocity is not directly proportional to the force and inversely proportional to the mass). Being able to habitually identify the p-prim on which the student based their reasoning is extremely important. If we can do this, we can use the existing p-prim and build on it by modifying the language or the context. We need to view p-prims as productive *resources* on which to build students' new knowledge (Smith *et al* 1994).

David Hammer and his collaborators developed the idea of resources further (see, e.g., Hammer (2000), Hammer and Elby (2003), Hammer *et al* (2005)). They proposed that resources are bits of prior knowledge that can be activated alone or with other resources as a student is reasoning about a physics topic. Resources are often context-dependent and may not be robust in their activation, i.e. a student may abandon a resource or change which resources they are activating rather quickly. Richards, Jones, and Etkina in the paper 'How students combine resources to make conceptual breakthroughs' (Richards *et al* 2018) describe resources using the definitions of different researchers as 'cognitive elements at various grain sizes that may be in different states of activation at any given moment' (Conlin *et al* 2010, pp 19–24); they can range from small, basic elements such as diSessa's p-prims, to more complex conceptual structures such as 'coherent theories about physical phenomena' (Harrer *et al* 2013, p 23101).

Hammer and colleagues discuss conceptual and epistemological resources. Conceptual resources are similar to p-prims but differ from them in size and scope and many of them are physics related. For example, a conceptual resource of energy as a substance can help a student successfully explain how a battery powers a lightbulb, but when used to analyse what happens to electric current in a circuit, it might lead to the incorrect answer that current is used up in a circuit. Epistemological resources relate to the nature of knowledge and learning (Hammer and Elby 2003). An epistemological resource of knowledge as fabricated stuff can help students think of developing their own explanations but if they have a resource that knowledge comes from authority, they will want their teacher to 'give them an answer'. Like p-prims, conceptual and epistemological resources are activated when we ask students questions or when they are interpreting reading materials. While these p-prims and resources sometimes lead the students to give answers that are 'wrong', we should try to 'diagnose' the source of their ideas and channel this source in a productive direction.

Here is an anecdote might help. In the Rutgers Physics Teacher Preparation Program in the course 'Development of Ideas in Physical Science' that we discussed in chapter 4, there is an assignment when the students need to interview a novice about a specific physics topic, transcribe the interview, and then interpret what the subject said. One of the pre-service teachers had to interview a novice about electric charge. This was his comment at the end of the discussion of the interview: 'It looks like my interviewee did not know anything about electric charge, but if you replace the word 'energy' with the words 'electric charge' in his answers, most of them are absolutely correct'. This comment shows how important it is to listen to the students carefully and think of what resources they activate when responding to our questions. Without a doubt, we both consider the frameworks of p-prims and resources much more productive in helping students learn than the framework of misconceptions (or alternative or naïve conceptions), and we try to avoid the term 'misconceptions' when talking about student ideas.

Naturally, within the resource framework, we still appreciate that students often need to overcome difficulties when constructing physics ideas. There are several well-documented student difficulties (most created by instruction, or by the confusing language, or by the context of our questions that is unfamiliar to the students) that we acknowledge and list them in every chapter of the *Instructor Guide* (IG).

Here is an example from IG chapter 3, Newtonian mechanics:

> The most difficult is the meaning of the word 'force' as a quantity that characterizes an interaction between two objects as opposed to the motion of an object. The reason for this difficulty is the language we use in everyday life. The difficulty that stems from our teaching is thinking that *ma* is a force and using *ma* to calculate any force. Other common difficulties include thinking that objects move in the direction of forces, and that any two forces that are the same in magnitude and opposite in direction are Newton's third law forces. When drawing force diagrams for an object of interest, students mistakenly put forces exerted by the object of interest on some other object.

This example shows the causes of the first two difficulties but does not address the causes of others. We can think of the 'moving in the direction of the force' difficulty as arising from focusing on the experience when any motion starts—the object always starts moving in the direction of the sum of the forces. Therefore, this difficulty stems from generalizing from some of our real-life experiences and forgetting about others. If we start from this experience and then ask students to analyse their experiences when the forces are exerted on an already moving object (for example, an object upward, under the condition that the students apply the correct definition of force), they will see that their rule only applies for the beginning of the motion, and not when an object is already in flight on the way up. The difficulty with Newton's third law stems from teachers' focus on 'equal in magnitude and opposite in direction' with less emphasis on the fact that these two forces characterize the same interaction. Here, student thinking is very productive for applications of Newton's second law and all we need to do is ask on what systems those equal and opposite forces are exerted and what interactions they describe. Again, there is nothing wrong with students' thinking here, it is simply misapplied. And the last difficulty can again be caused by teaching—through not identifying the system and the environment before drawing the forces. Here it is interesting that combining student reasoning related to Newton's third law (above) with this difficulty can help students with the application of the third law when they are trying to put the forces that the system exerts on an object in the environment on their force diagram. Therefore, none of these difficulties is a firm wrong concept that needs to be removed from students' minds but a productive resource that can be used to develop conceptual understanding.

5.3.3 The ISLE approach and student ideas

At this point in the chapter, you may be wondering: how does the ISLE approach address students' ideas, including p-prims and resources? As we do not ask students to make predictions before initial observational experiments, they are free to observe the phenomenon(a) without any expectations. When they describe what they observe using simple language, again, we are trying not to tap into their resources (yet). But when they have to come up with 'wild ideas' explaining the observational experiments, this is when their resources and p-prims come into play. For example, students conduct an observational experiment described in chapter 1 with a lightbulb illuminating the walls and the ceiling of the room. The students need to represent how the bulb's light rays reach all the points of the room. Their first model is the following (see figure 5.8(a); we met this model in chapter 1).

Why would they come up with this model? The resource here that they tap is a commonly used drawing of the Sun (see figure 5.8(b)). While this will turn out to be a wrong model, the beauty of this model is that it is easily testable. When the students themselves devise the testing experiments, make predictions of their outcomes using their model, run the experiments and find that the outcome do not match the predictions, they do not get upset at all. In fact, they get intrigued and think how they can tweak the model to account for the outcomes. Therefore, you can see that in ISLE the students bring their ideas at the stage of the testing experiments and reflect on why those ideas may be rejected sometimes. These actions become epistemological resources—e.g. every idea needs to be experimentally testable—which help them navigate new physics knowledge.

Figure 5.8. (a) The ray model of an extended light source drawn by a student. (b) The Sun drawn by children.

As another example of the ISLE approach to address student ideas, we can use observational experiments to help students create a 'correct' model when the model is so counterintuitive that they do not believe it. This is how one of the users of the ISLE approach, Allison Daubert, describes this experience:

> The biggest place where I think we bring in student ideas is during testing experiments. (Key here—not observational experiments at the

beginning). So, as an example, for Newton's third law, students observe that the readings of $F_{1 \text{ on } 2}$ is always the same as $F_{2 \text{ on } 1}$, but really, they don't believe it yet. We let them test all of these ideas about how motion or mass might affect the sizes of these forces in the testing experiments where they themselves design the experiments and they themselves get to disprove these ideas. I often discuss with students at the end of the testing experiments about how their initial ideas just contained slightly wrong language—that's all. While yes, the objects exert the same magnitude forces on each other their accelerations are drastically different, and that's the experience that they have in life, and that's the value in their idea. (Daubert 2022)

From the above two examples emerges a very big idea. One of the mains goals of the ISLE approach is to help students create new epistemological resources which resemble the epistemological resources of scientists! In other words, one of our main goals is to help our students develop scientific epistemology (the word epistemology means the study of the structure and development of knowledge). Having scientific epistemology, i.e. the approach to the nature and development of knowledge that scientists have, will allow our students to learn how to question claims that people make, how to view scientific knowledge as continuously improving (this is something that lay people have a hard time accepting), how to separate experimental facts from hypotheses or opinions, and so forth. If you are truly implementing the ISLE approach, your students will develop all these resources and they will be using them for the rest of their lives.

5.3.4 Unexpected treasure: student learning resources

While we discuss conceptual resources that students bring into learning and epistemological resources that we wish them to develop, it is good to remember that our students have other resources that they bring into learning. We can call them learning resources. Normally when we talk about those, people mean resources that the students can use to learn—textbooks, websites, etc. Here, we mean a completely different thing: *resources that the students bring into learning* that relate to how people learn. These are the resources that our learning system should build on (and ISLE does).

Think of all the stuff that our students learn without our help. What skills do they develop doing it and how can those skills help them learn physics through the ISLE approach?

1. Our students know how to persevere and to take time to learn something (think of multiple lives in computer games or repeating the same trick on a skateboard). In ISLE, this helps them when they design their own experiments, have to come up with a new explanation after rejecting the previous one, and when they are given an opportunity to improve their work by submitting a new version. Here their perseverance is rewarded when the experiment that they have designed works, the new explanation fails to be

rejected, or the resubmitted work allows them to learn more and get a better grade. ISLE offers a system that rewards perseverance and respects that different people may need different numbers of attempts or different amounts of time to learn something. It is the final outcome of this learning, not the speed with which it is achieved, that is rewarded.
2. Our students know how to work in a group and learn from group mates (e.g. they play with other people online where there is a ton of learning going on; they practice their skills in skateparks where more experienced skateboarders give support and advice to younger people). This helps them with group work in ISLE where most of the learning is through collaboration. This collaboration is rewarded when the students work together designing their own experiments or solving problems and submit one report for the whole group. As members receive the same grade for their work, they are motivated to work together. The same is true for any group activity that is organized well.
3. Our students know how to fail and how to learn from their mistakes (think about those who like to cook, garden, and, of course, gamers, skateboarders, musicians, athletes, *all* of them!) This helps them in ISLE when it is difficult to figure out stuff and you need to try again. We build on this resource with resubmissions of work and by encouraging students and giving them time to rethink their models and redo their experiments. In fact, rejection of a model through an experiment is a good thing in the ISLE approach, not a bad thing.
4. Our students know how to look for feedback and how to deal with feedback (think again about all the activities in which they engage). In the ISLE approach, a lot of feedback is provided through scientific abilities rubrics which allow the students to engage in self-assessment—and this self-assessment is crucial when they need to improve their work in the classroom or on a chess board or on the computer screen to survive.
5. They are creative (watch them play Minecraft!). ISLE builds on their creativity through engaging them to develop their own explanations, design their own experiments to find patterns or test ideas, and, most importantly, pose and answer their own questions.

Bottom line—our students are expert learners. And then they come into our traditional education where they are rewarded for individual problem solving and speedy answers, where grades are given once and forever for an assignment, and they are required to follow detailed instructions in the labs, which mainly verify what they have heard in lectures. All these elements (and many others) serve to reject the learning resources that students bring with them and teach them that failure is not an option, perseverance is only good until the first try, speed is more important than understanding, helping others does not improve your learning or your grade, and there is no room or no time for creativity. Slowly, day by day in such an environment, they stop brining their wealth to the table and start developing apathy, boredom, and lack of interest to struggle. We all are familiar with these issues and often blame the students for their lack of motivation, perseverance, etc. But really it is the result of our educational system. Now, when you get the students who have been in this system for many years and have stopped applying their real learning resources in school a long time ago, how do you remind them that they have these resources? We need to remind them because otherwise they will not be successful in the ISLE classroom.

To help your students remember all the learning resources that they possess, you can engage your students in the Expertise Activity, brilliantly designed by David Brookes and Yuhfen Lin (Brookes and Lin 2012). David Brookes described it and other methods to help students activate their learning resources in one of the interludes in our first ISLE book (Etkina *et al* 2019, pp 3–9):

> In this activity, we ask students to identify a hobby they are accomplished at and divide them into thematic groups based on their responses. For example, there is often a cooking group, a sports group, a board and computer games group, etc. We task each group with drawing up a learning cycle on their whiteboard that explains to the rest of the class how one can move from becoming a novice to an expert in their chosen field of expertise. The important point is that they must draw a repeatable cycle, not a list of 'what does it take to become good at something'. Having them construct a learning cycle draws out all the keys features of real-life learning: the need for motivation and persistence, the role of critical self-evaluation and seeking feedback from others, etc. These are all features of the ISLE classroom. At the end, students present their learning cycles to each other and I draw the discussion together at the end, highlighting common features, connecting those features to how the ISLE class is set up, and sometimes have them watch Dr Tae's TED talk.

5.3.5 Developing teacher habits in recognizing and building on resources

How do we help future teachers develop habits of recognizing p-prims and all types of resources when they interact with the students? First, as part of their education, they have to read the research papers cited above so that they are aware of different approaches to the interpretation of students' 'wrong answers'. But this is only the first step.

Second, pre-service teachers need to reflect on their own ideas when they are proposing explanations for the observational experiments or trying to answer questions. Where do their own ideas come from? What resources or p-prims did they use?

The next step is to provide pre-service and in-service teachers with examples of student answers to physics questions and their reasoning and to ask them to identify productive and unproductive elements and explain how to arrive at the 'correct' answer. In the textbook CP: E&A, among the end-of-chapter problems for every chapter there are questions and problems of those types. We provide students with several answers to a problem and ask them to explain what correct physics ideas could have led to the incorrect answers. Here is an example (for similar problems see the problems in chapter 4 of CP: E&A):

Problem Trisha, Gaurang, and Tyron are sitting on a sled on a slope covered with a hard snow. The sled is stationary. The friends have different suggestions for how to make the sled start moving:
Trisha: If one of us gets off, the sled will start moving.
Gaurang: We should invite another person to join us, and then the sled will start moving.
Tyron: We should get off the sled, polish the bottom of the sled to make it smoother, and sit back down on it. The sled will then start moving.
Discuss the students' suggestions and decide whose reasoning is correct and explain why you think this. In your explanation use physics concepts, including appropriate diagrams, why those who were incorrect said what they did, and indicate what part of their reasoning was correct.

We also provide teachers with students' incorrect answers to questions and problems and ask them to identify p-prims and resources that led to the construction of the answers. Many examples of such can be found in the distractors for traditional multiple-choice instruments such as the Force Concept Inventory (FCI) (Hestenes *et al* 1992) and the Conceptual Survey of Electricity and Magnetism (CSEM) (Maloney *et al* 2001). When those surveys were created, the authors used student interviews. They posed open-ended questions to the interviewees and recorded their answers (they were interested in the incorrect answers as they wanted to use them as distractors). For example, these are the answer choices to one of the FCI questions:
A stone dropped from the roof of a single-story building to the surface of Earth:

1. Reaches a maximum speed quite soon after release and then falls as constant speed thereafter.
2. Speeds up as it falls because the gravitational attraction gets considerably stronger as the stone gets closer to Earth.
3. Speeds up because of an almost constant force of gravity acting upon it.
4. Falls because of the natural tendency of all objects to rest on the surface of Earth.
5. Falls because of combined effects of the force of gravity pushing it downward and the force of air pushing it downward.

We can find productive resources in every answer choice and some of them can also be explained using the model of p-prims. For example, the first answer would be correct if the object that we dropped were a large piece of paper (think of a coffee filter) or, in another context, a parachutist. In this case, it would reach terminal velocity very quickly. This answer is an excellent resource for learning terminal velocity. The second answer would also be correct if the stone was dropped from a very large height. It follows from the p-prim 'the closer—the stronger' that we discussed above. At the same time, it is a great resource to use when the students learn how gravitational force depends on the distance between two interacting objects. Answer choice four is also based on a p-prim of a natural tendency of objects to do certain things and observations that everyone makes of falling objects (disregarding helium balloons for example), but as a resource, it is very useful when studying energy. Finally, the last choice is a resource for understanding that there are two objects interacting with the falling rock—Earth and air. While most students do not have a problem with Earth pulling down, the direction in which air pulls or pushes on the object is more complicated. In fact, when the rock is thrown upward, air and Earth both exert a downwards force on it. From the above analysis, you can see how on one hand, we can always find a context in which 'wrong' answers would be right, and on the other hand, we can figure out what prompts the students choose the wrong answer choices and how to build on their ideas.

In addition, over the years we have accumulated a large library of questions for pre- and in-service teachers to interpret student reasoning. Examples of such questions and possible correct answers are in appendix A for this chapter.

Finally, playing the Expertise Activity with teachers allows them to see how their own hobbies and interests help them develop habits that are essential for their work as a teacher and for the learning of their students.

5.4 Recognizing experimentally testable student ideas and knowing how to test student ideas

This section starts with a very old memory. One of us (Eugenia) was teaching a science methods course for future elementary school teachers. She put a streak of rubbing alcohol on the board and asked the students to observe it. Then she asked the students to tell her what happened to the streak. They all agreed that the streak disappeared. After that Eugenia asked her students to explain how the streak disappeared. She expected them to say that the particles of alcohol left the streak as they were moving randomly and some had enough energy to leave. But instead, she heard two completely different ideas.

Idea 1: Alcohol was 'sucked into the board'.
Idea 2: Air absorbed the alcohol.

Eugenia was surprised that a question that she thought to be very easy for the students turned out to be so difficult. She did not know how to respond to those ideas. She put students in groups and asked them to think of possible experiments whose outcomes they could predict using those two ideas. The students came up with the following experiments: (A) take a piece of paper weigh it and then wet it in alcohol, weigh it again and wait for it to dry; (B) wet a piece of paper in alcohol and put it in a vacuum. Their reasoning for these experiments is described in table 5.2.

The first experiment is easy to conduct, and Eugenia could do it right away as she had a sensitive scale and plenty of rubbing alcohol. See the experiment here: https://mediaplayer.pearsoncmg.com/assets/_frames.true/sci-phys-egv2e-alg-12–1–3a.

The second experiment while theoretically possible, is not feasible in the classroom. We can use a vacuum pump and a glass jar (see figure 5.9) but even with the best pump we cannot achieve a total vacuum. After several discussions with D T Brookes, they came up with an experiment that uses student ideas but allows for regular classroom experiments. Watch the video and think why they needed two pieces of paper (one in the glass jar at lower pressure and one in the room) and not one https://mediaplayer.pearsoncmg.com/assets/_frames.true/sci-phys-egv2e-alg-12-1-3b.

Table 5.2. Student reasoning process for making predictions of the outcomes of testing experiments using two different ideas.

Testing experiment	Prediction based on idea 1	Prediction based on idea 2
(A) Take a piece of paper, weigh it and then wet it in rubbing alcohol. Weigh the paper. Then wait for it to dry and then weight it again.	The weight of the paper after the alcohol dried should be the same as the weight when it was wet because the alcohol is still in the paper.	The weight of the paper after alcohol dried should be the same as the weight of the paper before it was wetted with alcohol because air would take all the alcohol from the paper.
(B) Wet a piece of paper in alcohol and put it in a vacuum.	The vacuum should make no difference assuming that it does not change the paper's ability to dry.	The paper should not dry as there is no air to absorb alcohol.

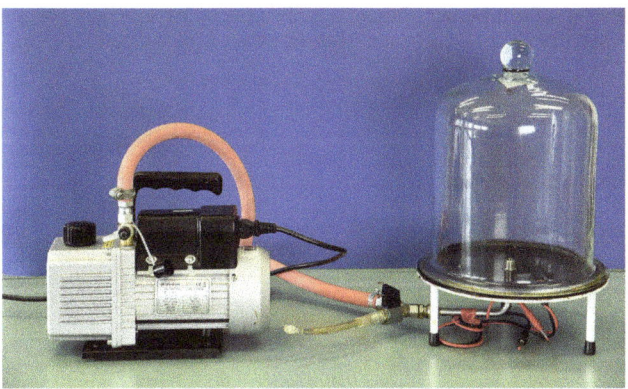

Figure 5.9. A typical school lab vacuum pump and a glass jar.

If you are thinking that they used the strip in the air as a control, you are right! They knew that the pump only pumps some air from under the jar, making it rarefied but not a complete vacuum. Therefore, they needed a control with unrarefied air. However, there was another complication with the experiment. While the pump was working, there was airflow across the piece of paper under the jar, therefore it was important to focus students' attention on what was happening *after* the pump was turned off and there was no movement of air inside the jar. With this 'improved' version of the testing experiment that the students proposed, the reasoning looks like that in table 5.3.

This example shows how sometimes student ideas for testing experiments, while technically correct, cannot be implemented in the classroom and need to be modified using the knowledge of the teacher (unless the students come up with the control idea themselves). Such thinking is the habit of mind that we need to strive to develop. We habitually think not only which students' ideas are testable and how to test them, but also how to modify students' ideas that seem impossible to implement into something that we can do.

Table 5.3. Improved testing experiments, predictions, and outcomes.

Testing experiment	Prediction based on idea 1	Prediction based on idea 2	Outcome
Take a strip of paper, weigh it and then wet it with rubbing alcohol. Weigh the paper. Then wait for it to dry and then weigh it again.	The weight of the paper after the alcohol dried should be the same as the weight when it was wet because the alcohol is still in the paper.	The weight of the paper after the alcohol dried should be the same as the weight of the paper before it was wetted with alcohol because air would take all alcohol from the paper.	The weight of the paper is the same as it was before it was wetted with alcohol. The outcome matches the prediction based on idea 2. We can reject idea 1.
Cut two identical strips of paper and dip them in rubbing alcohol. Put one under the vacuum jar and leave the other one outside. Pump out the air from the jar and then compare how fast the two strips dry.	Both papers should dry at the same rate, assuming air is not affecting how fast the paper absorbs the alcohol.	The paper left outside should dry faster because air will be taking alcohol away from it. Under the vacuum jar there is much less air (per unit volume), so the process should take longer.	The paper under the jar dries faster. We can reject idea 2.

In addition, you might be asking a question—what should I do if I do not have a vacuum jar and a pump at my disposal? This is where our developed materials come in handy. Over the years we listened to students' explanations and made videos of their proposed experiments. The videos are available for free at https://media.pearsoncmg.com/aw/aw_etkina_cp_2/videos/, http://islevideos.net/, the ISLE YouTube channel, and at https://www.youtube.com/@gorazdplaninsicfmful3516. If your students propose something and you do not have equipment, check out these resources and you might find something that fits.

Here are two other possible ways to find suitable testing experiments. The first one is to ask your students to come up with a substitution or a different experiment that you can do (you can tell the students which equipment is available). The second is to discuss the issue with the physics teachers or physics researchers in the community. For the latter we recommend joining the Facebook group 'Exploring and Applying Physics' (https://www.facebook.com/groups/320431092109343343)—this is a group of over 2400 physics teachers using the ISLE approach all over the world—they will help you!

Here follows a story about the former way—asking your students to come up with a different experiment.

When our students study wave optics, they learn that light can be modeled as a wave. When they study electromagnetic waves, they learn that those light 'waves' can be modeled as waves of changing E and B fields. The next step is to help students learn that light can be modeled as a stream of quanta. One of the first observational experiments that students observe in this unit (see chapter 27 in the textbook) is a negatively charged electroscope illuminated by regular and UV light sources. When illuminated by UV light, the electroscope discharges (see video https://youtu.be/X7EQJU9bxV4). The students need to use the electromagnetic wave model of light to explain this observation. One of their explanations sometimes is that the UV light ionizes the surrounding air and ionized air discharges the electroscope. How to test this idea? The students immediately suggest putting the electroscope in a vacuum—to get rid of the air. But this is not a feasible experiment. In this case we ask students to come up with a 'doable' experiment whose outcome they can predict using their explanation. After group discussions, they suggest charging the electroscope positively. If it is ionized air that discharges the electroscope, then there should be no difference in the behavior of the electroscope. To their surprise, the outcome of the experiment does not match their prediction (see video https://youtu.be/EgxVXOnsFx0) so they can reject the ionized air explanation and they need to come up with a new one.

In this example, we used qualitative testing experiments to test mechanistic explanations. We can also design testing experiments to test causal explanations or mathematical models.

When students study rotational motion, students observe that an object's rotational acceleration is affected by the sum of the torques exerted on it. Figure 5.10(a) shows one simple experimental set-up that allows students to observe this, consisting of a metal arm that can rotate freely around a vertical axis and two fans that can be

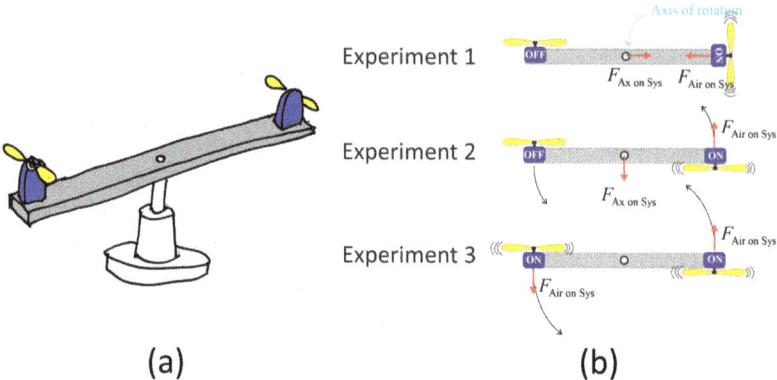

Figure 5.10. (a) Experimental set-up for studying rotational motion. (b) Sketches of three observational experiments (top view).

attached at different places along the arm and turned on and off (a similar activity can be found in chapter 9 of the textbook CP: E&A).

The initial observational experiments (figure 5.10(b)) allow the students to identify the following patterns:

- An external force that produces zero torque on the arm does not change the arm's rotational velocity.
- External forces that produce a nonzero net torque on the arm cause rotational acceleration. Doubling the net torque doubles the rotational acceleration of the arm.

The students might think that the mass of the object affects the rotational acceleration in a similar way as in Newton's second law that they learned earlier. This is an experimentally testable idea. In table 5.4 we show example testing experiments, but before you read on, try to come up with your own experiments and then compare them to ours.

Table 5.4. Testing whether the rotational acceleration depends on the mass of the rotating object (adapted from CP: E&A, chapter 9). The the videos of the experiments are at https://mediaplayer.pearsoncmg.com/assets/_frames.true/secs-egv2e-testing-the-hypothesis-that-mass-affects-rotational-acceleration.

Testing experiment	Prediction	Outcome
	If the rotational acceleration depends on the external torques and the mass of the system and we have the same fans as in experiment 2, then changing the location of the turned-off fan should not change the rotational	The rotational acceleration of the arm is greater than in experiment 2.

Repeat experiment 2 (figure 5.10(b)), but this time move the turned-off fan closer to the axis of rotation.

acceleration of the system.

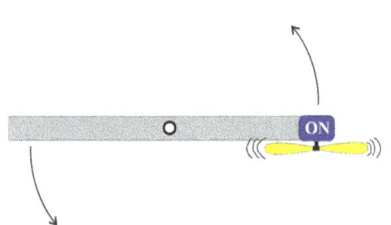

Because the mass of the system decreases, the rotational acceleration should increase.

The rotational acceleration of the arm is greater than in experiment 2 and greater than in experiment 1 (figure 5.10(b)).

Remove the turned-off fan from the arm and repeat experiment 2 (figure 5.10(b)).

The bottom line is that testing student ideas is not only important for 'correct' physics but also for helping the students develop the habits of mind needed to question the claims that bombard them all the time. Over the years, we have accumulated a huge library of testing experiments for different ideas and we have videoed many of them—the links are in CP: E&A, ALG and OALG.

5.5 Applying systems reasoning to the analysis of physical phenomena

The concept of a system in physics is different from the concept of a system in other disciplines, for example in biology. In physics, a system is an object or a group of objects (macroscopic or microscopic) that we choose, and the rest of the world is the environment. In physics, the objects outside the system are as important as the objects inside the system. Below, we show several examples of how we can choose a system and analyse the same situation using different systems.

To apply Newton's second law to solve problems, we first need to choose our system. We then look at the external objects that interact with our system. They can either touch it or interact at a distance (such as Earth, magnets, or electrically charged objects). We represent the forces that those external objects exert on the system with arrows on force diagrams. We then use the force diagrams to write Newton's second law. The choice of the system determines the analysis. The following exercise shows the difference.

> A dynamics cart is held at rest on a level track. A spring scale is attached to the cart and a string is attached to the spring scale. The string passes

over a pulley at the end of the track and a hanging object is attached (see the figure below). Predict what will happen to the reading on the spring scale when the cart is released.

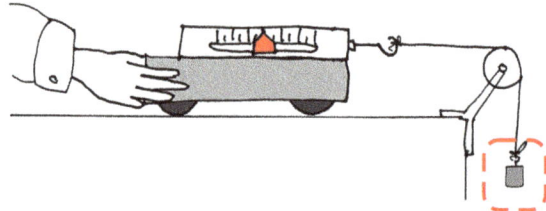

Here the choice of the system is crucial for making the prediction. We start with choosing the hanging object as the system (indicated by the red dashed line on the figure above). When the cart is held by the person, the sum of the forces exerted on the hanging object is zero (see the force diagram in figure 5.11(a)). The scale reads the force exerted by the string on the cart and on the hanging object. When the car is released, the object starts pulling it to the right and accelerating down itself. As it accelerates down, the sum of the forces exerted on it should point down (see figure 5.11(b)), this means that the force exerted on the object by the string decreases, and so does the reading of the scale.

See the video for confirmation (https://youtu.be/pznrUVWFNgU). While the problem is relatively easy to solve when we choose the hanging object as a system, it becomes impossible if we choose the cart as the system.

When we choose a system for force analysis, the forces internal to the system are not shown on the force diagram as their sum according to Newton's third law is always zero. A good illustration of this idea is the following exercise adapted from chapter 3 of the ALG.

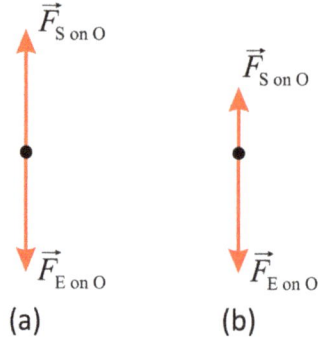

Figure 5.11. (a) Force diagram for the hanging object when it is at rest and is not accelerating. (b) Force diagram for the hanging object when it starts moving down, accelerating downward.

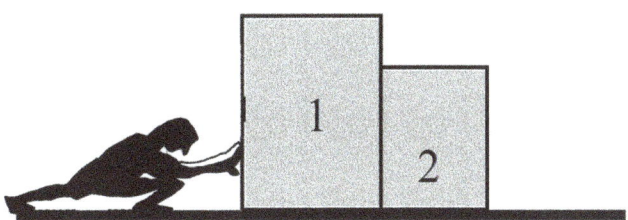

5.5.1 Reason

Block 1 has a mass of 6 kg, block 2 has a mass of 3 kg. The person pushing the blocks exerts a force of 90 N on block 1. Assume the surface is smooth.

a. Draw a force diagram for the combined two-block system. Do not forget to mark the perpendicular x- and y-axes. Use your force diagram to apply Newton's second law to determine the acceleration of the system of both blocks.

b. Now choose each block as a separate system and draw separate force diagrams for block 1 and block 2. Apply Newton's second law to block 1 and use the acceleration you found previously to determine the magnitude of $\vec{F}_{2 \text{ on } 1}$.

c. Evaluate the answer you found for $\vec{F}_{2 \text{ on } 1}$ by applying Newton's second law to block 2 to find the acceleration of block 2. (If you know $\vec{F}_{2 \text{ on } 1}$ you know $\vec{F}_{1 \text{ on } 2}$) Make sure the acceleration you find is consistent with the other accelerations you found in earlier parts. If not, try to figure out where you went wrong.

SOLUTION:
(a)

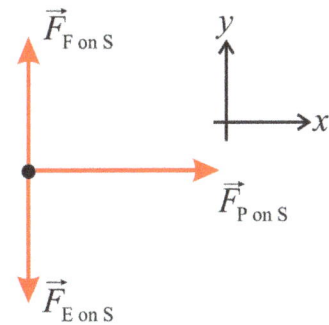

$$a_x = \frac{\Sigma F_{\text{on S } x}}{m_1 + m_2} = \frac{90 \text{ N}}{6 \text{ kg} + 3 \text{ kg}} = 10 \text{ m s}^{-2}.$$

(b)

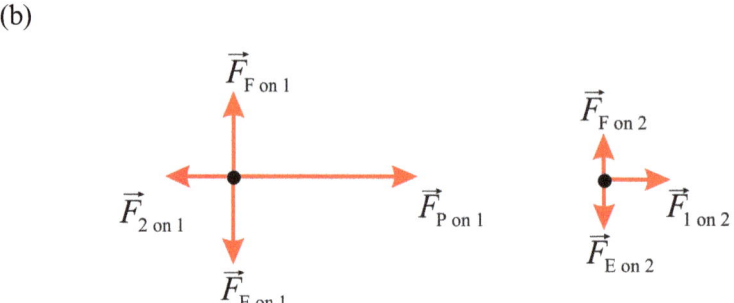

$$a_x = \frac{\sum F_{\text{on S }x}}{m_1} = \frac{F_{\text{P on 1}x} + F_{\text{2 on 1}x}}{m_1} \Rightarrow$$

$$F_{\text{2 on 1}x} = a_x m_1 - F_{\text{P on 1}x} = 10 \text{ m s}^{-2} \times 6 \text{ kg} - 90 \text{ N} = -30 \text{ N}.$$

(c)

$$F_{\text{1 on 2}x} = -F_{\text{2 on 1}x} = 30 \text{ N}$$

$$a_x = \frac{\sum F_{\text{on 2}x}}{m_2} = \frac{F_{\text{1 on 2}x}}{m_2} = \frac{30 \text{ N}}{3 \text{ kg}} = 10 \text{ m s}^{-2}.$$

Choosing the system for analysis habitually when analysing situations involving forces is extremely important. Here is a classical problem of a horse pulling a sled:

> When a horse pulls a sled, it exerts the same force in magnitude and opposite in direction on the sled as the sled exerts on the horse. As the forces are the same in magnitude and opposite in direction, they add to zero. This means that the horse should have zero acceleration. How can the horse start moving?

To resolve this seemingly unresolvable issue we need to ask ourselves: what is the system for analysis? If it is the horse (see figure 5.12(a)), then the forces exerted on the horse are by the sled, Earth, and snow. Drawing a force diagram one can see that the sum of these forces points forward. The force that the horse exerts on the sled is not a force exerted on the horse and thus should not need to be on the force diagram for the horse. If the system is the sled, then the objects interacting with it are the horse, Earth, and snow. Again, we can see from the force diagram that the sled (and the horse) accelerates forward. If we choose the horse and the sled as the system, then the forces that are exerted on the system are the forces exerted by the

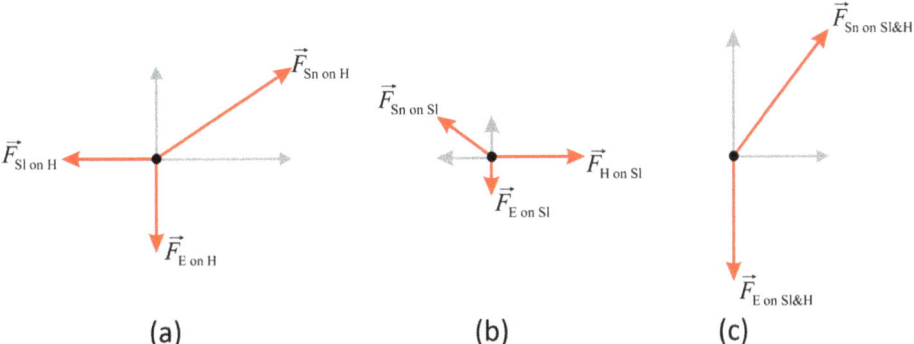

Figure 5.12. Force diagrams for the following choices of a system: (a) horse, (b) sled, and (c) horse and sled. We assumed that the rope connecting the horse and the sled is parallel to the ground. Gray arrows indicate the parallel-to-the-surface and perpendicular-to-the-surface components of the forces that snow exerts on the objects (friction and normal force).

snow and Earth. The forces exerted by the horse on the sled and sled on the horse are internal and add to zero. Any system we choose for analysis allows us to draw a force diagram that shows that every system accelerates forward (see figures 5.12(a)–(c)).

While using a system approach in dynamics is relatively straightforward, the situation is different when we use a system approach in situations involving conserved quantities such as momentum or energy. Here the choice of a system affects the analysis much more. First, we need to remember that conserved quantities (such as mass in classical physics, total energy, and linear or rotational momentum) are always conserved but this does not mean that they are constant for a system. The term constant means that the quantity does not change in time or in space. We can have motion of an object with constant acceleration, which means that the acceleration of the object at time t_1 is the same as at time t_2. This constancy does not mean that acceleration of an object is a conserved quantity. When the object changes its acceleration, let's say its acceleration increases by 5 ms^{-2}, this does not mean that the acceleration of some other object will decrease by exactly the same amount and if we redefine the system to include both objects, the total change of acceleration would be zero. Basically, the gain of acceleration by one object does not mean that some other object will lose the exact same amount of acceleration. Acceleration is not a conserved quantity. However, if a system gains 5 J of energy because some external object did work on it, this means that some other system lost exactly 5 J of energy. If we include all interacting objects in the system, the total change of energy would be zero. Energy is a conserved quantity. The same is true for momentum. This fact creates three issues when using a system analysis with energy (or momentum). First, the total energy of any system is always conserved (but not necessarily constant). The energy of a system changes when work is done on it or when energy is provided though the heating process or some other process (light, for example). Second, only the energy of an isolated

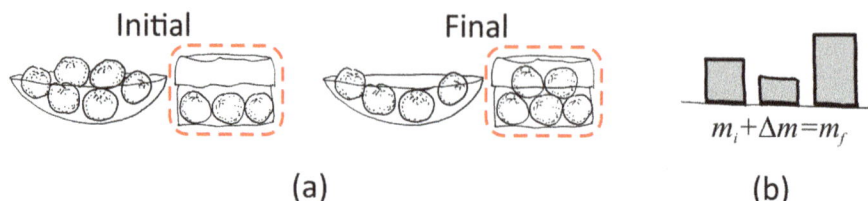

Figure 5.13. (a) Moving some oranges from the bowl to the bag (the process). (b) Bar chart representing the process for the system of oranges in the bag only. The red dashed line indicates the choice of system.

system (no interactions with the environment) is always constant. Third, the work done by objects inside the system provides a mechanism for the conversion of one form of energy into another form but it cannot change the total energy of the system. This idea of a conserved quantity allows us to represent processes with conserved quantities using bar charts (Van Heuvelen and Zou 2001). The example below shows how to draw a bar chart (for details see the textbook CP: E&A and the IG).

To understand how a bar chart works, let's take first a very simple situation involving the physical quantity of mass. Imagine you have a system (a bag of oranges) that has a total mass $m_i = 2$ kg. You have oranges in a fruit bowl too. You take some oranges $\Delta m = 0.5$ kg from the fruit bowl and add them to the bag. The final mass of the system (oranges in the bag) equals the sum of the initial mass and the added mass: $m_i + \Delta m = m_f$ or 2 kg + 0.5 kg = 2.5 kg (figure 5.13(a)). We see that while the number of oranges in the bag is not constant, it changes by $\Delta m = 0.5$ kg, if we redefine the system to be the oranges in the bag + the oranges in the bowl, the mass of this system remains constant as no oranges left the system or were added to it.

We can represent this process with a bar chart (figure 5.13(b)) for the system 'oranges in the bag'. The bar on the left represents the initial mass of the system, the central bar represents the mass added or taken away, and the bar on the right represents the mass of the system in the final situation. As a result, the height of the left bar plus the height of the central bar equals the height of the right bar. The bar chart allows us to keep track of the changes in mass of a system even if the system is not isolated (as shown in this simple case).

The same idea applies to momentum, except that momentum is a vector quantity and has direction. The change in momentum of a system is equal to impulse. Mathematically these changes are described by the following equations.

The x- and y-component forms of the generalized impulse–momentum principle are

$$(m_1 v_{1ix} + m_2 v_{2ix} + \ldots) + \sum F_{\text{on Sys } x}(t_f - t_i) = (m_1 v_{1fx} + m_2 v_{2fx} + \ldots)$$

$$(m_1 v_{1iy} + m_2 v_{2iy} + \ldots) + \sum F_{\text{on Sys } y}(t_f - t_i) = (m_1 v_{1fy} + m_2 v_{2fy} + \ldots).$$

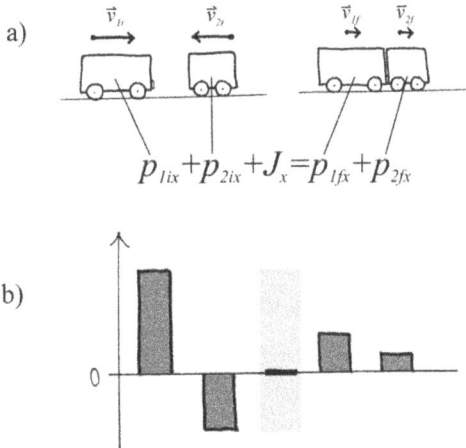

Figure 5.14. (a) Sketch of the process (*x*-axis points to the right). (b) Impulse–momentum bar chart representing the process (both carts are in the system).

Let us now represent the following process with a bar chart. A system of two carts with Velcro pads on the fronts move along a track at the same speed in the opposite directions. The mass of cart 1 is twice the mass of cart 2. The carts collide, get stuck together and start moving in the direction of velocity of cart 1 that originally had a larger magnitude of momentum. We will assume that the forces on the system due to friction and air drag are negligible.

Before constructing the bar chart, we represent the process using an initial–final sketch (figure 5.14(a)).

Identifying the initial and final states of the process is very important (in this case the initial state is before the collision and the final right after the collision). We then use the sketch to help construct the impulse–momentum bar chart (figure 5.14(b)). The lengths of the bars are qualitative indicators of the relative magnitudes of the momenta components. In the final state in the example shown, the carts are stuck together and are moving in the positive *x*-direction. Since they have the same velocity, the cart with the larger mass (1) has twice the final momentum compared to cart 2.

The middle, shaded column in the bar chart represents the net external impulse exerted on the system objects during the time interval ($t_f - t_i$). There is no impulse for the process shown because the sum of the forces exerted on the system during this time interval are zero (note that we neglected friction and air drag). The shading reminds us that impulse does not reside in the system; it is the influence of the external objects on the momentum of the system. Notice that the sum of the heights (note that height can be negative) of the bars on the left plus the height of the shaded impulse bar should equal the sum of the heights of the bars on the right. This 'conservation of bar heights' reflects the conservation of momentum.

We can use the bar chart to apply the generalized impulse–momentum equation. Each nonzero bar corresponds to a term in the equation; the sign of the term

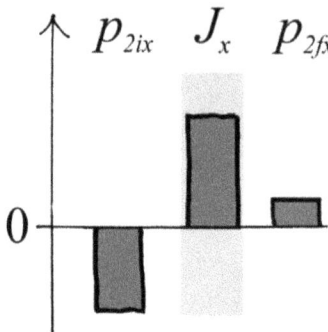

Figure 5.15. Impulse–momentum bar chart for the collision of two carts (cart 2 is the system).

depends on the orientation of the bar: $(p_{1ix} + p_{2ix}) + \Sigma F_{\text{on Sys }x}(t_f - t_i) = (p_{1fx} + p_{2fx})$. Here the x-subscript indicates the x-component of momentum (the direction of the momentum of the first cart is assumed to be positive). As we know the momenta of the carts before and after the collision: $p_{1ix} = -2p_{2ix}$; $\Sigma F_{\text{on Sys }x} = 0$ and $p_{1fx} = 2p_{2fx}$, we can for example find out what happened to the momentum of cart 2 as the result of the collision:

$$(-2p_{2ix}) + p_{2ix} = 2p_{2fx} + p_{1fx} \Rightarrow$$

$$p_{2fx} = -\frac{1}{3}p_{2ix}.$$

Cart 2 has momentum in the opposite direction and 1/3 of its initial magnitude.

We can now reanalyze the process using bar charts for a system that consists only of cart 2. In this case cart 1 is not a part of the system so it does not appear as a momentum bar, but it exerts an impulse on the system. The bar chart in figure 5.15 represents the process.

We can analyse the process using mathematics: $p_{2ix} + \Sigma F_{\text{on Sys }x}(t_f - t_i) = p_{2fx}$. From this expression we see that if we know the final momentum of the cart from the first approach, we can determine the impulse that the first cart exerted on the second:

$$p_{2ix} + F_{1 \text{ on } 2x}(t_f - t_i) = -\frac{1}{3}p_{2ix} \Rightarrow$$

$$F_{1 \text{ on } 2x}(t_f - t_i) = -\frac{4}{3}p_{2ix}.$$

Note that the impulse is negative—it means that the force exerted on cart 2 points in the direction opposite to its original velocity. If the time of the collision is known, we can determine the average force, and if the average force is known, we can find the time interval for the collision. This example shows how the choice of a system allows us to analyse the same situation from different angles.

A similar approach allows us to represent processes involving energy and work. Mathematically, work–energy processes are represented using the following equations.

Generalized work–energy principle: The sum of the initial energies E_i of a system plus the work W done on the system by external forces equals the sum of the final energies of the system E_f:

$$E_i + W = E_f \quad \text{or in mechanics}$$

$$(K_i + U_{gi} + U_{si}) + W = (K_f + U_{gf} + U_{sf} + \Delta U_{int}),$$

where K is the symbol for kinetic energy, U_g is the symbol for the gravitational potential energy; U_s is the symbol for the spring potential energy (or elastic energy) and U_{int} is the symbol for the internal energy.

Note that we have moved $U_{int\,i}$ to the right-hand side ($\Delta U_{int} = U_{int\,f} - U_{int\,i}$) since the values of internal energy are rarely known, while internal energy changes are.

Imagine that we wish to represent the following process. A low friction cart, which is moving along the horizontal track hits the sponge and bounces off it (figure 5.16(a); the video of the experiment is at https://youtu.be/EsbAYXjc8JI). Using the sensors in the cart we can record the graphs that show the time dependence of the velocity of the cart and the force exerted by the sponge on the cart (figure 5.16(b)). We notice that the speed of the cart after the collision is smaller than the speed before the collision. From the slow-motion video we can also see that the sponge compresses during the impact but resumes its original shape after the collision.

We will choose the system that consists of the cart and the sponge. The initial state will be right before the cart hits the sponge and the final state right after the cart bounces off it. We will assume that the friction and drag forces on the system are negligible. We can now construct the bar chart to represent the process (figure 5.17(a)).

The system starts with some kinetic energy and ends with smaller kinetic energy. As we included the sponge in the system, there is no work done. There is no elastic potential energy in the initial or final state, because the final shape of the sponge is the same as the initial shape, so the only other form of energy that can account for the decreased kinetic energy is the internal energy. We can now represent the process mathematically

$$K_i + 0 = K_f + \Delta U_{int} \Rightarrow$$

$$U_{int} = K_i - K_f = \frac{1}{2}mv_i^2 - \frac{1}{2}mv_f^2.$$

Knowing the initial and final velocity ($v_i = -0.50$ m s^{-1}, $v_f = 0.30$ s^{-1}, see figure 5.16(b)) and the mass of the cart (0.296 g in our case) we calculate $\Delta U_{int} = 0.024$ J. Can we test the hypotheses that the internal energy of the system increased? Since the change of the internal energy is positive, we expect the temperature of the sponge and the bumper to increase. An ideal device to detect

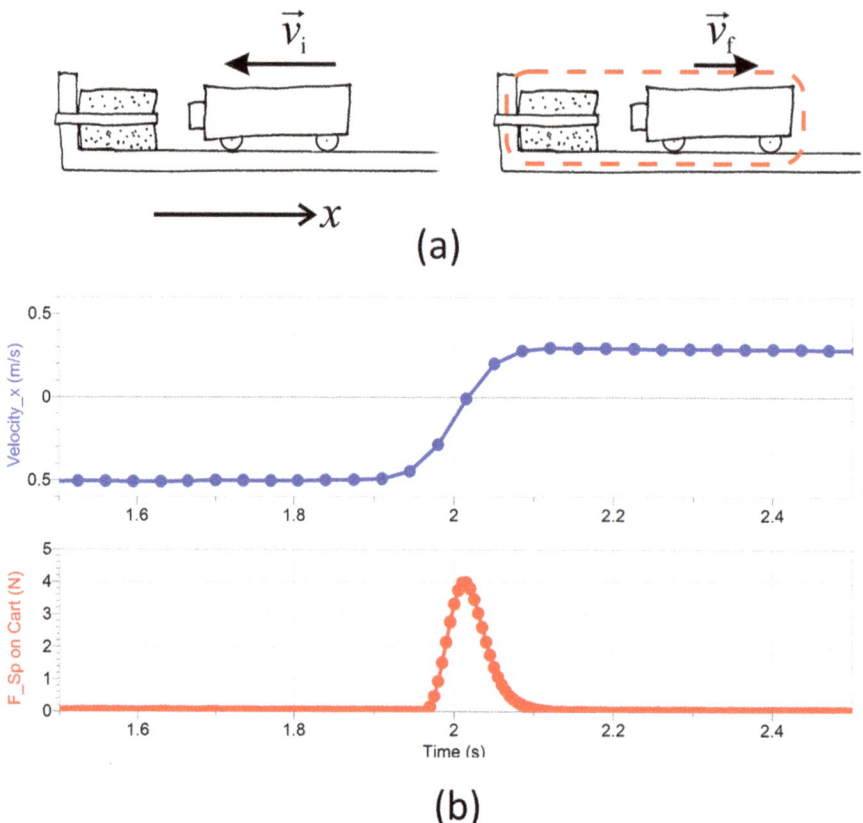

Figure 5.16. (a) A sketch of the collision of the cart with the sponge. (b) Velocity-versus-time and force-versus-time graphs recorded by motion and force sensors.

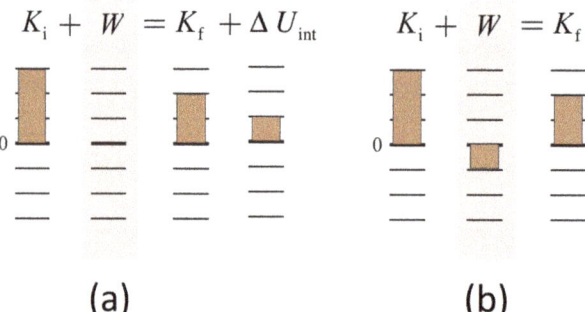

Figure 5.17. Work energy bard charts representing the collision. (a) The system is the cart and the sponge. (b) The system is the cart only.

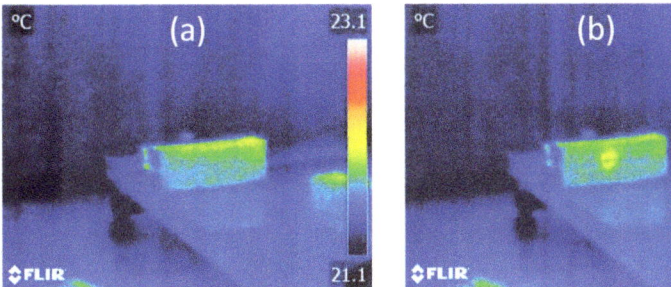

Figure 5.18. IR image of the sponge (a) before and (b) after the collision with the cart.

small temperature changes of the surfaces is an infrared camera (due to their relatively low cost IR cameras are becoming a common piece of school equipment). Two IR images of the sponge, before and after the cart bounced off it, show that the surface temperature of the sponge indeed increased by about 0.5 °C at the place where the bumper hit the sponge (see figure 5.18).

What if we repeat the analysis for a different system? Let us consider the system to be the cart only (remember, we neglected any friction and interaction with air). The initial state is right before the cart hits the sponge and the final state is right after it bounces off the sponge. The bumper of the cart is much harder than the sponge, so, practically, it does not deform during the collision. Thus, the change of the internal energy of the cart is negligible and the only candidate for balancing the bar chart is work, which should be negative (figure 5.17(b)). The sponge is the object outside the system so it can do work on the system, but how do we know that this work is negative? In order to answer this question, we need to examine how the force exerted by the sponge on the cart changes with the position during the collision (recall that the area under the curve $F(x)$ is equal to the work done by the force F on the system). Unfortunately, the limitations in the sampling rate of the position does not allow us to do this directly from the measurements shown in figure x. However, we can devise the following alternative experiment that helps students understand why the total work done by the sponge is indeed negative (the video of the experiment is at https://youtu.be/uIKF77sVknQ).

Place the cart next to the sponge so that its bumper is barely touching the sponge (see figure 5.19(a)). While measuring the force and the position start slowly pushing the cart with your hand. When the force reading reaches the value that was typically obtained during the collisions, start reducing the force exerted by your hand until the cart moves away from the sponge. The results of the measurements are shown in figure 5.19(b).

During the compression the force exerted by the sponge points to the right while the displacement is to the left, therefore the work is negative. During the loosening period the displacement changes the direction, but the direction of the force remains, therefore the work is positive. Because the magnitude of the work in the first part is larger than in the second part, the total work done by the sponge is negative. As you can see, the analysis of the situation when the sponge is not in the system is much

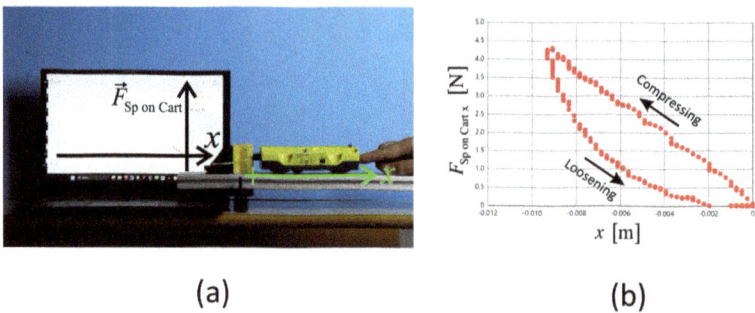

Figure 5.19. The explanation of why the work done by the sponge on the cart is negative (a) the force and the displacement; (b) the change in the force exerted by the sponge on the cart during the collision when the sponge is compressing and relaxing.

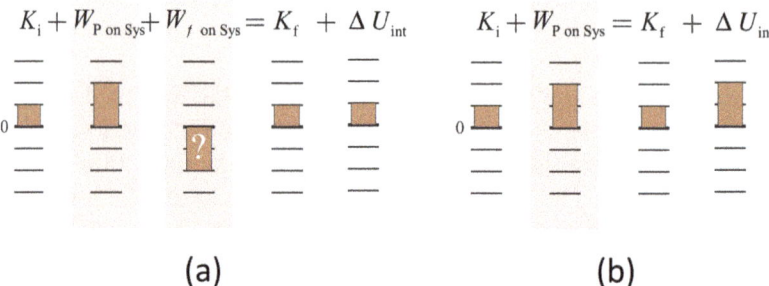

Figure 5.20. Pulling a box at constant speed along a rough horizontal floor: (a) the box is the system; (b) the box and floor are the system.

more complicated compared to when the sponge is a part of the system. Choosing the system wisely might simplify the problem situation and solution path.

Another example of the importance of system choice is the situation when a person is pulling a box across the floor at constant velocity. As the velocity is constant, the pulling force exerted by the person on the box must be equal in magnitude to the friction force on the box. If the box is the system, then it seems like the positive work done on the box by the person is equal in magnitude to the negative work done by the friction force and thus the total work done on the box is zero (see figure 5.20(a)). However, the bottom of the box gets warm, thus the internal energy of the box increases. If we draw a bar chart for the process, we realize that it is not balanced: the total final energy of the system is larger than the total initial energy plus work. The discrepancy is due to a mistake we made in reasoning about the work done by the friction. The work done by the friction is actually smaller in magnitude than the work done by the person. In addition, the box and the floor are exchanging material while rubbing against each other, which means that the box as a system is ill-defined. The correct analysis requires understanding of the processes at microscopic level and is therefore too complex most introductory classes

(Sherwood and Bernard 1984). To avoid this complication, it is much easier to include the floor in the system and consider internal energy change of both surfaces (see figure 5.20(b)). Note that the change in internal energy is larger with this choice of system, because we include the changes of the internal energy of both the box and the floor.

Finally, in most energy analyses, we exclude people from the processes. What would happen if we did not? Imagine a person lowering a heavy ball slowly at constant speed from height h_1 to height h_2. If the ball, the person, and Earth are in the system, then the energy of the system is constant, and no external objects do work on it (we disregard the outside air again). The initial state is when the ball is at h_1 and the final when it is at h_2. The gravitational potential energy of the system decreases. This means that some other energy should increase. What energy is that? Since the increase of the thermal energy of air and the ball due to air drag is negligible, this leaves us with an increase of the internal energy of the person. We can further separate the internal energy change of the person into a thermal energy change (change of the body temperature) and a chemical energy change (chemical changes in the body due to metabolism). As for every living organism, the person reduces their chemical energy ($\Delta U_{ch} < 0$) to perform this physical task (strictly speaking, the chemical energy of person plus oxygen in the air decreases.) In our case the increase in thermal energy of a person comes partly from the chemical energy change but also from the change of the gravitational potential energy of the ball (see bar chart in figure 5.21). We can see the process in the following way: the person is preventing the ball from speeding up by continuously transforming any increase in the kinetic energy of the ball into thermal energy of the muscles and joints.

From the above examples, it is clear how important it is to choose a system before starting an analysis of a process. Some systems make analysis easier, as in the example of a cart bouncing off the sponge or moving a box across a rough surface at constant speed. How do we develop a habit of choosing a system productively? First, when we model problem solving for teachers, we can make sure that we always start by identifying the system. Next, when the teachers work on force, energy, and momentum related activities, ask them to start by deciding what system they would use for the analysis. Then, ask them to analyse the processes using several different

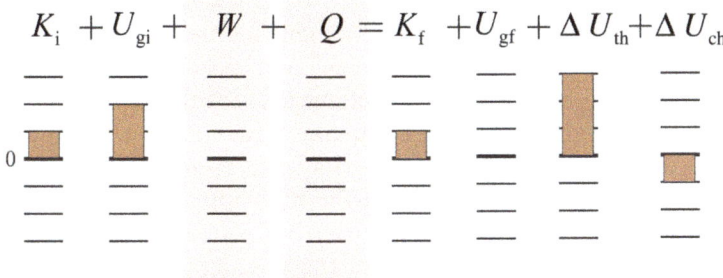

Figure 5.21. Energy analysis of the system involving the ball, Earth, and person.

systems and discuss when a particular system makes the analysis easier, just as we did here. Research shows that physics teachers and physics students (undergraduate physics majors) have very weak knowledge of systems and their role in the analysis of physical processes (Seeley *et al* 2019). Therefore, it is crucial that teacher preparation and professional development activities help rectify the situation.

5.6 Choosing the right language

Language is one of the representations that we use to construct physics knowledge or any kind of knowledge. Think about how we learn new ideas. How do we know what an apple is? We need to see, smell, touch an apple, and we need someone to tell us that this object is an called an apple. In other words, the visual image of an apple and the word for it need to be coordinate. But then, we will not need a real apple to recognize it anymore. We could see a picture of an apple in a book and know exactly what it is. Now the pictorial representation of an apple is coordinated with the sound of the word or the written word. In another language the word for an apple would be different. For example, in Slovenian language the word is jabolko. If Slovenian were our native language, we would coordinate the real apple or a picture of an apple with jabolko.

This consistent coordination of different representations is needed for us to construct meaning (Lemke 1990). Therefore, when this coordination is broken, the meaning is lost. Below we will show different examples of such 'breaks'.

> **Example 1** The way physicists speak about a physics idea is not coordinated with the meaning of the idea that we wish our students to construct.

When students learn physics, they not only use language to describe and explain observed phenomena, but they also encounter words that are common in everyday language as the names of different physical quantities. Very often the language that we (experts) use contradicts the very meaning of what we wish our students to learn and thus the coordination that we spoke about above is broken. Here are a few examples.

Many physics textbooks use the term 'weight' for the force that Earth (or any other planet) exerts on an object of interest (our system). When we talk about weight, we say 'weight of an object', which means that the object has weight. And yet, weight is a force, which means it is a quantity that characterizes the interaction of two objects and therefore, by definition of a force, it *cannot* belong to one object. Therefore, when the textbook says 'the weight of an apple is 1 N,' this sentence does not make any sense unless we decode it as 'Earth is pulling on the apple, the apple is pulling on the Earth, and the magnitude of this interaction measured in newtons is equal to 1 N.' In other words, we when say 'weight of an object', we take a huge shortcut. Do students understand all these intricacies or do they start wondering what forces really are? As mentioned in chapter 4, when students talk about forces, they often think about them as being properties of moving objects as opposed to the external things that change the motion of objects (Brookes and Etkina 2009).

Another example is the term 'heat'. In physics, heat is not a thing: the term is reserved for a process through which energy is transferred between two systems that have different temperatures and neither of them does work on the other. By default, heat cannot reside in a system, similar to work. And yet, we speak of heat as a substance. We say heat is transferred to the system. If it is transferred to the system, it might reside in it, and yet it does not. Research shows that the students who talk about heat as a 'thing' are less able to reason successfully about thermodynamic processes that those who talk about heat as a process of energy transfer (Brookes and Etkina 2015).

The goal of these examples is to highlight how issues arising from physicists' careless use of language make the coordination of different representations of the same idea impossible for the students.

> **Example 2** The way lay people speak about concepts is not coordinated with the concepts we wish our students to construct.

There is more to the language issue than the use of the terms by the experts. First, the terms themselves are ambiguous. If we study the history of physics, we find that it takes the physics community a long time and a great deal of effort to figure out what to call a specific physical quantity such as force, or energy, or heat. These words are used widely in everyday life and very often their everyday meaning does not match the physics meaning of the same term. When in a movie the hero says: 'May the force be with you', we think that force is something that we can carry. When the news talks about wasted energy, we think that energy is something that can be lost. Both thoughts contradict the physics meaning of those quantities. As we said above, a force characterizes an interaction between objects and cannot belong to any one object, and energy is a conserved quantity, therefore it cannot be wasted or lost.

> **Example 3** The way physicists speak about some physics concepts is not coordinated with other physics concepts that we wish our students to construct.

While the use of 'weight', 'heat', and 'force' in ways that contradict their physics meaning is a problem, but the physics ideas behind those words are clear for the physicists, some physics language, while universally adopted, is fundamentally incorrect. An example of such 'incorrectness' is the use of 'conservative' and 'non-conservative' forces (see Seeley *et al* (2022) for details). As discussed above, momentum, energy, and electric charge are conserved quantities. They are very distinct from the rest of the quantities that might be constant in some processes. Therefore, the language of 'conservative' versus 'non-conservative' forces applied to energy analysis makes students (and teachers) think that there are some processes in which energy is not conserved. We need to remember that those terms for forces operate in the context of mechanical energy only, which is not a conserved quantity by default. Only total energy can be and is called a conserved quantity. Therefore, Seeley *et al* recommend calling those forces 'path independent' and 'path dependent'.

> **Example 4** The grammatical construction of our sentences is not coordinated with the physics concepts we wish our students to construct.

The grammatical constructions of our sentences also create issues for student learning. When we say 'tension in the rope' we again communicate that forces can reside inside objects. When we say 'electron in an atom falls from one energy level to another' or 'jumps', we communicate that (1) an electron by itself has quantized energies (which is incorrect, a free electron can have only kinetic energy and it is continuous), (2) that those energy states are physically at different distances from the nucleus, and (3) that the electron is engaged in physical movement covering some distance. All those ideas prevent our students from forming a correct understanding of the process of the change of energy states of an atom with an emission or an absorption of energy.

> **Example 5** Language and new ideas.

In addition to the issues with terminology and sentence structure, there is another issue related to language. Historical analysis of the development of quantum mechanics conducted by David Brookes (Brookes and Etkina 2007) and the analysis of physics textbooks and the discourse by professional physicists showed a contradiction between the language that physicists use when they are trying to make meaning of new ideas and what language students encounter when they are trying to make meaning of new ideas. When physicists are trying to understand what is going on in an unfamiliar territory (for Brookes and Etkina this territory is quantum mechanics) they use analogical reasoning thinking of new properties as being 'like' some other properties. However, as the time goes by and the new concept gets into textbooks, the word 'like' is dropped and what is left is just the word 'is'. An example of talking about electrons is helpful here. Originally, the founders of quantum mechanics talked about the behavior of electrons being like the behavior of waves. In other words, the wave model of an electron was an analogy. Later, the 'like' was dropped and the textbooks talk that an electron is a wave as it can be described by a wave equation, it diffracts, etc. In this language, the electron is spoken about as a wave metaphorically. For an unexperienced student this subtlety is lost and they start thinking of an electron literally being a wave. The analogical aspect of the comparison is lost (Brookes and Etkina 2007). Brookes and Etkina argue that the progress from an analogical model to metaphorical statements is encoded both in words and in the grammatical constructions of the sentences and often the grammatical constructions contradict the essence of the initial analogy. This is when student difficulties arise.

> For example, to comprehend a metaphor such as 'the electron is a smeared paste,' the reader has to come up with an *ad hoc* category shared by both entities. A physicist who understands the quantum mechanical behavior of an electron might suggest an *ad hoc* category of 'things that don't have a well-defined location.' There is no guarantee that a student will come up with the same classification. We hypothesize that students often come up with an *ad hoc* category that is inappropriate to a given situation. This inappropriate categorization is at the heart of their difficulties. These difficulties may manifest themselves as student difficulties. (Brookes and Etkina 2007, p 010105).

The details of their theoretical framework are provided in the paper that is cited here, but the bottom line is that the language that we use is simultaneously a productive and counterproductive tool in the learning process. Every teacher should habitually focus and reflect on the language that they use and remain continuously concerned that the language helps student instead of derailing them.

There are several important habits of mind and practice of ISLE teachers related to language:

1. We can think of language as one of the many messaging systems that we use to communicate. The problem with language is that, by itself, it does not carry messages unless the person who hears it has tools to decode the messages. Learning a foreign language is a good example here. When a person speaking a different language tells us something, unless we can decode their message by translating it into a language we know, there is no meaning in the message. Thus, learning physics often resembles learning a foreign language: a teacher uses words that the students cannot decode and thus, they cannot construct meaning of what the teacher is trying to communicate. The ISLE approach addresses this issue in a natural way. When students describe observational experiments, they are encouraged to only use simple words that a five-year-old can understand. When they come up with explanations, they naturally use analogies and

mechanisms that they are familiar with, therefore the new terms do not arise. If they are finding a pattern in the data and this pattern leads to the development of a new physical quantity, then the teacher helps them name this quantity *after* they have constructed it. If they find a relationship between the quantities that is knows as a physics law, the teacher shares the name for the law *after* the students have devised it. The bottom line is that in the ISLE process, the idea, the concept, the mechanism, the image come first and the name second. This approach avoids the situation when the students use words whose meaning is not shared by everyone. Once the meaning is agreed upon, the new term goes into a glossary that the students keep and return to it if they forget the agreed-upon meaning. Appendix B to this chapter shows the ISLE-specific terms whose meaning must be shared by all students so that they can decode messages using those terms.
2. We stay true to the physics meaning of the physics concepts and tweak our language to help our students avoid contradictions between the words, grammar and physics meaning. For example, we do not use the term 'weight' but use the words 'force exerted on the system by Earth (or another planet)'; we do not use the term 'heat' but instead use the term 'heating' to underscore that it is not a thing but a process; we do not say 'gravitational potential energy of an object' (unless it is the GPE of a star as one object) but 'gravitational potential energy of the object–Earth system'; we do not say 'energy level of an electron' but 'energy state of an atom'. All these small changes help maintain the consistency between our language and the physics concepts.
3. We try to maintain the 'analogical' nature of the language that we use, avoiding metaphors until later in learning a concept. For example, we talk about light behaving like a wave, not being a wave. We underscore the word model in the 'wave model of light' and 'photon model of light'.
4. We insist on the students using 'correct' language in their discourse. However, when they are first grappling with a new idea, this 'language policing' needs to be very careful as focusing too much on the words that they student is using we might break their train of thought when they are just figuring things out.

5.7 *Interlude* Eugenia on motivation

We often hear that motivation is the 'holy grail of education'. We know about the horse and water saying—we can bring a horse to water but we cannot make it drink. Talking to many teachers and university professors we hear often that the students are not succeeding because they are not motivated. Different researchers have written about intrinsic and extrinsic motivation and provide different suggestions on how to move students from being extrinsically motivated to being intrinsically motivated (Dehaene 2022). But in this interlude, we would like to talk about what motivates students in the ISLE environment and how these motivational approaches relate to research findings.

However, first let me share some personal experience. From my very first day of teaching physics in high school and until the very last day 40 years later, I noticed a very peculiar thing. Every lesson that I taught seemed too short. I never noticed how time passed and while I was watching the clock on the wall like a hawk to make sure that the students do everything that I planned, the clock seemed to move too fast. I would take a look and then, boom! 20 min would pass and I thought it was only five. How was it possible? The answer to this question came from the work of a person with a very long and hard to pronounce name—Mihaly Csikszentmihalyi. Try to pronounce his name!

It was the idea of 'flow—the psychological experience', the term coined by Mihaly Csikszentmihalyi in his 1990 book *Flow, The Psychology of optimal Experience* (New York: Harper and Row) that explained this very strange passage (or I should say 'flow') of time when I was teaching. Csikszentmihalyi interviewed dozens of composers, artists, chess players, computer game designers, etc, and found that all of them sometimes experienced a very special mental state, that he called the flow state. Being in the flow state is like '… being completely involved in an activity for its own sake. The ego falls away. Time flies. Every action, movement, and thought follows inevitably from the previous one…'. The flow state is the state of concentration and engagement. When people experience flow, they are intrinsically motivated. What conditions are necessary for us to be in the flow state?

We can summarize Csikszentmihalyi's findings in four points:

- The person needs to be skilled in doing the task but the task still needs to be challenging (this is the balance between perceived challenge and skill).
- The person needs to have a clear goal for what they are doing and how to proceed.
- The person needs to be focused on the doing not on the reward that comes.
- The task should have immediate feedback built into it so that the person knows how they are doing.

Now, if you think about these conditions, it is clear that the time passed without me noticing it when I was teaching:

- I was skilled in teaching, but, as you know, every lesson brings a new challenge.
- My goal was to help students learn.
- I did not think of any rewards other than seeing my students engaged.
- Learning and their responses provided immediate feedback to me of whether my methods were working.

Needless to say, I loved teaching and was extremely motivated to become better at it.

Now, if we think about traditional learning environments, most of their features destroy potential flow in our students. We either make students work on the problems that they are not yet skilled at or give them repetitive assignments that hold no challenge. We create goals that are attached to extrinsic rewards (grades) and we wait for a long time to give feedback on their final tests. Therefore, one of the major goals of the ISLE approach is to help our students experience the state of flow in our classrooms, as this state is the ultimate motivator (you can read more on this subject in Karelina *et al* 2022). You all probably experienced a moment when one of your students would say: 'Oh, wow, it is the end of the class? I did not even notice how time passed!' This is the testament to the fact that you helped that student to be in the flow state during the lesson.

Using the concept of flow as a guideline for creating motivation, we group 'motivating' activities into the elements necessary for creating a flow state (see table 5.5).

In addition to the motivation built into ISLE-based learning due to its two intentionalities, we have additional recommendations for increasing student motivation:

Table 5.5. How the ISLE approach addresses flow conditions.

Flow condition	What we do in the ISLE approach	Why it helps
The person needs to be skilled in doing the task but the task still needs to be challenging (this is the balance between perceived challenge and skill)	Students work in groups and use multiple representations to analyse phenomena and solve problems. Students do not have cook book instructions when they design experiments, they need to work collaboratively to devise their own explanations of the observed phenomena. The teacher continuously informs the students about the skills that they have been developing and that they are getting better.	Team work helps the students see that the team can solve the problems that one person cannot. It increases the feeling of perceived skill. MRs help develop skills that are useful in any area of physics. Design and invention help maintain challenge. It also extends their 'zone of proximal development' (Vigotsky 1934). Helps create the feeling of being skilled, helps created the balance between perceived challenge and skill.
The person needs to have a clear goal for what they are doing and how to proceed.	Before every unit we create the 'need to know' for the students. This can be a 'cool' experiment or a video that the students cannot explain yet, but can at the end of the unit. This experiment is the motivation for learning during the unit. Students work with observational experiments that deal with phenomena familiar for them from everyday life but that they have not questioned. The students have self-assessment scientific abilities rubrics (Etkina *et al* 2006).	The 'need to know' experiment is the motivation for learning during the unit. Connection to everyday phenomena helps see physics as relevant. Self-assessment rubrics help the students see the goals and simultaneously how to proceed when they conduct different types of experiments, collect and analyse data, etc. See more about the rubrics in the first ISLE book (Etkina *et al* 2019) and on scientific abilities website.

The person needs to be focused on the doing not on the reward that comes.	Allowing and encouraging resubmission of work helps the students focus on the process of learning not the grade.	Downplaying the importance of grades and praising effort not only helps the students to focus on the task, not the coming reward, but also helps them develop growth mindset (Dweck 2006).
The task should have immediate feedback built into it so that the person knows how they are doing	Designing and conducting testing experiments for which the students make predictions based on their own ideas under test. Using self-assessment rubrics when writing lab reports and solving problems. Sharing solutions on group whiteboards and encouraging the students to revise their solutions based on the work and presentations of their peers.	The outcome of a testing experiment provides immediate feedback about the hypothesis under test. Self-assessment rubrics help create self-feedback. Sharing whiteboards after completing the assignment also provides immediate feedback.

- Let them explore something that they are interested in. Maybe sports they play or a vocation they plan to choose, or something that their parents do.
- Let them know that they are getting better at something in learning physics. If they do not see that they are getting better, they are not motivated to struggle.
- Be 'cool'. Your coolness and creativity motivate them. Make them feel that learning is 'cool'.
- Make work a reward, not punishment. If somebody did not do their work, say: 'Sorry, you did not do your homework, and that is why you cannot work on this problem. Next time do your homework and you will be rewarded'. When we reward our students by decreasing their work, this sends a message that work is punishment and no work is a reward. Turn this around!
- Encourage students to take photos or videos of everyday objects or situations that relate to a physics principle or a concept that they will learn about (for example: drop a small stone into a pond or lake and take a video of the waves that form—to begin learning about waves) or that they have just learned about (for example: take photos of everyday devices or situations that employ torque, friction (static/kinetic/viscous), photos of projectile motion, refraction, and reflection, and videos that show energy conversion in everyday life…)—and then post their work on the class website.
- Move the lessons at a fast pace but not too fast. If you move slowly, all students are bored. If you move too fast, students are lost. It is important to move 'border line fast' so that the students can still maintain the pace but they need to work hard to do it. We call it 'creating the sense of urgency' (a term we learned from Dan McIsaac).

- Stop lecturing. It destroys all motivation.

Appendix A

In this appendix we share several questions and possible correct answers for pre- and in-service teachers to interpret student reasoning.

1. Students were solving the following problem. A wooden block (base 10 cm × 10 cm, height $H = 20$ cm, density 0.8 cm^{-3}) is attached to a thin, light stick. Holding the stick, you are lowering the block very slowly, with constant speed into water that is in a large container (see the figure below). Draw a graph that shows how the force exerted by the hand on the block $F_{\text{H on B}}$ depends on the position of the bottom surface of the block z, for $0 \leqslant z \leqslant 30$ cm. We choose $z = 0$ to be at the water surface and the z-axis pointing down

Kayla, Hera, and Stavros presented the following graphs:

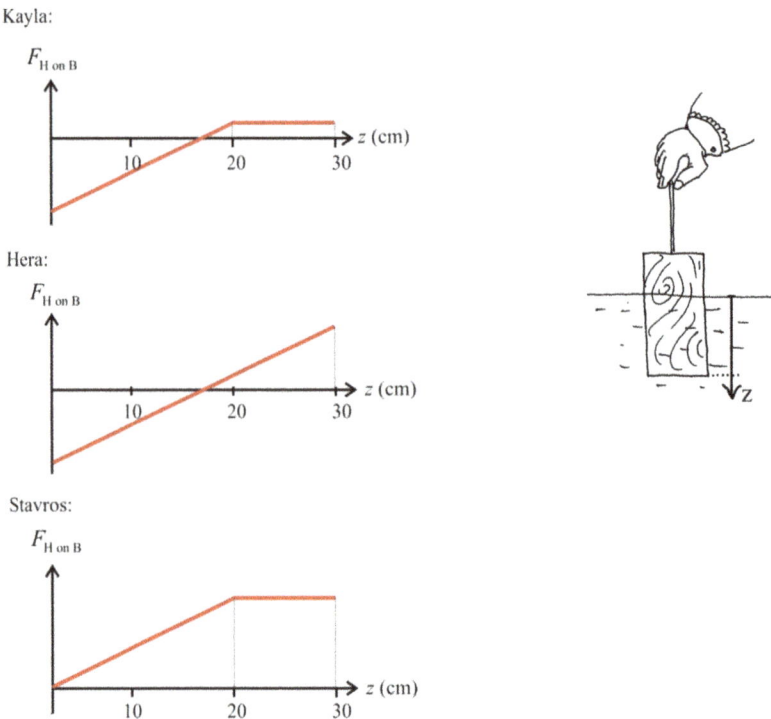

a. Which student presented the correct solution? Explain his/her reasoning.
b. Which difficulties do the other two students most probably have? Describe what part of their reasoning was correct (if any).

The same three students were solving the next problem, as a continuation of the previous problem:

Next to your graph draw three separate force diagrams for the wooden block, when its bottom surface is at the following three positions: $z = 1$ cm, $z = 20$ cm, and $z = 30$ cm. Represent the block with a point and label each force with a double index, indicating which object is exerting a force on the block (for example—force exerted by the hand on the block).

Next to the graphs above, draw the force diagrams that Kayla, Hera, and Stavros most probably drew, assuming their reasoning is consistent with that in solving the first problem (drawing graphs).

Possible solution:
a. Kayla is correct. First, we have to exert an upward force on the block, to balance the force exerted by Earth on the block. When the block is descending, the force exerted by the water on the block (buoyant force) is increasing (and pointing upward) until the whole block is immersed under water. As a result, $F_{\text{H on B}}$ decreases, becomes zero (at the point when the block would float) and increases again in the opposite direction. After the whole block is immersed, the buoyant force remains constant and so does the $F_{\text{H on B}}$.
b. Hera correctly identified all three forces. However, she thinks that the buoyant force increases proportionally to the depth to which the block is immersed. Probably she is focusing only on the force exerted by the water on the bottom surface of the block, and she fails to see that once the block is totally immersed, the total force exerted on the block by the water remains constant.
c. Stavros most probably focused only on the force exerted by water on the block. He did not consider the force exerted by Earth on the block. However, (unlike Hera) he correctly understands that the buoyant force on the block depends on the immersed volume of the block and not on the depth to which the object is immersed.

Kayla:

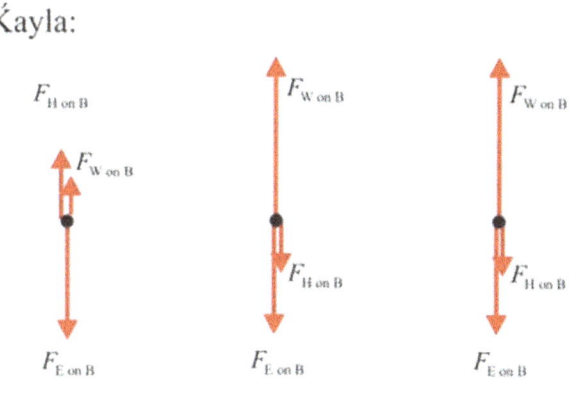

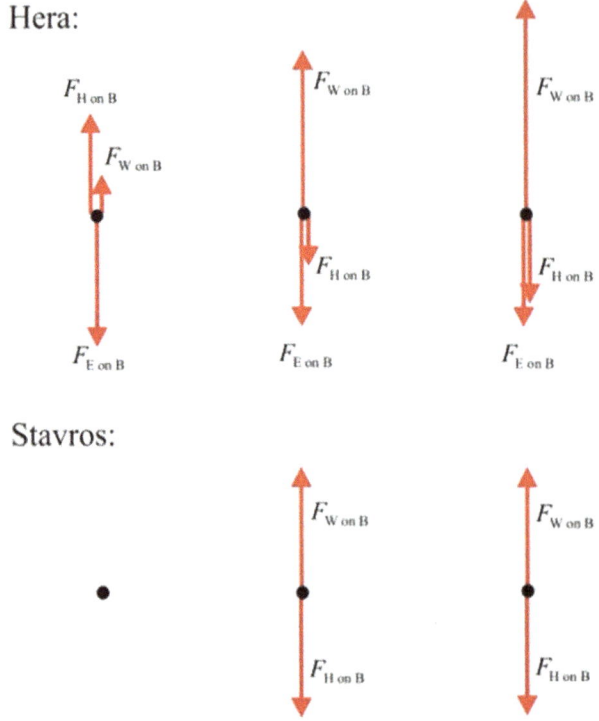

2. Students were solving the following problem: The figure shows two circuits that consist of equal bulbs, ideal batteries, and connecting wires with negligible resistance. Rank the bulbs according to their brightness and explain your answer.

Three students gave the following answers:
Harry: $A > B = C$, because the current in the second circuit divides between the bulbs B and C and therefore each get less current than A.
Fiona: $A = B = C$ because the potential differences across all three bulbs are equal.
Ron: $B = C > A$ because the total equivalent resistance of the second circuit is smaller than the resistance of the first circuit and therefore the current through the battery in the circuit is larger than in the first circuit.

Comment on all three student answers. For each explanation indicate, which productive and which problematic ideas it contains. Mark which answer you think is correct.

Possible solution:

	Productive ideas	Problematic ideas
Harry	Student knows that the sum of the currents into the junction equals the sum of the currents out of the junction (Kirchhoff's junction rule).	Student thinks that a battery is a constant current source.
Fiona (correct answer)	Student knows that a battery is a constant voltage source.	None.
Ron	Student knows that adding resistors in parallel decreases the effective total resistance of the circuit. Student knows that the smaller the resistance, the larger the current (Ohm's law).	Student does not take into account that the current in circuit 2 divides between the bulbs B and C (fails to do the quantitative analysis).

3. Students were solving the following problem: In a certain region we have a homogeneous electric field (see the first sketch below). Sketch the electric field lines in this part of the region after placing a point charge in the center of the area.

Four different sketches (A to D) drawn by the students are shown below.

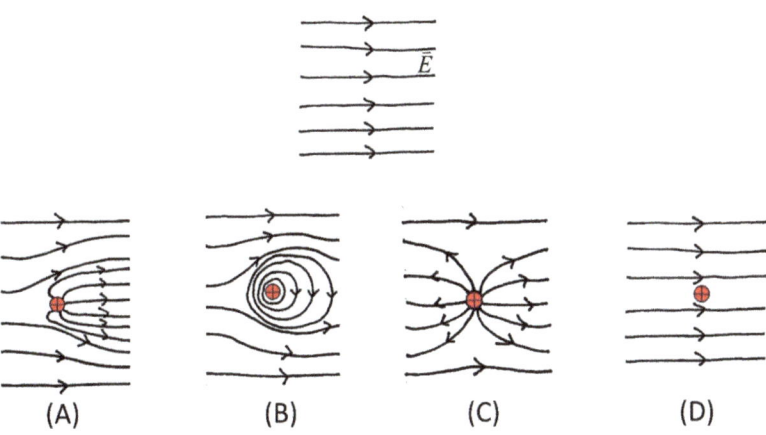

a. Comment on all four student solutions. For each solution, write which productive and which problematic ideas it contains. Mark which solution you think is the best.

b. Write down which important concepts the students must develop to successfully solve the above task.

Possible solution:
(a)
(A) Correct solution. Student knows that the total E field at any point is a vector sum of the E field produced by the point charge and the uniform E field.
(B) The student realizes that the homogeneous E field will modify the E field of a point charge, but they confuse the field lines with equipotential surfaces.
(C) The student is aware that the E field lines near the point charged must be similar to those of a lone charge and far away to those of a homogeneous field, but is unable to correctly apply the principle of superposition of E fields.
(D) The student thinks that the charged particle does not change the properties that the space had before the particle was placed in it.

(b)
- The shape of the E field lines of a point charge.
- Superposition of E fields.
- General properties of field lines.

4. Students were solving the following problem: In a certain region we have a uniform magnetic field (see the first sketch below). Sketch the magnetic field lines in this part of the region after placing a long straight conductor in the center of the area so that it is perpendicular to the plane of the sheet. There is a current in the conductor, going into the plane of the sheet.

Four different sketches (A to D) drawn by the students are shown below.

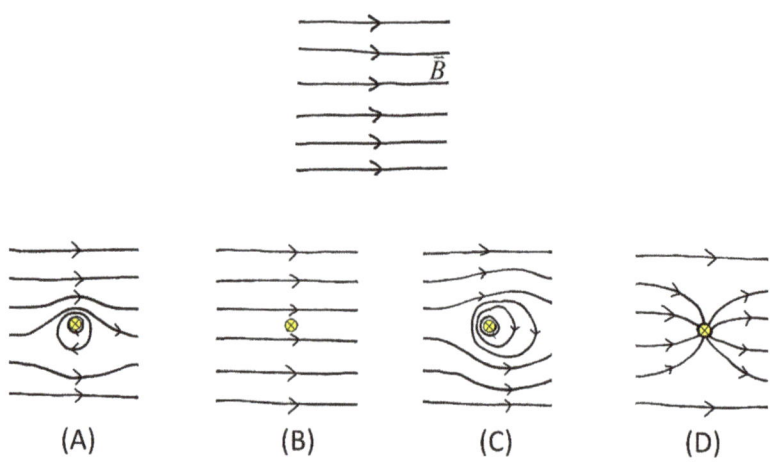

a. Comment on all four student solutions. For each solution, write which productive and which problematic ideas it contains. Mark which solution you think is the best.
b. Write down which important concepts the students must develop to successfully solve the above task.

Possible solution:
(a)
- (A) Correct solution. Student knows that the total B field at any point is a vector sum of the B field produced by the current carrying wire and the uniform B field.
- (B) The student thinks that the current carrying conductor does not change the properties that the space had before it was placed in it.
- (C) The student is aware that the magnetic field lines near the current carrying conductor must be concentric circles but cannot apply the principle of superposition of B fields or they think that a uniforms magnetic field 'carries away' the circular magnetic field lines in the direction of the field.
- (D) The student is aware that far from the wire the magnetic field must be uniform. The student does not know that the magnetic field lines of a current carrying conductor are concentric circular loops. The student's solution is probably a combination of ideas from electrostatics and magnetism.

(b)
- The shape of the magnetic field lines of a straight current carrying conductor.
- Superposition of B fields.
- General properties of B field lines.

5. Students were solving the following problem: You place the object in front of the convex lens so that the distance between the object and the lens is equal to 1/3 of the focal length of the lens. By drawing a ray diagram, determine the location where the image of the object is formed. What else can you say about the image based on the ray diagram?

Below is Maya's solution. Find and comment on productive and problematic ideas in Maya's solution. If you think there are errors in the solution, present an improved solution.
Maya's ray diagram

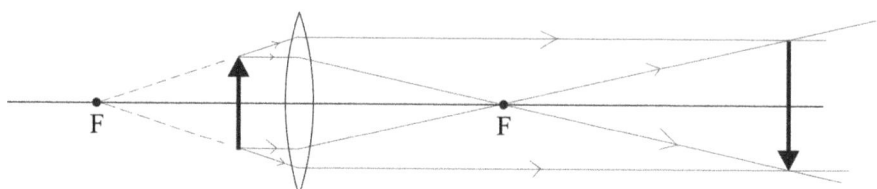

Based on the ray diagram, I can tell that the image is magnified, inverted and real. The magnification is about 1.4.

Comment on Maya's solution. Find productive and problematic ideas in Maya's solution. If you think there are errors in the solution, present an improved solution.

Possible solution
Productive ideas
- The student knows that a lens has two foci that are equidistant from the lens.
- The student knows that a ray that is parallel to the optical axis is refracted so that after passing through the lens it travels through the focal point.
- The student knows that the ray emerging from/passing through the focal point is refracted so that after passing through the lens it travels parallel to the optical axis.
- The student probably remembered that the real image is inverted when mirrored by a lens (although she does not understand how the image is created).

Problematic ideas
- The student does not understand how the image of the object is created. The student thinks that the image of a point on an object is formed at the intersection of any two rays that can emerge from different points.

Possible response to the student
> You correctly remembered that a parallel ray is refracted so that it travels through the focal point after passing the lens and that a ray coming from the direction of the focal point is parallel to the optical axis after passing the lens (*building on productive ideas*). But you have problems with the construction of the image (*addressing problematic ideas*).
>
> Let's take any point on the object, for example, the tip of the arrow (*folding back*) and try to construct the image of that point. Can rays that are emerging from other points of the object contribute to the image of this point? Think about what the picture would look like in this case. Would you even be able to get a sharp image of the object? (*Encouraging metacognition*).

6. Andrew, Jette, and Mary were solving the following problem: A student pulls a 5 kg block across a rough horizontal surface, exerting a constant force on the block. The magnitude of the force is initially 5 N, and the block moves at a constant velocity of 2 m s^{-1}. While the block is moving, the student instantly increases the force to 10 N. How will the block move now?

The answers of three students are described in the first column of the table below. For each student's explanation please describe what you think the students' strengths and weaknesses are (the second and the third column). We will discuss the last column together. (*Possible solutions are italicized*).

Student answer	Strengths	Weaknesses	How would you respond to this student if they were a student in your class?
Andrew: The block will move at a constant velocity of 4 m s^{-1} because it was initially moving with constant velocity and the final force is two times larger than the initial force.	None.	*The student thinks that the speed of the object is directly proportional to the sum of the forces exerted on the body. The student does not realize that a constant friction force is exerted on the block. He may not be able to apply Newton's second law.*	*Can you tell me, in your own words, what Newton's second law says? Can you tell me in which situations does a body move with constant velocity?* *Try to think why the block moves with constant velocity in the first place in this situation.* *Please draw a force diagram for the first situation ... and now for the second situation. Is your reasoning consistent with force diagrams? How does this affect your answer?*
Jette: The block will move at a constant acceleration of 2 m s^{-2}, because the	*The student knows that the acceleration of the body is*	*The student overlooks that a constant force of friction is exerted by the*	*You are on the right track, I see that you understand the content of the second NL, but*

(*Continued*)

(Continued)

Student answer	Strengths	Weaknesses	How would you respond to this student if they were a student in your class?
sum of the forces exerted on the block is 10 N and the mass of the block is 5 kg.	*proportional to the sum of the forces, and she remembers Newton's second law.*	*surface on the body in the opposite direction of motion.*	you missed something. First, explain to me how that initially the block was moving with constant speed, even though we exerted on it a force of 5 N? Please draw a force diagram for the first situation … and now for the second situation. Is your reasoning consistent with force diagrams? How does this affect your answer?
Mary: The block will move at a constant acceleration of 1 m s^{-2}, because the sum of the forces exerted on the block is 5 N and the mass of the block is 5 kg.	*The student is reasoning correctly and knows how to apply Newton's second law.*	None.	Great reasoning. You correctly determined the acceleration of the block based on the given data. Can you just explain how you know that the sum of the forces exerted on the block is 5 N?

7. Students were solving the following problem: A spherical street lamp accidentally explodes. Three equal pieces A, B, and C fly off the lamp holder with equal speeds but in different directions, as shown in the figure below. Compare the speeds with which each piece hits the ground. Indicate any assumptions that you made.

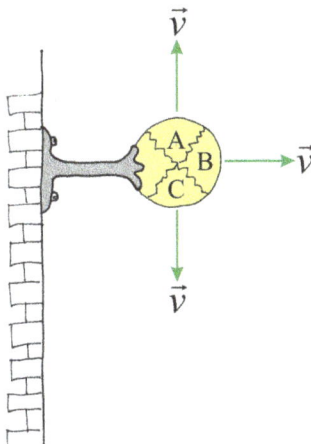

Meghan solved the problem as follows:

> *I assumed that the air resistance was negligible. As part A moves downwards past the explosion site, it will have the same speed as part C did at the start, which means that parts A and C will end up with the same speed. Part B will have the highest speed when it hits the ground because its velocity will have both an x- and a y-component.*

Comment on Meghan's solution. Identify and describe the productive and problematic ideas in Meghan's solution. How would you respond to Meghan if she were a student in your class? If you think the solution is wrong, present an improved solution.

Possible solution
Productive ideas:
The student has made a reasonable assumption and has applied it consistently in the solution. The student is correct when she says that as part A moves down past the explosion site, it will have the same velocity as part C initially had, meaning that parts A and C will end up with the same velocities. It is also correct that B will have both an x- and a y-component of velocity just before hitting the ground.

Problematic ideas:
The student probably thinks that all three parts have the same vertical velocity component when they hit the ground. Since particle B also has a horizontal component of velocity, the student thinks that its velocity is maximum.

Possible response to the student:
You have correctly determined that the velocities of parts A and C are equal when they hit the ground. It is also true that part B will have both an x- and a y-component of velocity, but what is the initial y-component of velocity for part B and what is the initial y-component of velocity for parts A and C? Does that change your answer? Do you think you could solve this problem using an energy approach?

Appendix B
Shared language in the ISLE approach
Credit to Eugenia Etkina and David Brookes

Observational experiment: An experiment where you investigate a phenomenon by collecting qualitative or quantitative data without specific expectations of the outcome. No predictions are made of the outcome of an observational experiment. We can call observational experiments 'hypotheses generating experiments'.

Description: A statement of what was observed in an experiment without explaining it (qualitatively or quantitatively). It answers the question, 'What happened?' You can describe with words, pictures, diagrams, etc.

Explanation: A statement of a possible reason for the reasons for something that happened in the experiment. It answers the questions 'why' or 'how'. An explanation might contain a hypothetical mechanism of how something happened. In this case it is a mechanistic explanation. For example, the mechanistic explanation for the drying of alcohol is the random motion of its particles. However, sometimes an explanation does not have a mechanism in it—it only explains the causal aspect of the phenomenon. In this case it is a causal explanation. For example, an object's acceleration is explained by the sum of the forces exerted on it and its mass. If you are collecting data, an explanation might be an inference from the data—why the data look the way they do.

Hypothesis A synonym for an explanation. There are multiple hypotheses that can explain what happened. A hypothesis should be experimentally testable. A hypothesis can be disproved by a series of testing experiments (see below). It can turn out to be wrong.

Model: A simplified version of an object, a system, a phenomenon or a process. A scientist creating the model decides what features to neglect. A particle model of an object neglects its size, a model of ideal gas neglects the sizes of its particles and the interactions at a distance between them, a model of a free fall neglects interactions of falling objects with air, a model of energy constancy of an isolated system neglects interactions of this system with the environment. Models can be conceptual or mathematical. In a way explanations, models, and hypothesis all belong to the same group of concepts—mental constructions describing or explaining physical phenomena. In many cases the terms models/hypotheses/explanations are synonyms.

Prediction: A statement of the outcome of a particular experiment (before you conduct it) based on the hypothesis being tested. It says what should happen in a particular experiment if the hypothesis under test is correct. Prediction is not a guess. Without knowing what the experiment is, one cannot make a prediction. A prediction is not equivalent to a hypothesis but should be based on the hypothesis being tested.

Testing experiment: An experiment whose outcome you should be able to predict using the hypothesis being tested. We can call these experiments 'hypotheses testing experiments'. A testing experiment test the hypothesis, not the prediction. A testing experiment cannot prove the hypothesis to be correct (if its outcome matches the prediction) but might disprove it (if the outcome does not match the prediction).

Here an important note is in order: A hypothesis can be disproved by a series of testing experiments. It can turn out to be wrong. A prediction, however, is only wrong when it does not follow from the hypothesis being tested. If the outcome of the testing experiment does not match the prediction, it does not mean that the prediction is wrong. It only means that the hypothesis on which the prediction was based is wrong or some assumptions were overlooked. In this case the prediction is said not to match the outcome of the testing experiment.

Assumption: An assumption is some factor in the physical situation you choose to ignore or assume to be true, that simplifies a calculation or a model, or an experiment.

Application experiment: An experiment with the goal of solving a practical problem or determining the value of some physical quantity using the relations/models that have not been disproved by multiple testing experiments. We can call application experiments 'multiple hypothesis applying experiments'.

System: A system is the object (or objects) of interest that we choose to analyse. Make a sketch of the process that you are analysing. Then, make a light, pretend boundary (a closed, dashed loop) around the system object to emphasize your choice. Everything outside the system is called the environment and consists of objects that might interact with and affect the system's motion. These are external interactions. Interactions of the environment objects with the system cannot be neglected. External objects exert forces on the system, do work on the system, exert impulse and so forth. Internal objects cannot do any of these things.

The terms below were already defined in chapter 4, but we are repeating them here so that you have them al together.

Physical quantity: A physical quantity is a feature or characteristic of a physical phenomenon that can be measured in some unit. A measuring instrument is used to make a quantitative comparison of this characteristic with a unit of measure. Examples of physical quantities are your height, your body temperature, the speed of your car, or the temperature of air or water.

Operational definition: A rule that tells you what to do (what other quantities to measure and what mathematical operations to use) if you need to determine the value of a particular quantity. For example, for motion at constant velocity, $\frac{\Delta x}{\Delta t}$, is an operational definition of velocity.

Cause–effect relationship: A rule that tells you what will happen to a quantity when another quantity changes. For example, for motion at constant velocity, $\Delta x = v \cdot \Delta t$ is a cause–effect relationship that shows if the time interval of travel is doubled, the distance traveled is doubled. However, the operational definition of velocity is not a cause–effect relationship because if you double the distance that the object travels, the velocity will not change (since the time interval for the doubled distance will be doubled too).

References

Brookes D T and Etkina E 2007 Using conceptual metaphor and functional grammar to explore how language used in physics affects student learning *Phys. Rev. Spec. Top. Phys. Educ. Res.* **3** 010105

Brookes D T and Etkina E 2009 Force, ontology and language *Phys. Rev. Spec. Top. Phys. Edu. Res.* **5** 010110

Brookes D T and Etkina E 2015 The importance of language in students' reasoning about heat in thermodynamic processes *Int. J. Sci. Educ.* **37** 659–779

Brookes D T and Lin Y 2012 Designing a physics learning environment: a holistic approach *Proc. 2011 Physics Education Research Conf.* **vol 1413** ed N S Rebello, P V Engelhardt and C Singh pp 131–4

Brookes D, Etkina E and Planinsic G 2020 Implementing an epistemologically authentic approach to student-centered inquiry learning *Phys. Rev. Phys. Educ. Res.* **16** 020148

Clement J 1982 Students' preconceptions in introductory mechanics *Am. J. Phys.* **50** 66–71

Conlin L D, Gupta A and Hammer D 2010 Framing and resource activation: bridging the cognitive-situative divide using a dynamic unit of cognitive analysis *Proc. Cogn. Sci. Soc.* **32** 19–24

Daubert A 2022 personal communication, 27 December

Dehaene S 2022 *How We Learn: Why Brains Learn Better than Any Machine … For Now* (London: Penguin)

diSessa A A 1982 Unlearning Aristotelian physics: a study of knowledge-based learning *Cogn. Sci.* **6** 37–75

diSessa A A 1993 Toward an epistemology of physics *Cogn. Instr.* **10** 105–225

Dweck C S 2006 *Mindset: The New Psychology of Success* (New York: Random House)

Etkina E, Brookes D and Planinsic G 2019 *Investigative Science Learning Environment: When Learning Physics Mirrors Doing Physics* (IOP Concise Physics) (San Rafael, CA: Morgan and Claypool Publishers)

Etkina E, Van Heuvelen A, White-Brahmia S, Brookes D T, Gentile M, Murthy S and Warren A 2006 Scientific abilities and their assessment *Phys. Rev. Sp. Top. Phys. Educ. Res.* **2** 020103

Halloun I A and Hestenes D 1985 The initial knowledge state of college physics students *Am. J. Phys.* **53** 1043–55

Hammer D 1996 Misconceptions or p-prims: how may alternative perspectives of cognitive structure influence instructional perceptions and intentions *J. Learn. Sci.* **5** 97–127

Hammer D 2000 Student resources for learning introductory physics *Am. J. Phys.* **68** S52–9

Hammer D and Elby A 2003 Tapping epistemological resources for learning physics *J. Learn. Sci.* **12** 53–90

Hammer D, Elby A, Scherr R E and Redish E F 2005 Resources, framing, and transfer *Transfer of Learning from a Modern Multidisciplinary Perspective* ed J Mestre (Greenwich, CT: Information Age) pp 89–119

Harrer B W, Flood V J and Wittmann M C 2013 Productive resources in students' ideas about energy: an alternative analysis of Watts' original interview transcripts *Phys. Rev. Spec. Topic. -Phys. Edu. Res.* **9** 023101

Hestenes D, Wells M and Swackhamer G 1992 Force concept inventory *Phys. Teach.* **30** 141–58

Kaltakci-Gurel D, Eryilmaz A and McDermott L C 2016 Identifying pre-service physics teachers' misconceptions and conceptual difficulties about geometrical optics *Eur. J. Phys.* **37** 045705

Karelina A, Ektina E, Bohacek P, Vonk M, Kagan M, Warren A R and Brookes D T 2022 Comparing students' flow states during apparatus-based versus video-based lab activities *Eur. J. Phys.* **43** 045701

Lemke J L 1990 *Talking Science: Language, Learning, and Values* (Norwood, NJ: Ablex)

Maloney D P, O'Kuma T L, Hieggelke C J and Van Heuvelen A 2001 Surveying students' conceptual knowledge of electricity and magnetism *Am. J. Phys.* **69** S12–23

McCloskey M 1983 Intuitive physics *Sci. Am.* **248** 122–31

Richards A J, Jones D and Etkina E 2018 How students combine resources to make conceptual breakthroughs *Res. Sci. Edu.* **50** 1119–41

Schuster D 1993 Semantics and students' conceptions in science *Paper Presented at the Summer Meeting of the American Association of Physics Teachers (Boise, ID)*

Seeley L, Vokos S and Etkina E 2019 Examining physics teacher understanding of systems and the role it plays in supporting student energy reasoning *Am. J. Phys.* **87** 510–9

Seeley L, Vokos S and Etkina E 2022 Updating our language to help students learn: mechanical energy is not conserved but all forces conserve energy *Am. J. Phys.* **90** 251

Sherwood B A and Bernard W H 1984 Work and heat transfer in the presence of sliding friction *Am. J. Phys.* **52** 1001–7

Smith III J P, diSessa A A and Roschelle J 1994 Misconceptions reconceived: a constructivist analysis of knowledge in transition *J. Learn. Sci.* **3** 115–63

Van Heuvelen A and Zou X 2001 Multiple representations of work–energy processes *Am. J. Phys.* **69** 184–94

Vigotsky L S 1934 *Thought and Language* (rev. ed. A Kozulin trans) (Cambridge, MA: MIT Press)

Zull J E 2002 *The Art of Changing the Brain: Enriching Teaching by Exploring the Biology of Learning* 1st edn (Sterling, VA: Stylus)

IOP Publishing

The Investigative Science Learning Environment: A Guide for Teacher Preparation and Professional Development

Eugenia Etkina and Gorazd Planinsic

Chapter 6

ISLE and development of routines

In the previous chapters, we discussed several physics and physics teacher habits of mind and practice that relate to the tasks of teaching focused on the development of conceptual understanding and reasoning skills by their students. In this chapter, we talk about the routines that follow from the habits that are helpful to be implemented in a classroom that follows the ISLE approach. A reminder: a habit is *what* we are used to thinking about or doing and the routine is *how* we go about regularly doing the habit. For example, we all have a habit of brushing our teeth in the morning, but the routines that we use are different. Some do it with a mechanical brush, some with an electronic one, some do it before breakfast, and some do it after. Some squeeze the toothpaste from the bottom up and some just twist it to get the toothpaste as fast as possible. And so forth[1].

6.1 Positioning students, group work, developing accountability

Every classroom starts with ... the classroom. What do students see when they walk in? Do they see the neat rows of desks and tables facing the board, theater-style seating with the stage in the front, or a room with tables for 3–4 people with small whiteboards and markers on each table with no center stage? See the difference in figure 6.1.

These settings send completely different messages to the students. The students see what is expected of them—either to watch the person on the stage or to work themselves. As the ISLE approach assumes the latter, the first routine is clear—to set up your classroom in a way that sends a clear message to the students that learning physics is a collaborative enterprise with them at the center.

---

[1] We remind the reader that the abbreviation CP: E&A stands for the textbook *College Physics: Explore and Apply*, ALG stands for the *Active Learning Guide*, OALG stands for the *Online Active Learning Guide*, and IG stands for the *Instructor Guide*. Proper citations for these materials are in chapter 1.

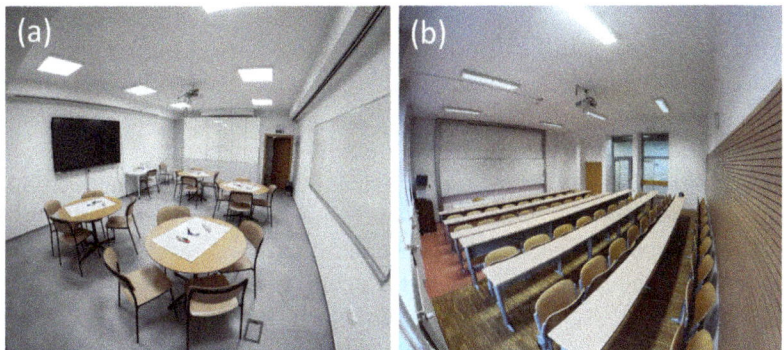

Figure 6.1. (a) Interactive engagement classroom; (b) traditional classroom.

How do we develop this routine in pre-service teachers and how do we help in-service teachers who come to our workshops to change their existing routines? As always, we lead by example.

Our routine is to first set up the whiteboards. As they last for many years with good care, there is a good chance that you will only need to do this once. Here are a few suggestions:

Size: We recommend the size of approximately 50 cm × 70 cm or larger, depending on the size of the desks where your students are working. Figure 6.2 shows a simple solution, where we used a 50 cm × 70 cm whiteboard and combined two traditional school desks to make a table for a group of four students.

Figure 6.2. A whiteboard for group work and a set of colored markers.

Material: The most important thing is that the white surface is really smooth. If you can't find panels with a suitably smooth surface, you can buy untreated panels and cover them with smooth white self-adhesive wallpaper.

Care: Insist that students clean the whiteboards after each lesson. Ink left on the boards for several days is difficult to remove. You can use a mild detergent solution or water for daily cleaning, but occasionally you will need to clean the boards with alcohol.

In addition to whiteboards, you need different color dry erase markers, an eraser or some cloth (old rags work best), and a cleaning solution for each group.

The next step is to organize the tables and chairs. It is a good routine to come to the classroom (if you are sharing it with some other teachers) a few minutes before class and organize the tables and chairs for group work. It is even better if some of your students and pre-service teachers come early too and help you do this. This way, the setting up becomes a routine for them that they will replicate in their own classrooms.

The next step is to put any necessary equipment on every desk. It is truly important to develop a habit of thinking about each activity as experimental. Even when the students solve paper-and-pencil problems, it is great for them to have equipment to immediately check their answer.

Your students will be working in groups at those tables, each holding a marker of a different color in their hands so that the contributions of each member are clear. How do you form groups? There are different approaches to this task. During the first class of a course, we let groups be formed naturally, as the students walk in to the classroom and choose seats. Once they are settled, it is good to check that there are no groups with one female and the rest males. You can ask students to move in this case. While some females make themselves heard in a group of males, many have difficulties doing so even when they have a great grasp of physics. Thus, at the beginning, it is a good routine to avoid such situations. Later, it is necessary to monitor the groups to make sure every voice is heard.

Is there a best size of a group for group work? If the activity has one right answer, two people are enough, but if it is a creative activity with multiple possible approaches, then three is a minimum number and an optimal number for all students to participate. Four is still OK, but one person might not participate. Starting with five members, only a few of them participate. As the semester progresses, it is great to change groups to make sure that everyone in the course works with everyone else.

As one of the goals of any ISLE-based course as well as of a teacher preparation program is to create a community, the first necessary condition is that people know each other well and experience working together in different contexts (more about this later). With the possibility of a large variability in the group member's preparation, we found it useful to have groups of mixed ability. Thus, as you get to know your students, try to organize the students so that the groups in each course meeting have students who are high achieving, medium achieving, and those who struggle. If the atmosphere in the course is supportive (and this depends on you), then the struggling members will grow and improve quickly.

Working in groups is important in the ISLE approach. Do you remember the two intentionalities of the ISLE approach? The first is to help students learn physics by constructing knowledge that follows the processes that mirror those of practicing physicists. The second is to help students stay motivated, feeling that they belong in physics and develop growth mindset. Group work fosters both intentionalities as scientists work in teams creating knowledge and group work can help develop the sense of community, belonging, and extend students' zone of proximal development. The problem is that not all groups are functional. What does it mean to have a functional group? A functional group is a group where group members work together, listen to each other, and support each other to solve a problem. By working *together*, they extend each other's zone of proximal development. This sounds good in theory, but how do you help them learn to work this way? And does belonging to a functional group make a difference? It turns out that it does!

From the above follows that it is important that the group work is really collaborative, not led by one dominant person while everyone else is either passively listening or is 'checked out' until the answer is provided. How can we help group members collaborate? Research by David Brookes, Yuehai Yang and colleagues (Brookes *et al* 2021) found that in the groups where the more knowledgeable person 'hedges' the answers (makes them sound a little uncertain, for example: 'What do you think of this idea?' or 'I could be wrong but these are my thoughts…') instead of declaring them authoritatively, other group members participate and collaborate more equitably. Such hedging opens space for other members of the group to contribute. The consequence of this more equitable engagement is that these groups make far more progress on challenging activities (the activities that require the students to leap into their zone of proximal development). If there is no equitable engagement, the other group members do not challenge the statements of the person who is perceived as more knowledgeable. It seems that hedging creates a feeling of some kind of psychological safety that is necessary for effective collaboration.

As with everything else, working in a group is a skill that needs to be learned. Using hedging in your own speech to model desired behavior of your students is the first step. The next step is openly talking to your students about what makes groups effective—this knowledge will serve them long after they finish the physics class. Sharing with the students that being socially aware of other people in the group, making sure that everyone has a chance to speak, and understanding that everyone's contributions are important is crucial for productive collaboration. People have different strengths and building on these different strengths is what makes the intellectual value of a group much higher than the sum of intellectual values of its members. When students work in groups, it is helpful to stand behind a group and listen to the tone of their conversations and watching if all members have a chance to speak and if they are trying to hedge. If you notice a problem, either talk to the whole group ('I noticed that not everyone has a chance to speak in your group, please make sure that everyone does'), or asking the most dominant group members to meet you outside of class and talk to them separately about the importance of being aware of other people in the group ('Why do you think we hedge when we make statements? Why do you think listening to other people might help you learn more?').

After the groups finish their activity, they will need to report it to the rest of the class. Who will do the reporting? A good practice is for each group to choose a 'spokesperson' for the day. This person will deliver the groups solution to the class during that lesson. You will need to monitor the list of spokespeople so that everyone gets to play this role before the new round starts. Choosing a spokesperson for the day can be the first assignment for each group at the beginning of each class.

When the groups start working on the assignment, it is a good routine to announce the time that they will have. If you do not do it, the start will be slow and about 4–5 min at the beginning will be lost. It is good to have a timer or some other means to remind students how much time is left. This time monitoring creates 'the sense of urgency' in the lesson that prevents it from dragging. How do you know how much time will be needed for a particular activity? First, time yourself doing it when you are prepping the lesson and then multiple it by two. This would be the minimum time. You can always extend it in class (indistinctly), but having the limit is crucial for a quick start of group work.

To have accountability for student work, the students need to put everything on their whiteboards and then present what they found to the rest of the class. But while such an approach helps keep the students accountable, it might lead to drudgery in the lesson and a lack of a 'sense of urgency'.

Therefore, a key step in maintaining the sense of urgency is deciding when to cut off the time for the activity and how to organize groups' presentations. It is tempting to wait until all groups finish and then let them all present their findings by taking turns. There are several dangers in this routine. Those who finish first get bored waiting for the rest of the groups and when all the groups have the same solution to the problem, it is boring to listen to the same thing again and again. Here are possible alternative routines:

1. Notice when the first group finishes and if their work is correct, stop the rest of the groups. Let this group present and then ask representatives of the groups that did not finish to ask questions or repeat what the first group said and then give them a few minutes to finish their boards using the work of the finished group.
2. Invite the members of the finished group to visit the groups that are not done and help them. Then the first group presents.
3. Give an additional activity to the finished group and let the rest of the class finish. Then, if the solutions are the same, any group can present. However, if the solutions are different, invite the groups to visit each other, talk and then share the differences that they found without repeating the things that all groups did the same.

It is important to *not* be the first to validate the results and solutions yourself, but, rather, let students discuss them. However, at the end it is a good practice to summarize the results of the group activity and clearly state why the students did it and what they were expected to learn from it. A good routine is to keep these summaries short on the class board and let the students take photos for their journals at the end of the lesson.

There are a few other important routines to keep in mind specifically for group work in the ISLE environment:

1. Before each group activity, ask the students where it belongs in the ISLE process —are they working on an observational experiment, on the patterns, on the testing experiment, etc. For example, if it is a testing experiment, it is helpful to put the hypothesis that the students will be testing on the class board for clarity.
2. After every group activity, summarize what the students found so that they can proceed to the next one being 'on the same page'. For example, if it is an observational experiment, it is helpful to put the patterns that students found on the class board; if it is an application experiment, it is helpful to put the results on the board and ask how we know that they make sense. The bottom line is that developing an epistemological aspect of reasoning is as important as doing the activities.

The 'group work' routines described above work for the lessons when students learn new material and when they do long labs (if you are teaching in college and the course is run in a traditional mode the labs are separate from other activities).

If you are teaching a large enrollment physic course using the ISLE approach and have 'lecture time' (we call it a 'large room meeting') when all 200–300 students or more are in a theater-sitting environment, it might feel that no group work is possible. But team work is always possible. A student needs to turn to their neighbor to discuss the activity. Their consensus can either come from direct sharing with the rest of the class, or by choosing an answer among the choices that you provide using a student-response system. Even if you cannot organize team work in a large room meeting, the students can still work in groups in the labs or when doing problem solving activities. Then the routines described above are relevant.

At the beginning of this section, we asked the question: How do we help pre-service teachers develop these routines? The answer is very similar to the answers we gave concerning the development of habits. Fist, let them participate as students in group work and reflect on how the groups are formed, how the whiteboards are used and so forth. Then have a discussion with them as to WHY you used those routines and how they helped them feel comfortable and accountable at the same time, and how they facilitated learning. Finally, invite them to set up the classroom before each lesson and let them practice these routines during microteaching when they teach physics lessons to their peers in physics teaching methods courses (see more about such courses and microteaching in chapter 7).

6.2 Setting up experimental work for the students

Traditionally, student experimental work was thought to be carried out when they are participating in instructional laboratories or labs. As the start of learning of any concept (and the rest of the progression) in the ISLE approach is experiment-based, leaving experiments only to the labs would be contradictory to the ISLE philosophy. Therefore, we wish to develop teacher's habits of mind and practices that would engage their students in experimental work (however simple) at every stage of their learning. Simple qualitative observational and testing experiments are always possible. Many are described in the textbook CP: E&A, the ALG, and the OALG (if you do not have the right equipment, then you can use the videos, see them at https://media.pearsoncmg.com/aw/aw_etkina_cp_2/videos/). Students can do many experiments at home and video them for sharing during class sessions—dropping small rocks in water to create waves, walking and running on sand to see the deformation of the road, making simple pendula to study periods, freezing water in a plastic bottle to observe its expansion, drinking with a straw, and many, many others.

In addition, one can see the role of experiments in our traditional paper-and-pencil problem solving using the ISLE lens. You can think of a problem statement as a description of a phenomenon for the students to analyse. The mathematical models for analysis are the hypotheses that the students use, as well as simplifying assumptions for what to ignore. The answer that they arrive at is the prediction that they make. If you view the problem solving process from this point of view, then you see how important it is for the students to compare the outcome to the prediction, i.e. to conduct the experiment replicating the phenomenon in the problem and collect data. While it is not possible for every single problem, many problems can be treated as 'testable' especially in mechanics (including hydrostatics), DC circuits (a quick testing experiment can be conducted with simulations such as PhET), geometrical and wave optics, and magnetism. The topic that is most difficult from the point of view of quantitative experiments is electrostatics.

While we strive to develop a habit of having equipment on every group's desk for every activity (not really possible, but good to strive towards), in this section we discuss routines that are helpful when you do set up equipment for every group, let it be in a studio classroom setting or a traditional high school classroom, or a lab. When preparing future teachers, it is very important to help them develop this routine when they engage in microteaching (see chapter 7 for more details). They need to go through the process described below multiple times before it becomes routine enough for them to replicate it on their own when they start teaching.

6.3 Reflecting

The word reflection is common in educational practice. We can think of reflection as an act of looking back in order to process experiences, or a way of thinking about one's own thinking in order to grow. A habit of reflection is crucial for one's development as a teacher. After every lesson we ask ourselves: How did it go? What went well? What should be improved? Without such thoughts, we will repeat the same mistakes next year and will not emphasize the aspect of the lesson that went well. But it is even more important to ask the questions: Why did it go well? Why didn't it go well? What made certain experiments trigger productive thinking in my students and other experiments not? How could I have better responded to that comment, or that question? From the above examples, it is clear that without habitual reflection there is no progress. How do we teach our students and future teachers to reflect and what are productive routines for doing reflections in a physics classroom?

At the end of each class meeting in our physics teacher preparation program lessons, we ask our students to answer the following questions: What did I learn as a student of physics today? What did I learn as a teacher today? How did I learn those things? What were the teaching methods that helped me learn … and …?

Such reflection can be done individually. At the end of the class the teacher says: 'Close your eyes and make a mental list of all the things you learned today. Group them into two categories: what you learned as a physics student and what you learned as a teacher. Think of what happened in class that helped you learn those things.' After 1–2 min, the students open their eyes and you ask them to raise their hands when they are ready to share. The rule is that one person can only say one thing from each category, and they cannot repeat what was already said. This means that the students who go first have an easier time choosing what to say. The last people will find it difficult to add to what was already said, so next time they will raise their hands first.

If you have too many students in the course to have individual reflections, you could have group reflections. Give 2–3 min to your students who have been working in groups during class to write what they have learned (same questions) on their whiteboard and then ask them to share. Each group says one thing from each category on their whiteboard and the reflection goes from group to group until all new things are said. Group reflections are also very helpful when you have students that are (for various reasons) shy or not accustomed to reflecting. Implementing regular group reflections helps them to break the ice and learn how to reflect, so that you can later proceed with individual reflections.

Is there an explanation as to why reflection is important? *Yes!* Indeed, our knowledge of how the brain works provides two mechanisms for the importance of reflection for learning. First, when a person who is asked to reflect on what they have learned searches through their mind for the answers, they activate the electric circuits that have just been formed. This activation makes the circuits stronger and thus more easily accessible later (reflection improves memory). But there is more to the role of reflection, oral or written, in learning. According to Kolb's brain cycle (Kolb 1984), when the brain learns, it goes through the following steps: sensory experience, reflective observation, hypothesis formation, and active testing. The active testing involves motor function—some kind of movement in our body. For example, you walk into a room at a party and see a person by the window (sensory experience). You start 'searching' in your memory where you could have seen that person before (reflective observations) and what her name is. You hypothesize that you saw her at the hostess wedding and her name is Jill (hypothesis). You approach her and say with a question mark in your voice: 'Jill? Nice to see you again' (active testing). If you never spoke her name, you would not know whether your hypothesis was correct; you had to test it!

You can think of oral reflections on learning new material or anything else that happened during the lesson as one form of active testing (another one is performing the actual testing experiments that are a part of the ISLE process). Therefore, if the reflections are useful for teachers, they should be useful for the students too. All the above practices concerning reflection are good for high school students too. You can see the reflection at the end of a lesson as instant non-threatening formative assessment technique. Did the students mention everything that you wanted them to learn? Could they explain how they learned it?

But there is more here. In addition to oral reflections, you can ask your students to reflect as a part of the homework (Etkina 2000). Asking them to answer the following prompts:

- What did I learn this week?
- How did I learn it?
- What remained unclear?
- What questions would I ask if I were the teacher to find out whether my students understood the material?

This homework does not only illuminate what and how students learned by providing feedback to us as teachers, but also helps the students recompile and reconcile all of the week's experiences and ask themselves how they learned what they did. Our research shows that when the students focus on the ideas and relationships instead of definitions and when they can articulate how they learned something by connecting experiments and reasoning, they have higher learning gains compared to those students who think that they have learned from listening to the teacher of from pure observations of experiments (May and Etkina 2002). Similarly, when they asked higher level questions ('how to?' and 'why?' instead of 'what?') they have higher learning gains (Harper *et al* 2003).

Another important reflection routine is the 'reflection on the solution' of a problem or an experimental result. The habits of mind that are used in reflections on the solutions to the problems involve checking the units, doing extreme (or limiting) case analysis (White *et al* 2023), considering how reasonable the result is and how consistent it is with different representations used to analyse and solve the problem. The important next step is to go back and fix the solution if any of these techniques show a mistake. How can we help future teachers develop such habits and what routines can we use in this process? Again, as in all previous cases, we, as educators, need to model it and to reflect on when we assess students' solutions. How do you catch mistakes that students make? While every case is unique and there is no unique recipe, one good routine is to use students themselves to find those mistakes when evaluating the solutions of another peer group.

Reflection on experimental results is also extremely important. You probably experienced a student asking you to validate their experimental result: 'We did the experiment and found the value of g to be 9.6 ± 0.2. Is it correct?' Although the students wrote the findings with experimental uncertainty, it does not help them learn whether it is a 'good' result. They need to compare it to something. This something is often an 'accepted value'.

But what if there is no accepted value? Imagine, the students need to find the coefficient of static friction between their shoe and the floor tile. They design an experiment in which they attach a force meter to the shoe and pull it until the shoe starts moving. They repeat the experiment several times and determine the random uncertainty of the result (as shown above). How do they reflect on the value that they found? You probably realized that this experiment belongs to the group of application experiments in the ISLE process. These experiments require the students to use multiple models to determine some unknown quantity or build a device to achieve a specific goal. To determine some unknown quantity, students are required to design two independent experiments. For those experiments, reflection is crucial as it helps the students compare the results obtained from two different experiments. Only by comparing the findings from two experiments including experimental uncertainties can they decide whether they are 'correct'. To help students learn how to reflect on the experimental result in application experiments, we guide them with self-assessment rubrics. All rubrics are in our first ISLE book, but here we show rubrics relevant for the reflection moment in student work (see table 6.1).

Table 6.1. Selected rubrics for designing and conducting an application experiment.

Rubric D: Ability to design and conduct an application experiment

Scientific ability	Missing	Inadequate	Needs improvement	Adequate
D4 Is able to make a judgment about the results of the experiment	No discussion is presented about the results of the experiment.	A judgment is made about the results, but it is not reasonable or coherent.	An acceptable judgment is made about the result, but the reasoning is flawed or incomplete. Or uncertainties are not taken into account. Or assumptions are not discussed. The result is written as a single number.	An acceptable judgment is made about the result, with clear reasoning. The effects of assumptions and experimental uncertainties are considered. The result is written as an interval.
D5 Is able to evaluate the results by means of an independent method	No attempt is made to evaluate the consistency of the result using an independent method.	A second independent method is used to evaluate the results. However, there is little or no discussion about the differences in the results due to the two methods.	A second independent method is used to evaluate the results. The results of the two methods are compared correctly using experimental uncertainties. But	A second independent method is used to evaluate the results and the evaluation is correctly done with the experimental uncertainties. The discrepancy

there is little or no discussion of the possible reasons for the differences when the results are different.	between the results of the two methods, and possible reasons are discussed.

6.4 Questioning techniques

If you ever attended a physics research seminar or a conference, you probably noticed that physicists never take any statement for granted. They question. How did you know? What were the uncertainties in your experiment? How does it compare to the experiment done by XX? Such questions are common, and they show that authority is not the reason to believe in anything. However, there is more. Without asking a question about something that they observe, there is no development of new ideas of physics. While observations of the real world always come first, the next step in the physics progress is asking a good question. In the US Next Generation Science Standards, asking questions is one of the important science practices that students are required to master. The question is (no pun intended here), how do we teach our students to ask good physics questions?

As always, we need to start modeling this practice ourselves first. This means developing a habit of asking our students good questions when they are learning. We can think of all questions that we ask our students as belonging to two big groups. We will call one group 'closed questions' and the other group 'open questions'. Open questions assume multiple correct answers, or they even do not care about the correctness, just students' ideas. Multiple students are welcome to answer.

Closed questions assume one right answer and one person who knows it. For example, you are interested in how your students understood the concept of acceleration. You might ask:

- Who knows what acceleration is?
- What is acceleration?
- What is the definition of acceleration?
- What is the unit of acceleration?
- What does it mean that the acceleration is negative/positive?

All these questions assume confidence in those who answer and the existence of one right answer. These are closed questions.

How can we turn them into open questions? Here are some examples:
- Please tell me what you 'see' when I say the word 'acceleration'.
- Please give me two examples of real objects that move with acceleration. How will you know?

- Think of a few differences between velocity and acceleration.
- Eugenia says: 'Acceleration is the change in velocity'. Why would she say this? Do you agree with her? If you disagree, how can you help her agree with your point of view?
- David says that an object with an acceleration of 5 m s^{-1} s^{-1} speeds up and an object with an acceleration of -5 m s^{-1} s^{-1} slows down. Eugenia disagrees. What can be possible reasons for her disagreement?
- How would you explain the idea of acceleration to somebody who has never studied physics?
- Please give me an example of an object that has a positive acceleration and is slowing down and another example of an object that has a negative acceleration and is speeding up.
- How do you know if an object is accelerating?
- What are your thoughts about acceleration?
- How would you represent an object slowing down with a motion diagram?

If you compare the first set of questions to the second, you will find that in the second set each question assumes the existence of multiple answers, and no one needs to know all of them. Therefore, the 'fear of the wrong answer' barrier is reduced and many more students can (and will) participate.

In general, it is a good rule of thumb to avoid starting questions with: What is …? Who knows …?

The next step is how to elicit answers to the questions. Here, the routines are important. Sometimes we think that a question is easy and everyone should be able to answer it. Then we ask the whole class and wait for volunteers. Try not to call on the same person the second time before all others have a chance to participate and try not to miss girls holding up their hands (as you know, they are often invisible). Give students 10–15 s to respond, let them see that you really mean it. But if after 10–15 s no hands rise up, say: OK, let's have 2 min in your groups to come up with ideas and then we will share them. This routine reduces the need for an immediate personal right answer even more.

In general, to invite more students to answer your questions, it is useful to start them with: Please share your thoughts about…. What are your ideas? How can we explain…? How can we test…? How would you approach…? How do you know…? What is your image of…? How can we convince A in…? Tell me more about… Who can add to…? Any ideas about how we can explain…? Any ideas how we can test…?

We often ask a question in a whole class discussion or when students are reporting on their group work, and a student answers with one word. If this word is correct, we often validate and move on. But does it always mean that the student really understands what they said? A good routine is to make sure to ask this question: 'What do you mean?'. This is a simple way to elicit their real understanding. The next step, even if they provide a good answer is to *not* validate it, but to toss it back to the class—do you agree? Sometimes, when one student answers a question, the others do not listen or do not understand. Asking 'Do you agree?' to the whole class with the expectation that somebody else would answer makes everyone focused on what the first student was saying. Habitually asking the rest of the class to evaluate each other's answers will become routine for them. The goal is to communicate the message: 'I am not the final authority; you need to figure it out yourselves.' ('What do you think?' response is called a reflective toss—the name coined by Jim Minstrell a long time ago.) When somebody in the class responds, stay back until the discussion between the students starts. Open questions encourage or trigger discussions, closed questions do not.

How do we help teachers learn to develop a habit of asking good questions? Here, again, we use the cognitive apprenticeship approach. First, when you, as a course instructor or a leader of a PD program teach a lesson on modeling a high school physics lesson, it is important to focus students' attention on the type of questions that you ask. After they see the pattern described above (of course, you try to ask open questions), you give the name for those questions and give examples of closed questions. When the students are practicing microteaching (see more about this in chapter 7), being teachers in the classroom where their peers play the role of students, you, again, need to focus on the questions that they ask. It is great to do it in advance when they are planning the lesson. In fact, it is good to write possible questions on flash cards first and to flip them during the lesson in order not to 'slip' into the mode of closed questions.

In our experience, beginning teachers' unsuccessful questions can be categorized into three groups:

- Too easy for the target students.
- Require only repeating/recalling what the teacher said, triggering low-level cognitive processes.
- Too complicated/not clear…. causes students to start guessing what the teacher had in mind. Thus, it is always better to prepare in advance. During the microteaching experience (see chapter 7), it is important to watch how much time pre-service teachers wait for an answer to their question before they send the class to groups to discuss it or even answer it themselves. Helping them develop a habit of asking good questions and allowing their students to be successful with answering those is important and can only truly be achieved during microteaching.

Now, the most difficult step arrives. How do we encourage high school students to ask good questions when they are learning physics? Although 'asking questions' is the #1 science practice in the Next Generation Science Standards, in teaching practice we only have tools and routines to reward students for good answers—but not for good questions. Have you ever given a grade to your student for asking a good question or just extra points? And what are good questions?

If you look at the history of physics, the scientists who we all know are those who dared to ask a question about something that everyone else accepted as true (dogma in a way). Galileo asked whether it was true that all objects fall at constant speed with that speed proportional to their mass (Aristotelian dogma). Newton asked how the Moon orbits Earth. Einstein asked how we can figure out what would happen if we could travel on a light ray or be placed in a closed elevator in free fall—would you know that you are not standing on Earth? So, bottom line, asking good questions is as important, if not *more* important than giving good answers.

When I (EE) was teaching high school, on the first day of class, I would tell my students that questions were very important and if anyone asked a great question (I would be the person to judge that), this person would receive the same number of points as they would on a perfectly correct test. And during my teaching about one person per term would get these points. And everyone clapped when I would say: 'This question is a great question, such and such scientist asked it too and this is what happened after—I would tell a short story—and the student who asked it would receive the points.' Of course, it is a subjective decision but in all my years of teaching, no student ever argued that some question that I found worthy of the points was not.

Often, a good question would change my lesson plan and we would continue our investigations to answer it. This is true even for questions that were not that remarkable—I tried to show my students that almost every question they asked was important to follow as it created a natural 'need to know' and led to more learning. However, we all know, that sometimes the questions are irrelevant or distracting. For those, we need all our tact to show the student that we respect the question, but it is out of the field of our studies or that it will be answered later.

Now, how do we teach students to ask good questions? First, the students should feel safe to ask. Neither we, nor other students should ever comment or make fun of a person asking a question. But this is not enough. The same way as we teach students to reason like physicists, or to read the textbook as experts (the interrogation method that is described in the first chapter of our textbook and in the activities in the first chapters of the ALG and OALG), we need to teach them to ask good questions. The whole ISLE approach is conducive to generating or producing insightful and valuable questions.

Below we give a few examples of good questions that the students learning through the ISLE process might ask.

Observational experiment questions:
  How do we infer a pattern from these data?
  How do we best represent the data?
  Which variable is independent and which is dependent?

Model/explanation/hypothesis development questions:
  How do I start thinking about making a mathematical model for the pattern?
  How do I know if it is a good idea to linearize data to find the pattern?
  How do I go about finding a mechanism?
  How do I know that my explanation is correct?

Testing experiment questions:
How do I design an experiment to test the model/explanation/hypothesis?
How do I know if this is a good testing experiment?
How do I make a prediction of the outcome of the testing experiment using the hypothesis under test?
How do I know if my experiment will give me the outcome that will allow me to differentiate between the two hypotheses that I have?
Were there any additional assumptions that I made when I made the prediction? How can I validate them?
How do I determine the uncertainty of my result?
How do I know if my experiment ruled out the hypothesis/model/explanation?

We can go on to make a list of good questions for application experiments and for different multiple representations, but you probably already see the pattern here. Almost all good questions start with 'how', ('How do I know …?', (the best question ever) or 'How do I do such and such?') and *not* with the word 'what'. Note also that we did not list any good questions that start with the word 'why'. Why is that? While the students often ask questions starting with 'why', those are in fact questions that have the 'how' in them, however (and this is the reason to avoid 'why' questions) 'why' questions question the purpose of the phenomenon, not how it happens. And often the answer is anthropomorphic (anthropomorphism is the attribution of human traits, emotions, or intentions to non-human entities. It is an innate tendency of the human mind). For example: Why do objects fall down on Earth? Answer: Because they want to be in a state with the smallest gravitational potential energy. In fact, the objects do not want anything, and we just gave them human characteristics with our answer. The most famous answer to the question 'why' was given by Newton who was asked 'why' gravity exists. He said that he did not care why, he only cared 'how' to describe it.

To teach your students to ask good questions, the teacher needs to model such questions (see the above discussion) and to explain to the student why a specific question is good. And of course, to reward them, as we described above. We have examples of such questions in our materials, for specific elements of the ISLE process—in the labs at (see ISLE-based labs at https://sites.google.com/site/scientificabilities/isle-based-labs).

A long time ago we did a study correlating the quality of the questions that students asked about the material once a week as a homework assignment and their learning gains (Harper *et al* 2003). Those who asked questions that we described above as 'good' had significantly higher learning gains than those who focused on 'what' type of questions.

How to translate the above discussion of teaching high school physics students to ask good questions into the development of habits and routines of the pre-service teachers? First, it is to use these very same strategies during the lessons that you lead in the physics teacher preparation program and then let them reflect on how you motivated them to ask good questions. The next step is to help them encourage their peers in asking good questions during microteaching. Reflecting at the end of each class on the 'best student questions' is an excellent practice too.

6.5 Responding (or not) to students' questions

How to decide when to answer a student's question, when to toss it back to the class, when to provide a hint, and when to return it with a 'what do you think' reply? Of course, the choice depends on the question, on the situation in which it was asked, on the timing, and on many other factors. Here we consider several common types of questions that a student or a group of students might ask during a lesson or a lab and possible routines in response to these types of questions. Although the questions and responses are different, they all have a common approach: to engage the person asking the question with some intellectual effort related to the answer. This is needed because we want the students themselves to make the connections in their brains—listening to our answer will not help them remember it. The routines described below are pure suggestions as it is impossible to provide the exact advice for a specific situation without knowing your students and the context in which the question was asked.

1. *Questions* about a definition of a quantity or a mathematical expression the answer to which can be looked up easily and the answer that the student forgot (What is 'a prediction'?). *Possible response:* Answer the question directly to save time. In particular if the definition that you wish the student to use is different from the ones that they can look up online. For example, when searching online for the definition of prediction, you find in Wikipedia:

   A prediction, or forecast, is a statement about a future event or data. They are often, but not always, based upon experience or knowledge. There is no universal agreement about the exact difference from 'estimation'; different authors and disciplines ascribe different connotations. (Wikipedia n.d.)

   However, this is not how we wish the students to think about a prediction if they are learning physics through the ISLE approach. We wish them to think that a prediction is the statement of the outcome of a testing experiment that follows from a hypothesis/explanation under test. Therefore, it is more useful to repeat the definition for the student and then ask back: Do you remember a situation when we made a prediction? How did we do it? The point is that even though you are 'giving away the answer' you are still engaging the student in some intellectual effort needed to remember what you said. This can only happen if the student connects the terms, relations, etc, to the knowledge in their brains.

2. *Questions* about a procedure that is content independent (How do we find a pattern?) *Possible response:* Show how to do it on a different example, explain every step, and then make the students find the pattern in their data themselves (these steps follow the apprenticeship approach steps: model

proficiency, reflect on what you did, let the novice try themselves with feedback and reflect on what they did). Here again, while it looks like you are 'giving away the answer', in fact, you are engaging the students in an active learning experience.

3. *Questions* about a procedure that is content dependent (How does the assumption that the system is isolated affect the prediction?) *Possible response*: Ask why they made this assumption and how the calculations would be different if they did not make it. What energy would be less and what energy would be larger? Here, there is no direct answer to the question as in the examples above, as your goal is to help the students reflect on their previous actions to be able to do it themselves.

4. *Questions* about the experimental result or an answer to a problem (We got 0.3 for the coefficient of friction, is it correct?) *Possible response:* Is this a reasonable value? What are the uncertainties in the value? Have you used the second independent method to get the same quantity? Did the results overlap within the uncertainties? If not, did you examine the assumptions in both methods? Here again, there is no direct answer as your goal is to help the students learn how to evaluate their result themselves.

5. *Questions* about steps in a derivation (How did you get from A to B?) *Possible response:* Ask the class if anyone can help. If no one volunteers, ask everyone to close their eyes and open them if they also did not understand the step. If you see that many students did not understand but they did not ask the question, repeat the step slowly while asking small questions to the class on how to proceed after each step. Then ask the first person if the whole derivation is clear and then ask to repeat the explanation of the step. Then repeat the trick with closing eyes and see that no one opens theirs. This trick is a metacognitive help for the students to ask themselves: Do I really understand? And then, if they realize that they don't, the question provides the 'need to know'.

6. *Questions* about help when stuck (We don't know what to do, can you help?) *Possible response:* Despite the variations in the context of the question (this can be a difficulty with starting to solve a problem or to design an experiment, or something else), it is tempting to help by asking leading questions that would take the students along the path of your reasoning. But it is probably not a very helpful approach as when you lead them through your reasoning path, you are utilizing connections in your brain that the students might not have. They can answer every small leading question, but the whole picture will escape them. There will be an illusion of help, but when you leave the group, they might have difficulties moving forward and will call for your help again. A better strategy in this case is to try to identify what difficulty they seem to have and what knowledge that *they* have that might help them to overcome this difficulty. Then, ask a question or give a hint that will help them fall back on this knowledge or skill so that they can make the conceptual leap themselves. Here is an example. The students are working on the horse-sled problem that comes after they learned how to

draw force diagrams and how to apply Newton's second law. This problem seems like a Newton's third law problem, but, in fact, it is still the second law problem. 'The horse is pulling on the sled exerting a force, and therefore the sled is pulling back on the horse exerting the force of the same magnitude in the opposite direction. The sum of these two forces is zero. How can the sled start to move?' The students are stuck and do not know what to do. A good 'fall back' question would be: 'What system did you choose for analysis?' Once the students decide that the sled is the system, the force that it exerts on the horse is irrelevant for the answer as it is not exerted on the sled, but the force exerted by the ground is important. Same for the horse. In this case, the students make the conceptual leap themselves and the teacher only helps them identify the knowledge that they need to be successful.

To summarize, there is no one routine for responding to students' questions. But there are some common habits of mind that might help you decide what to do. If the question is factual or about a routine procedure—answer it directly but then engage the students in some reflection and application. If the question is conceptual or involves a complex procedure, engage the students in an activity that will help them figure out what to do. Tossing the question back to the class ('What do you think?') is a good habit but it should not become a routine as not every question should be tossed back based on the examples above.

6.6 Setting up assessment routines

Much has been written about formative and summative assessment in the last 25 years. In a way, we can view any assessment as a way to provide feedback. The questions are: who receives the feedback and what is the response to this feedback?

We view summative assessment as assessment that gives the teacher feedback on what to change in the next unit or next year and gives the student feedback about their overall performance on a set of required tasks. Usually, the summative assessment is given at the end of the unit or a term and there is not much that a student can do with this feedback. However, the teacher should use the feedback to revise/improve the instruction in the future. The situation is different with formative assessment that happens as the students are learning. Here, assessment questions and activities provide instant feedback to both the teacher and the students, and both have an opportunity to change. Therefore, we can think of summative assessment as assessment 'after learning' and formative assessment as assessment 'during learning and for learning'.

Another important aspect of assessment is that any assessment provides feedback to the student about what is important. Therefore, assessment needs to match the goals of the course/unit/lesson. When planning a lesson/unit/course, we always recommend that the teacher first defines the goals. Then, design examples of assessment activities that show whether the students achieved the goals. Finally, plan the instructional sequence. Such a routine helps avoid the mismatch between what the teacher thinks the students would learn and what they think is important to focus on.

Below is an example that shows how to use formative assessment as assessment for learning. Imagine that the students are learning how to find the direction of acceleration of an object moving in a circle at constant speed. Below are two of the activities that they do. In the first activity and the 'time for telling' that follows, the students learn that the velocity of an object moving in a circle is tangent at every point and the reasoning tool that will help them determine the velocity change vector at a given point. The second is the activity where the students apply the tool and find that the acceleration points in the radial direction. This activity also informs the teacher as to whether they are able to use the technique of finding the direction of acceleration graphically.

Activity 1

You learned that the sum of the forces exerted on an object moving in a circle at constant speed is pointed toward the center of the circle. To explain, think of the motion of the object. While the speed is constant, the velocity is not. How can you find the direction of the velocity of such an object at every instant?

    a. Work together with your group members to draw the velocity vectors on a whiteboard for such an object at four different points of the circle. What is the direction of the velocity vector? What is its magnitude?

    b. What can you say about the motion of the object? Is it motion with constant velocity? If not, how can you determine the acceleration at each point in the motion? Think of the definition of acceleration $\left(\vec{a} = \frac{\Delta \vec{v}}{\Delta t}\right)$ and how you determined the direction of the acceleration for objects moving in a straight line.

    c. ('Time for telling') Follow the procedure shown below to learn the technique for determining the direction of acceleration of an object that is not moving along a straight line (figure 6.3).

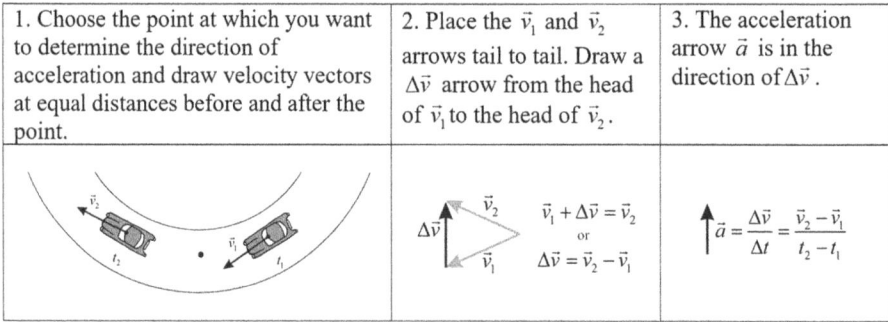

Figure 6.3. How to determine the direction of acceleration. (A similar procedure can be found in CP: E&A, chapter 5.)

Activity 2

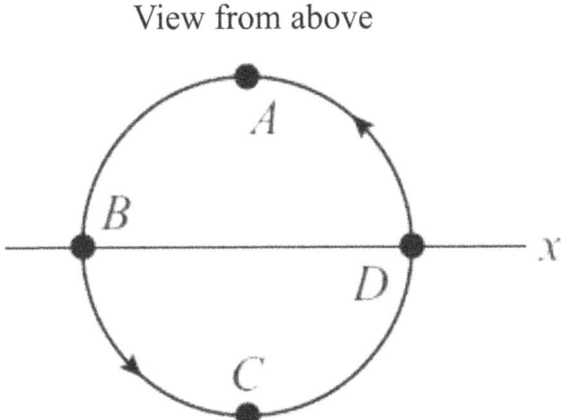

View from above

An object moves at constant speed in a circle.
  a. Your group needs to determine the direction of its acceleration at each of the four positions shown in the illustration. Split the work among group members so that each member is responsible for one point. Work on a shared whiteboard. Make sure you take a point right before the point of interest and right after. Use a ruler to make sure the lengths of the velocity vectors remain the same and their directions are tangent to the circle.
  b. Examine the findings of other members of the group. Do you have an agreement on a pattern in the directions of the acceleration vectors? If so, what is the pattern? Summarize your pattern on your group's whiteboard and compare what you found with the findings of another group. (Adapted from the ALG, chapter 5, pp 5–4.)

To discuss how to provide feedback to the students, we will go back to the tasks of teaching (ToTs). There, we find task III: Monitoring, interpreting, and acting on student thinking. The definition of the task is as follows:

> Teachers engage in an ongoing and multifaceted assessment, using a variety of tools. Teachers understand and recognize challenges and difficulties students experience in developing an understanding of key science concepts; understanding and applying mathematical models and manipulating equations; designing and conducting experiments, etc. Teachers also recognize productive developing ideas and know how to leverage them.

You are probably wondering where feedback is in this general statement. Let's look closer at the specific tasks of teaching:

Teachers:
- III.a Employ multiple strategies and tools to make student thinking visible.
- III.b Interpret productive and problematic aspects of student thinking and mathematical reasoning.
- III.c Identify specific cognitive and experiential needs or patterns of needs and build upon them through instruction.
- III.d Use interpretations of student thinking to support instructional choices both in lesson design and during the course of classroom instruction.
- III.e Provide students with descriptive feedback.
- III.f Engage students in metacognition and epistemic cognition.
- III.g Devise assessment activities that match their goals of instruction.

We can see from the above list that while items III.a and III.g speak directly to devising assessment, the rest of the items are related to providing feedback. To provide feedback, a teacher first needs to use the assessment activities to identify what students do well and what they miss. III.b Teachers then need to connect those to what the students need to work on and to what the teacher's next plans are III.d. But once those are identified, the teacher needs to provide descriptive feedback to the student III.e, which should include both what the students did well (going back to b) and what needs to be improved. However, to help the student do it on their own after feedback, the teacher needs to help the student reflect and question their own thinking. In other words, engage the student in metacognition III.f.

Our research shows (Dodlek *et al* 2023) that experienced teachers are successful in identifying student's strengths and weaknesses, and they often provide descriptive feedback for the faults but not for successes. While this seems like a reasonable thing to do if we wish our students to improve, if the students do not know what they did well, they will not build on this strength in the future. Moreover, observations of students show that they have trouble identifying their own strengths, and therefore, the teacher needs to explicitly tell them what these strengths are.

It looks like providing feedback is a complex and time-consuming task. One way to simplify it is to use clear rubrics that describe the process that the student needs to follow to be successful on a task. While every problem is different and it is impossible to provide content-based rubrics for every problem, the ISLE approach has a unique benefit here as we have tasks that repeat again and again. Those tasks can be self-assessed by the students using the same rubrics (all freely available on the scientific abilities website—https://sites.google.com/site/scientificabilities/) and used by the teacher without providing extensive comments, as the rubrics note the level of achievement that the student needs to reach. Research on the quality of student lab reports written in a project lab course shows that when the instructor provides individual feedback using rubrics compared to written comments, not only does the instructor spend less time providing the same level of feedback (65% of the time when using rubrics compared to writing comments), but also the quality of student work increases (Faletič and Planinsic 2020).

For example, the students need to test the hypothesis that in the motion of a projectile, vertical and horizontal motions occur independently of each other. They are offered the experimental set-up in which two coins are released simultaneously from a table. One coin is dropped and the other one is shot horizontally (see figure 6.4).

Figure 6.4. Set-up for the experiment.

They need to predict which coin lands first. Here, it is very tempting to predict that the coin that is shot horizontally will land last as it has a longer path to travel because of its horizontal motion. But this prediction is based on students' intuition, not the hypothesis under test. If the students making this prediction would use the rubric (see table 6.2), they would see that they made the prediction based on a source unrelated to the hypothesis (their intuition). Notice how the columns of the rubric allow the students to check whether their prediction is based on the hypothesis and describes the outcome of the experiment. They can use the rubric while constructing

Table 6.2. An example of a self-assessment rubric.

Scientific ability	Missing	Inadequate	Needs improvement	Adequate
Is able to make a reasonable prediction based on a hypothesis	No prediction is made. The experiment is not treated as a testing experiment.	A prediction is made but it is identical to the hypothesis, OR the prediction is made based on a source unrelated to the hypothesis being tested, or is completely inconsistent with the hypothesis being tested, OR The prediction is unrelated to the context of the designed experiment.	A prediction follows from the hypothesis but is flawed because: • relevant experimental assumptions are not considered and/or • the prediction is incomplete or somewhat inconsistent with the hypothesis and/or • the prediction is somewhat inconsistent with the experiment.	A prediction is made that: • follows from the hypothesis, • is distinct from the hypothesis, • accurately describes the expected outcome of the designed experiment, and • incorporates relevant assumptions if needed.

the prediction and immediately after. Using the rubric, they would go back and use the hypothesis to make the prediction: if the horizontal and vertical motions are independent, then the coins should land at the same time (neglecting air resistance—assumption) because in the vertical direction they both have the same motion (dropped with zero initial velocity). Here, the prediction follows from the hypothesis and is distinct from it, describes the outcome of the experiment, and incorporates a relevant assumption.

A full set of rubrics is described in the first ISLE book (Etkina *et al* 2019) and is available for free downloads at https://sites.google.com/site/scientificabilities/rubrics.

Now that we have discussed habits of practice related to the content of assessment and feedback, next it is important to talk about the routines. How often should students be assessed? Should we grade their homework? How do we set up a system that will allow the students to improve their work and resubmit it to you without you spending too much time assessing the resubmissions? While there are no single right answers to any of these questions, we will share the routines that we and other ISLE adopters use.

How often should students be assessed? Every time you analyse what your students did when either designing and conducting an experiment, solving a problem, or working on an activity, you assess them. Although this kind of assessment that does not communicate back to the student gives feedback to you, it lacks feedback to students. In order to have a closed feedback loop, it is necessary to devise assessment routines so that you and your students receive feedback as often as possible. A short quiz at the beginning of each lesson in a high school setting is an ideal set-up for you to see where the students are and for them to see where they are after you provide feedback. It also allows you to see the patterns of achievements and difficulties and to prepare for the next lesson. These quizzes need to be very short and if you assign grades, they should be subjected to improvements (see below). They need to be easily graded (about 30 seconds per person per quiz will make this part of your daily work possible). The easiest way to write those quizzes is to have questions that do not require long calculations but allow students to draw different representations and look for consistency. For example, right after your students learned that the acceleration of an object is in the same direction as the sum of the forces exerted on it and they know how to draw force diagrams, an excellent quiz the next day can be the following:

Below you see an elevator being pulled by a cable (figure 6.5). On the right of the sketch, you see three force diagrams representing the forces exerted on the elevator. Which diagram represents the elevator going up? Explain.

Here, the answer will tell you immediately whether the student understands that force diagrams predict the acceleration of an object, but not its velocity. If you score the quizzes after the class, for the next lesson you will have a good strategy of how to provide additional instruction if it is needed.

It is also useful to set up a very crude grading scheme for the quizzes (if you use grades). For example, it can be 0, 1, and 2. Two—the answer is mostly correct, 0 is

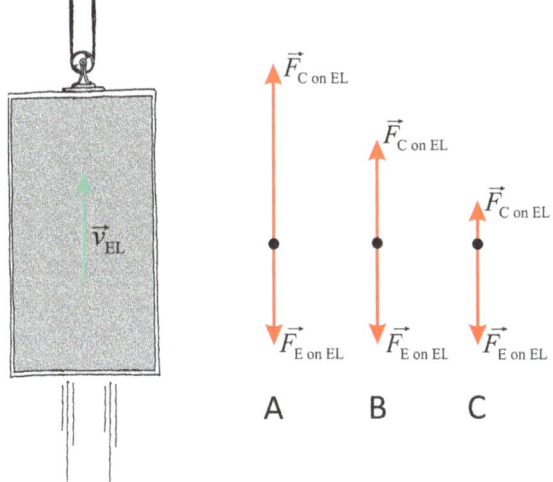

Figure 6.5. An elevator pulled by a cable and three force diagrams.

not correct and 1 is in between. This way you can grade the quizzes quickly by sorting them into piles. This way the students will not have any feedback from you, just a sign that they did not do well on the quiz. What happens next? They are free to find anyone who got a 2, talk to this person, figure out what they did wrong and then come to you during some specially set hours to improve their grade. You can decide what improvement involves. Some teachers let the students do a similar problem in front of them, some ask before-hand to write a note about what they did wrong in their first attempt and how they learned the 'right' stuff, and some teachers interview the student after they complete the resubmitted assignment. However, whatever approach you choose, if a student showed you that they mastered the material, their grade should reflect their state of knowledge and skill at that moment, without being reduced for multiple trials.

Another quick way is to ask your students to write electronic notes (using google docs or google forms) describing one most important thing that they learned during the lesson, how they learned it and what questions they have. While these responses should be graded only for pass/fail (pass if they gave it a significant effort and fail if they did not), the feedback the next day should help students learn how to reflect. The students should also be able to improve these notes too.

If you are designing a summative written assessment (a test), a good rule of thumb to estimate how long it will take your students to complete it, is to give this test to your colleague to solve (perfectly, with all the steps that you require from the students), ask them to time themselves, and multiply this time by two times or even three times. If you are solving the test that you have composed, multiple your time by three or even five times.

If your students write lab reports in groups, then it is very useful to do this activity by using an online platform such as google docs or similar. If each group has its own google doc, then you can track their work in real time, provide feedback right away and see their revisions and contributions of different students. For more details see (Buggé 2020).

Instead of writing a lab report, another approach is having the students write a letter to their parents, grandparents or any guardians about the lab explaining what they did and what they learned.

Standards-based assessment. For a long time, student achievement has been assessed by their performance on tests, quizzes, and other assignments. A new approach to assessment has emerged in the last ten years. This approach focuses on students' mastery of specific skills that are described at the beginning of the courses with the level of mastery outlined by rubrics. One can think of standards-based assessment as a way to create a map of the course for the students that they can follow as they progress through the course. All they need to do is to meet each standard at the highest level defined by the rubric. They can show their mastery on multiple assignments. An example of a standard can be: The student can represent situations involving forces using a force diagram. The rubric for the force diagram is shown in table 6.3.

The teacher creates similar rubrics for all the standards that they wish their students to meet and gives their students multiple opportunities to demonstrate that

Table 6.3. Force diagram self-assessment rubric.

Scientific ability	Missing	Inadequate	Needs improvement	Adequate
Force diagram	No representation is constructed.	FD is constructed but contains major errors such as incorrect mislabeled or not labeled force vectors, length of vectors, wrong direction, extra incorrect forces are added, or some forces are missing.	FD contains no errors in forces but lacks a key feature such as labels of forces with two subscripts or force vectors are not drawn from single point, or axes are missing.	The diagram contains no errors in the number of forces, and each force is labeled so that it is clearly understood what each force represents.

they meet the standards by working on specific assignments or activities. The IG outlines content-based standards for each chapter at the beginning of the chapter and process standards in the introductory chapter. The scientific abilities rubrics and the IG can help you make a list of standards and rubrics for your students.

While standards-based assessment encourages focus on mastery, it is rather tricky if a school follows traditional grading policy with numerical grades. How do you convert mastery into grades? Some teachers use additional tests at the end of the term/year to give grades to their students which puts an additional stress on the students. Some teachers assign numerical scores to the rubrics' descriptors of proficiency and then calculate the average to assign a final numerical grade. All these methods dilute the power of standards but are unavoidable in the current educational system with its numerical grades. However, in university courses, instructors have more freedom. Therefore, we recommend using standards-based assessment in physics teacher preparation courses where the focus on the grades is minimal and use it with caution with high school teachers. A list of standards for future physics teachers in the course Teaching and Assessment in Physical Science at Rutgers University is provided below[2].

Grading and activities. Your course final grade will be based on how you meet the standards listed below. Each standard will be assessed multiple times through assignments that you complete in class and at home (quizzes, homework, projects, microteaching, final exam). You have to convince me and your classmates that you meet the standard. If at any point you fail to meet the standard, you will have an

---

[2] Some standards refer to the Next Generation Science Standards (NGSS), adopted by many US states as the guide to their curriculum development.

opportunity to be assessed again. Each assignment can be improved. I encourage you to try as many times as you need to make the assignment perfect.

General standards (GS)

GS1: Is familiar with NJ Model Curriculum, New Jersey Student Learning Science Standards, and Next Generation Science Standards and can use them when planning, instructing, and assessing student learning.

GS2: Is able to formulate the goals of the instructional unit that reflect the Disciplinary Core Ideas, Science Practices, and Cross Cutting Concepts (see the NGSS) relevant to this unit and can be assessed.

GS3: Is able to interpret student responses (oral or written) and revise planned instruction based on the responses during microteaching.

GS4: Is able to collect (or to describe) evidence that will indicate that students achieved a proposed goal.

GS5: Is able to write a lesson plan that has all required elements and implement the lesson in practice.

GS6: Is able to write a unit plan that has all required elements. Is able to explain how the unit can be modified for students of different abilities and different cultural backgrounds.

GS7: Is able to devise a beginning of a lesson that builds on student ideas and engages them in meaningful exploration of physics ideas during micro-teaching ('need to know').

GS8: Is able to solve (or explain why the solution is not possible) selected physics problems at the level of algebra-based physics in the areas that are addressed in the course.

Content specific standards (selected topics):

Area of physics	Subject matter knowledge (SMK)	Pedagogical content knowledge (PCK)
Kinematics	SMK1—Can make connections between physical quantities used in kinematics and concrete graphical representations, knows how to derive $x(t)$ functions for different motions and is able to articulate the connections between the concepts and science practices in the unit, including what concepts yield better to specific practices. Is able to demonstrate an understanding of students' ideas in kinematics (productive and	PCK1—Is able to outline the sequence of kinematics topics during the unit. Is able to create an overall storyline for the unit. Is able to articulate what the most important ideas in the kinematics unit are so that the students can move forward. Is able to provide a list of needed equipment and the list of resources to teach kinematics. Is able to discuss research findings in the area of kinematics and give

(Continued)

(Continued)

Area of physics	Subject matter knowledge (SMK)	Pedagogical content knowledge (PCK)
	unproductive), is able to interpret student work on graphs and provide examples of formative and summative assessment in kinematics. Is able to use the concept of index to help students write linear functions.	examples of TPT articles relevant to kinematics. Is able to give examples of formative assessment for the most important ideas and provide examples of possible student responses and feedback.
1D dynamics	SMK2—Is able to articulate the relationships between Newton's laws and explain why particular representations are important.	PCK2—Additional: Is able to provide an example of how to set the goals for one lesson on Newton's laws and show the evidence that the goals are achieved.
2D dynamics	SMK3—Is able to explain the relationship between normal and friction forces and demonstrate how to approach multiple-objects problems.	PCK3—Additional: Is able to interpret student work on forces and suggest instructional sequences to address student difficulties.
Circular motion	SMK4—Is able to identify productive and unproductive language in circular motion and derive the expression for centripetal acceleration without calculus.	PCK4—Additional: Is able to design a two hour laboratory for circular motion where students develop specific scientific abilities while constructing, testing or applying physics concepts.
Energy	SMK5—Is able to demonstrate fluency with the system approach to energy and productive representations of work–energy processes; Is able to explain why the energy of a bound system is negative.	PCK5—Additional: Is able to show how to help students develop mathematical expressions for four types of energy used in mechanics and work. Can demonstrate understanding of student difficulties in this area. Can articulate the curriculum sequence for teaching work–energy unit.
Fluids	SMK6—Is able to explain where the expressions for the fluid pressure and buoyant force come from	PCK6—Additional: Is able to show how research-based questions related to student learning of this material help design instructional units.
Vibrations	SMK7—Can explain the difference between periodic motion and simple harmonic motion; can describe and explain SHM; Is able to demonstrate familiarity with useful representations in this area.	PCK7—Additional: Is able to demonstrate an understanding of students' ideas in the area of vibrations, can design an instructional progression for the unit and final assessment.

Electric field	SMK8—Can explain the difference between the concept of electric field and the physical quantities characterizing it; Is able to use multiple representations to explain the behavior of conductors and dielectrics in electric field.	PCK8—Additional: Is able to address common difficulties that students have with the concept of electric potential in a lesson.
DC current	SMK9—Is able to reason through complex problems in electric circuits (including power) using the language of potential difference	PCK9—Additional: Is able to design a lesson in which the students learn to reason through complex problems in electric circuits (including power) using the language of potential difference and connect this material to their everyday experience

6.7 Homework—to assign or not to assign?

Homework is probably one of the most contentious issues in education. Some studies show that assigning homework does not improve student learning, some argue that the students need to rest at home, some say that even if they assign homework, very few students do it. There are lots of arguments against assigning homework. What are the arguments in favor of assigning homework? We list them below.

Argument 1: Homework teaches people to plan their work. They need to decide where and when they will do it. They need to rely on themselves to do it. They need to figure out how to communicate with other people if they cannot accomplish the homework on their own. This also requires planning. Planning intellectual work is one of the aspects of metacognition. Therefore, the mere need to do the homework develops metacognition. In class, all aspects of the work are planned by the teacher.

Argument 2: Homework helps people remember what they just learned. A long time ago, in the nineteenth century Hermann Ebbinghaus used himself as a study subject to learn how he remembered some information that he just learned. From his limited research came a 'forgetting curve' that shows that within the first day (or more precisely, during the first 10 h) a person forgets about 70% of what they learned. However, if they review the material within this time, their memory brings up the new knowledge to the same level. Repeated review drastically reduces the amount of forgotten information.

While now we know more about memory and how to boost it in the first encounter with the new information, the main idea remains: we forget new stuff very quickly. Therefore, having an opportunity to work with new ideas within 10 h of the first encounter and then again in class increases the chances that the new ideas will stick in memory.

Argument 3: Homework helps people 'catch up'. If something was not clear in class, working on the homework will bring these issues to light and encourage the person to see answers—either with their friends or with the teacher.

Argument 4: Homework prepares people to learn in class next time. Sometimes, it is useful to work on a problem or an experiment before seeing the material in class to create 'the need to know' or a question that will be answered later. If an experiment is videoed and data collection is time consuming, the students can collect data at home and prepare for discussion.

Argument 5: Homework helps learn to interrogate scientific text. There is a great deal of research (Podolefsky and Finkelstein 2006) that shows that our students do not read textbooks. They find the material in the textbooks not helpful and if they open a textbook, then they mostly look for worked examples and mathematical representations. This is unfortunate, as being able to learn from a scientific text is an important skill for current and future education. We have developed a strategy that teaches students to read scientific text in ways similar to how physicists do it (called interrogation strategy and described in the next section). However, to practice this reading strategy, the students need different amounts of time as everyone reads at their own pace.

Based on the above, we have enough arguments in favor of assigning homework. Assuming that you decide to assign it, new issues arise: What to assign? How to provide feedback? To grade or not to grade?

What to assign? What to assign for homework depends on your goals. Do you wish the students to 'strengthen' the new brain connections that they just developed? Then assign several interesting problems, experiments to perform or activities to complete which the students need to use the material that they just learned. You can also assign them to read the textbook, but this is only if they have learned how to do it (see the next section in this chapter). Do you wish to motivate them for the next lesson? Then assign them to observe some experiments (real or videos), collect data and try to find patterns that you will discuss during the next lesson. Do you wish for them to prepare for a test of the whole unit? Then assign them to make a list of the most important things that they think they have learned in the unit and make a test for the unit with the explanations of how each problem assesses those important things. Note that the latter assignment requires team work, thus it should be assigned to teams of students (two or three) to work together and they should have ample time to complete it (3–5 days).

How to provide feedback? While there is always an option to collect homework (or see them online if the students submit online), reading every single one of them and providing feedback is a time and effort consuming approach. How can we save time providing feedback? There are multiple options.

For the homework, which serves the purpose of strengthening the material of the previous lesson, one way is to start the class with a short quiz with one or two questions that are based on the homework. The quiz should be every day at the beginning of class and not take more than 5 min. If a person did the homework, the time should be enough to complete the quiz. But if the person did not do the homework, then 5 min will not be enough. We discussed above how to grade these quizzes and how to provide feedback. Another way to provide feedback is to have a group activity in class the next day that mirrors the homework and then provide feedback in that moment to the class as a whole. You could also post the solutions to the assigned problems a day after the homework is due and ask those who have questions about the problems to come after class (or at any designated time) to talk to you. The main idea is that you should not discuss how to solve homework problems in class the next day as it sends the message to those who did not do them that they will learn in class regardless of whether they complete the homework. So, why bother do it at home?

For the homework that prepares the students for the next lesson, the feedback is provided when you continue the activities through sharing in class. But in this case, there is no accountability—if the homework was to collect data from an experiment, those who did the experiments and collected data will benefit, but those who did not do the experiments would wait passively and not learn. What to do? Here it is good to check the completion of the homework (not correctness) and assign some points for this work to motivate the students to do it. In addition, if you have time and space, you can offer them to come to class after and do the experiments there. It is not the best option, but it is better than letting those who did not do the work lose the learning opportunities that your assignment provided.

To grade or not to grade? Homework is work in progress. If you consider it learning, then providing feedback is necessary but grading is not. You could grade for effort, for completion, for clarity, but not for correctness. The bottom line is that the grade should not be the motivation for a student to do the homework, but the real goals that the homework has. Therefore, it is important to have a conversation with the students about the goals of the homework that we discussed above so that they know *why* they are doing it and how it helps them today and most importantly, in the future. Intrinsic motivation is always better than the extrinsic one. And having exciting homework helps for motivation too!

How do future teachers develop homework routines that they will implement later with their own students? To answer this question, we post the excerpt from a syllabus for one of our physics teaching methods courses (Rutgers University) that shows a possible approach towards helping future teachers develop habits and even routines for homework. The syllabus is for the course, Teaching and Assessment in Physical Science, which will be discussed in detail in the next chapter.

Homework: Every week after class you will
1. For the first eight weeks of classes, you will write a lesson plan or another assignment for one of the concepts discussed in class that will show that you met a specific standard. Follow the outline at the end of the syllabus. The homework should be submitted by Thursday night. I will read the reports on Friday and provide feedback typed in the same document but in a different color. You will then make revisions in a third color and send the revised document back to me. You can revise the homework as many times as you wish to improve your grade, but you have to do it before the next class. For the last six weeks of classes, you will write a reflective journal answering three

questions as if you were a student in that class (answers need to relate to physics, not teaching):
- What did I learn in class?
- How did I learn it?
- What remained unclear?
- If I were the teacher, what questions would I ask to find out whether my students understood the material?

Or you will complete another assignment if necessary.
2. Work with chapters/sections of CP: E&A and the ALG to analyse the structure of the cycles and complete problem solving tasks—these are your responsibility. Every week you will be assigned a problem to solve. The solution needs to be submitted with your homework. To get help with the problems, attend a problem solving help session on Tuesdays at 2 pm.
3. Read assigned articles and be prepared for class discussions.

The above text in the syllabus shows several aspects of the homework that are important to discuss:
1. The homework is based on what happened in class, but it involves creative transfer and application of the ideas. The transfer and application are important as although we wish our future teachers to learn how to teach every topic in the ISLE approach, it is not truly possible as each school district or university has their own requirements for the curriculum or available equipment or class setting. Therefore, it is important when preparing teachers that we let them practice designing their own lessons matching their course set-up. The more ISLE-based lesson plan writing they experience, the easier it would be for them to navigate their own course.
2. While future teachers experience the ISLE approach in their own methods courses and reflect on their experiences (see more about it in chapter 7), their proposed lessons might still need improvements. Therefore, building into the course structure the expectation that they would need to revise their lesson plans is crucial for their intellectual growth.
3. Including reflection questions in the homework helps future teachers develop a habit of reflection and sets them up for using this habit when they become teachers.
4. Practicing problem solving in the ISLE approach also helps develop habits of mind and practice problem solving routines. They will only help their students develop productive problem solving routines if they possess those themselves.
5. Reading research articles is a productive habit but this reading needs to be connected to practice.

6.8 Reading textbooks

As we discussed above, one of the goals of the ISLE approach is to help students experience learning physics similar to how physicists construct knowledge. A crucial part of functioning as a physicist or any scientist is reading scientific texts. Although evidence suggests that the ability to effectively read science texts is important, students enrolled in STEM courses do not regularly read science texts (Podolefsky and Finkelstein 2006). In one survey of life science and engineering majors (Stelzer *et al* 2009), the researchers found that 70% of students never or rarely read texts. A study of students enrolled in introductory physics courses found that while 97% of the students report buying the text, fewer than 40% of those students regularly read it. Studies in other content areas have yielded similar results, finding that less than 30% of students regularly complete reading assignments. This lack of consistent reading can be detrimental not only to the students' learning in school, but also to their success out of school.

Why don't students read the textbook even if they spent money on it? We can see three reasons here. The first one is the textbooks are written in a way that does not help students learn. The second one is that the students do not know how to *read* textbooks. They usually approach a textbook as they approach fiction, thinking that just by reading a sentence after a sentence they can learn new ideas and solve problems. But reading is just one part of the brain learning cycle (Kolb 1984, Zull 2002). According to brain research, the learning process starts with sensory input. It then proceeds to the reflection of this input and the subsequent formulation of a hypothesis that explains the input by connecting it to existing knowledge. Finally, the hypothesis explaining the input needs to be tested through the engagement of the learner in active testing of the hypothesis which involves motor functions (it can be talking, writing, performing an experiment, etc). Based on this process, reading involves sensory input. However, if after this input the reader does not reflect on the read sentence and does not try to place it in the set of knowledge that they already possess by making connections, does not hypothesize what this read sentence or paragraph can mean, and does not talk to other people about it, or write it, then the read information does not become 'knowledge' or 'understanding'. Therefore, if we wish the students to learn something from reading the textbook, they need to learn to go through the above processes.

The third reason for the students not reading textbooks might lie in the use of internet and social media. Research points to the reduced attention span and inability to concentrate of those who spend a lot of time on online social networks (Paul *et al* 2012). To benefit from textbook reading, one needs to invest significant time and mental energy. The lack of this time and inability to focus on the same content for a prolonged time might also contribute to the lack of textbook reading. However, this reason might not be that important as even before the expansion of social media our students did not read textbooks much.

To help students develop abilities to comprehend and think critically about their reading, we need to teach them to read scientific text the same way we teach them how to design experiments, collect and analyse data, solve problems, etc. One method that was found effective to achieve this goal is called elaborative interrogation (Smith *et al* 2009). Elaborative interrogation is a reading strategy in which students are prompted to read the text and then answer a 'Why is this true?' question based on the reading. The results of interrogation studies are encouraging, as the method shows increased comprehension over more traditional comprehension techniques such as rereading. Two studies examined the reason for the effectiveness of this method and both came to a similar conclusion. The 'why' questions help focus students' attention on the relevant information within the text, which reduces the cognitive load of irrelevant information. But the reduction of cognitive load is not the only benefit of the elaborative interrogation approach. If you think of what is happening in our brain when we ask ourselves 'Why is this true?', you will see that the steps resemble Kolb's learning cycle. To answer this question, the reader needs to first think of what this 'this' is—reflective observation. Then they need to figure out how this new piece of information relates to what they already know to answer the question why this is true—making a hypothesis. Finally, if the assignment requires the student to write an answer to the question as a part of the homework assignment or class activity, then they engage in active testing as motor function is involved.

While it might seem that elaborative interrogation is just another technique that we, as teachers, need to learn, in fact, there is nothing new to it. All experts, when reading scientific papers, interrogate them. They ask themselves the following questions: How do the authors know this? What is the evidence? What is the uncertainty in the evidence? How does this new idea fit into my previous knowledge? How can we test this new idea? And so forth. These are interrogation questions and when we ask them, we go through the Kolb's cycle again and again. How can we teach our students to do the same?

Here is an example. We call it 'reading aloud with the students'. Ask the students to open a page in their textbook (it needs to be the same page for the whole class) and read a paragraph. Then you model how to interrogate each sentence in the paragraph. For example, here is a paragraph from the textbook CP: E&A (chapter 17, p 506):

> We now understand why rubbed objects acquire opposite charges. Two objects start as neutral—the total electric charge of each is zero. During rubbing, one object gains electrons and becomes negatively charged. The other loses an equal number of electrons and with this deficiency of electrons becomes positively charged. Sometimes when you rub two objects against each other, no transfer of electrons occurs. When the electrons in both materials are bound equally strongly to their respective atoms, no transfer occurs during rubbing.

Below we show how to 'read it aloud' with your students to help them learn how to interrogate this text. We will show the original text in Italics and the interrogation progression in a regular font.

We now understand why rubbed objects acquire opposite charges. When did we learn that? Oh, right, we did experiments when we saw that when you rub the foam stick with fur, this fur attracts that stick and repels the stick rubbed with plastic. Assuming that the stick rubbed with fur is negatively charged, and the stick rubbed with plastic is positively charged, then the fur should be positively charged. I guess the sentence makes sense.

Two objects start as neutral—the total electric charge of each is zero. That is right, zero does not mean the absence of charge as I remember though. Maybe a neutral object already has both charges inside, but they cancel each other?

During rubbing, one object gains electrons and becomes negatively charged. The other loses an equal number of electrons and with this deficiency of electrons becomes positively charged. We just studied that atoms are made of positive nuclei and negative electrons and only electrons are mobile. When we charge objects by rubbing, then it is possible that some of the electrons of one object jump onto the second object. Then the first object that lost electrons is now positive and the object on which the electrons jumped is negative. But the total charge is still zero as we did not create or destroy any electrons.

Sometimes when you rub two objects against each other, no transfer of electrons occurs. Hmm, this does not make any sense. Does this sentence contradict the previous sentence?

When the electrons in both materials are bound equally strongly to their respective atoms, no transfer occurs during rubbing. Oh, I see now. If the electrons are bound equally strongly in both objects, then nobody can pull electrons from this other object. Then it means that for charging by rubbing to work, the electrons in the two rubbing objects need to be bound differently to their nuclei. However, if I think about it, it does not make sense. It is very hard to imagine that all electrons are bound equally strong to an object. So, maybe, if the electrons are bound approximately equally strongly in both objects, each object takes about the same number of electrons from the other object. Therefore, the total number of electrons on each object practically does not change. I wonder which one is more correct...

The process described above takes about 5–7 min in class, but it is extremely important as it shows how you think. When we talk about cognitive apprenticeship, the most difficult part of it is to make the thinking of an expert visible to a novice so that the novice can use it as an example. Thinking aloud when reading the text achieves this goal. The next step is to engage the students in the same activity for a couple of examples and then to assign reading and interrogation for homework. But how do you know that the students did indeed interrogate the text at home? One way to do it is to explicitly assign interrogation questions as homework and then put them on quizzes and tests. In the textbook CP: E&A, each section ends with an interrogation question, the answer to which can be found in the text of the section.

6.9 *Interlude*: Gorazd about recognizing multiple levels of complexity in a specific physics experiment

It was a hot Sunday in August. We decided to take the kids to a park with a stream running through it. I thought that I might come across some interesting physical phenomena on the trip, so I took my GoPro camera with me. GoPro works underwater and I can use it to film underwater (of course, the camera could also come in handy for

(a) (b)

Figure 6.6. A ball rolling on the bottom.

some family shots). Just before we left, I had thrown a green lacrosse ball in my backpack. It might come in handy for playing in the water, or maybe for an experiment.

After we had cooled down in the shallow water, Finn and I spent some time throwing the lacrosse ball around. The ball is made of hard rubber, and it sinks when dropped in water. As we play, I notice that when the ball hits the sandy bottom, it starts rolling along the bottom in the direction of the current, and moves forward at a roughly constant speed. Hmm, interesting! I immediately start thinking about how I can explain the phenomenon and which chapter of physics I could apply it to. This was my thinking: 'Looks like a constant velocity motion … so the sum of the forces on the ball is zero … drag force due to the motion of the fluid …and rolling friction… more than enough reasons to take some videos'. First, I choose the underwater footage. I ask Finn to drop the ball into the water, a few meters away from me, and I take a video as the ball rolls past me (see photo in figure 6.6(a) and video at https://youtu.be/0xdX6OKC2as).

As I watch the footage, I think about a possible problem for my students. What is a unit where I could use it? What other information will they need to explain the phenomenon? The mass and the size of the ball are certainly important data, but I can determine this at home. But I definitely need information about how fast the water moves. Since I cannot tell this from the underwater video, I decided to take a video of the same phenomenon but from a different angle. I hold the camera above the water and point it down to get the top view of the ball's motion. I ask Finn to throw some dry leaves into the water, which float on the surface and are carried along by the water current (see figure 6.6(b)). While Finn and I are doing this new experiment, the wind blows and shakes the leaves off the trees by the stream. Now we have plenty of leaves that are carried by the stream. I quickly take the opportunity to take five more videos of the same experiment so that I can choose the best one when I am at home.

When I got home that evening, I downloaded the videos from the camera to my computer and had a quick look at them (I haven't deleted the videos on the camera yet, as they can serve as a backup). First, I looked at the video that shows the ball's

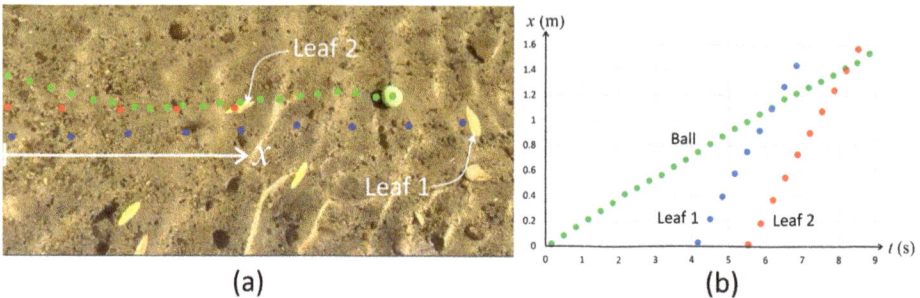

Figure 6.7. Rolling ball and leaves.

motion from the side, which I had taken underwater. The video clearly shows that the ball is rolling. It might come in handy later as evidence of how the ball is moving. Next, I looked at the videos, which showed the top view and the movement of the dry leaves on the surface of the water. I chose the videos where there were more leaves and where the ball and the leaves were moving roughly parallel to the frame of the image, as this would allow me to set the coordinate axis horizontally and thus make the analysis simpler.

I was ready to import the video into a video analysis software. If I wanted to determine the position of the objects in a video in meters, I needed to know the size of an object that was in the video. In my case the best choice was the ball, so I measured its diameter, which turned out to be 6 cm. First, I walked through the video and selected the objects whose motion I planned to observe. In addition to the sphere, I chose two leaves that traveled as close to the trajectory of the ball as possible, since I wanted to know the speed of the water close to the ball. Then, by clicking on successive frames, I manually determined the position of the objects to be observed (the ball and the two leaves) at successive moments. As I did this, a motion diagram was produced in front of my eyes (figure 6.7(a)), which I might use later in activities for students. Based on the position and time data, the program drew a graph showing the time dependence of the positions of the observed objects. Since the motion in my case was predominantly in one direction, I plotted a graph of the time variation of the x-coordinate (see figure 6.7(b)).

I am fascinated by the graph. The ball movement seemed uniform when playing in the creek, but I didn't expect such a nice linear dependency. When I look at the graph, I keep thinking how it is consistent with what I can see on the video and what new information I can learn from the graph. The slopes of the lines tell me that the ball is moving slower than the leaves (we noticed this at the stream already), but I also learn that the leaves are moving at almost the same speed. How can I use this material in the student activities and in which chapters?

In kinematics I can build on the following questions:
- What is the speed of the ball and what is the speed of the leaves?
- What is the relative velocity of the ball with respect to leaf 1?

- How does an observer (e.g. an ant) sitting on leaf 1 see the motion of the ball/leaf 2/the stone at the bottom of the stream? Describe in as much detail as possible.
- Where was leaf 1/leaf 2 at time $t = 0$?
- At approximately what time (relative to $t = 0$) was the image in figure 6.5(b) taken? Explain.
- Based on the graphs $x(t)$, draw graphs $v(t)$ for the three objects. Which graphs provide more information in this case? Explain.

Then I go deeper. What can I say about the forces exerted on the ball? If I ignore the mild curling of the ball, I can say that the ball is moving at a constant speed in the x-direction and, therefore, the sum of the forces exerted on the ball is zero. But which bodies are exerting forces on the ball? These are Earth, the water, and the sand. Students should have no problem with Earth and the sand. The force exerted by Earth points downwards and the force exerted by the sand can be broken down into two components. The first one is the friction force, which points in the direction opposite to the direction of motion. The second component is the normal force, which points perpendicular to the sand's surface, and thus, vertically upwards. It is not so easy to figure out how water exerts a force on the ball. I realize that water interacts with the ball in two ways. Water exerts a buoyant force (because the pressure in water increases with depth) and a drag force (because the ball is moving relative to the water). Can I determine the magnitudes of these forces? Buoyant force is easy. Knowing the diameter of the ball (6 cm) and the density of water, the students can determine the buoyant force. If I give the students the mass of the ball (140 g in my case), they can also determine the density of the ball and confirm that it is denser than water.

What about the drag force? Since the ball is moving slower than the water, the water is pushing the ball with the drag force that points in the direction of the ball's motion. Wait ... that's true if the water is flowing evenly around the ball. In my case, the water is practically not moving at the bottom. But the further I go from the bottom, the water speed increases. This is a difficult problem, so I decided to first look at a simple case where I assume that the water is flowing uniformly around the ball. In this case, the drag force indeed points horizontally which is also in the direction of the ball's motion. Using the results from the previous activity, students can determine the relative speed of the ball with respect to the water. Since they know the diameter of the ball and the density of water, they can estimate the magnitude of the drag force by assuming that the water flow around the ball is uniform, as mentioned earlier. Now students should be able to draw a force diagram for the ball (such as the one in figure 6.8). This problem will be suitable for the students who have already completed the chapter on hydrodynamics. But even for those students this is not an easy task, so maybe I will have to add some hints in the activity.

Can I go one step further? As noted earlier, the water does not flow evenly around the ball. The speed of the water relative to the ball is greatest at the top of the ball and is zero where the ball touches the sand. Although the problem is too complex even for me to address quantitatively, I can nevertheless say something about the forces at a qualitative level, and so can students. Since the relative velocity of the water with

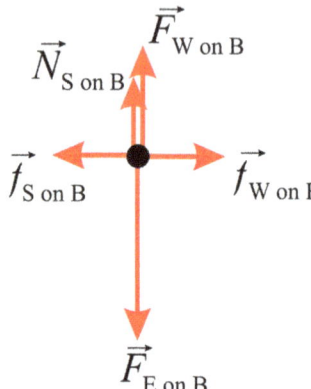

Figure 6.8. Force diagram for the ball that is rolling along the stream, assuming that water flows uniformly around the ball. B: ball, E: Earth, S: sand, W: water. The drag force and the friction force are labeled with a lower-case letter f. N is the normal force.

respect to the ball increases towards the top of the ball, the sum of the forces with which the water current pushes on the ball should point at some angle below the horizontal. If this hypothesis is correct, then the drag force has both a horizontal and a vertical component. As the vertical component pushes the ball towards the ground, this should also result in an increase of the friction force. This reminds me of a case when you are pressing at an angle on a book, making it move at constant speed along the horizontal surface. So, how can I test my hypothesis about the drag force…? Hmm…I think this problem can make for a challenging project for the students…

References

Brookes D T, Yang Y and Nainabasti B 2021 Social positioning in small group interactions in an investigative science learning environment physics class *Phys. Rev. Phys. Educ. Res.* **17** 010103

Buggé D A 2020 The short and long-term effects of the isle approach on high school physics students' attitudes and development of science-process abilities *Doctoral Dissertation* Rutgers, The State University of New Jersey, School of Graduate Studies

Dodlek D, Planinsic G and Etkina E 2023 How to help students learn: an investigation of how in- and pre-service physics teachers respond to students' explanations *Phys. Rev. Phys. Educ. Res.* **20** 010120

Etkina E 2000 Weekly reports: a two-way feedback tool *Sci. Educ.* **84** 594–605

Etkina E, Brookes D and Planinsic G 2019 *Investigative Science Learning Environment: When Learning Physics Mirrors Doing Physics* (IOP Concise Physics) (San Rafael, CA: Morgan and Claypool Publishers)

Faletič S and Planinsic G 2020 How the introduction of self-assessment rubrics helped students and teachers in a project laboratory course *Phys. Rev. Phys. Educ. Res.* **16** 020136

Harper K, Etkina E and Lin Y 2003 Encouraging and analyzing student questions in a large physics course: meaningful patterns for instructors *J. Res. Sci. Teach.* **40** 776–91

Kolb D A 1984 *Experiential Learning: Experience as the Source of Learning and Development* (Eaglewood Cliffs, NJ: Prentice-Hall)

May D and Etkina E 2002 College physics students' epistemological self-reflection and its relationship to conceptual learning *Am. J. Phys.* **70** 1249–58

Paul A P, Baker H M and Cochran J D 2012 Effect of online social networking on student academic performance *Comput. Human Behav.* **28** 2117–27

Podolefsky N and Finkelstein N 2006 The perceived value of college physics textbooks: students and instructors may not see eye to eye *Phys. Teach.* **44** 338–42

Smith B, Holliday W and Austin H 2009 Student's comprehension of science textbooks using a question-based reading strategy *J. Res. Sci. Teach.* **47** 363–79

Stelzer T, Gladding G, Mestre J P and Brookes D 2009 Comparing the efficacy of multimedia modules with traditional textbooks for learning introductory physics content *Am. J. Phys.* **77** 184–90

White G, Sikorski T R, Landay J and Ahmed M 2023 Limiting case analysis in an electricity and magnetism course *Phys. Rev. Phys. Edu. Res.* **19** 010125

Wikipedia n.d. Prediction *Wikipedia* https://en.wikipedia.org/wiki/Prediction

Zull J E 2002 *The Art of Changing the Brain: Enriching Teaching by Exploring the Biology of Learning* 1st edn (Sterling, VA: Stylus)

IOP Publishing

The Investigative Science Learning Environment: A Guide for Teacher Preparation and Professional Development

Eugenia Etkina and Gorazd Planinsic

Chapter 7

Organizing ISLE-based teacher preparation programs for the development of habits of mind and practice

In this chapter, we will discuss the organization, structure, and details of physics teacher preparation programs. These programs develop productive habits of mind, practice, maintenance, and improvement, which help teachers implement the ISLE approach in the classroom[1].

7.1 The importance of coherence and duration in the program

Research on the best teacher preparation programs finds that the best programs provide their candidates with prolonged coherent experiences where course work is coupled with clinical practice (Darling-Hammond *et al* 2005; Feiman-Nemser 2001; Hammerness *et al* 2005). Why are the length and coherence of the messages that teacher candidates receive important? Now that we have discussed the habits and routines of good teaching, the answer is clear—these two conditions are necessary for the formation of the habits and routines (repeated contexts and practice). The goal of this chapter is to provide examples of how to create coherent and prolonged physics teacher preparation programs that help future physics teachers develop habits of mind and practice necessary to implement the ISLE approach.

Both authors of this book were or still are the leaders of physics teacher preparation programs focused on the ISLE approach. We will use our own experiences and examples to describe how to create and maintain such programs.

---

[1] We remind the reader that the abbreviation CP: E&A stands for the textbook *College Physics: Explore and Apply*, ALG stands for the Active Learning Guide, OALG stands for the Online Active Learning Guide, and IG stands for the Instructor Guide. Proper citations for these materials are in chapter 1.

Below we show a table that provides an overview of both programs and shows how they achieve coherence. Both programs operate on semester teaching schedules and each semester is 15 weeks of instruction. The Rutgers program is in New Jersey, USA and the University of Ljubljana program is in Slovenia, Europe (table 7.1).

Note that here we are only reporting on the aspects of the programs related to the development of ISLE-based physics teaching habits of mind, practice, maintenance, and improvement. We are not reporting on general education courses that future physics teachers take with other preservice teachers (history of education, educational psychology, educational law, and so forth).

In what follows, we focus on the details of the Rutgers program. All physics teaching related activities in the Rutgers program described in this chapter have four goals. The first goal relates to the development of preservice teachers' dispositions necessary to teach physics through the ISLE approach. The second goal relates to some fundamental understating of teaching of all physics concepts. The third goal relates to the teaching of specific topics. Finally, the last and most important goal of the course work and clinical practice is helping preservice teachers develop productive habits of mind, practice, maintenance, and improvement discussed in the previous chapters in this book. These goals are achieved through all activities in the program that form a coherent progression of experiences for preservice teachers. To provide the details for the reader, we first start with the course work. While each course contributes to the achievement of all four goals of the program, the table 7.2 shows how each course addresses the fundamental and topic-specific goals.

In all of those courses, the ISLE approach is the lens through which the students examine and interact with the material. They also use the same textbook CP: E&A, the IG, and the ALG throughout all of the courses and thus get to know the materials very well. (We already shared the syllabus for the first course, Development of Ideas in Physical Science, in chapter 4. In appendix A of this chapter, we share the syllabus for the Teaching and Assessment in Physical Science course and in appendix B we share the syllabus for the Multiple Representations in Physical Science course.) The logical progression of the courses is important.

In the first course, 'Development of Ideas in Physical Science', the students learn *how* the major concepts, models, relations, etc, that they will be teaching in high school were developed. They learn what the original experiments were, how physicists came up with different hypotheses and how they ruled them out, what tools they used and so forth. This knowledge is not only important for the development of future teachers' scientific epistemology and physics habits of mind, but also for the learning of struggles of physicists constructing those concepts as these struggles will be similar to the struggles of their future students. For example, it took physicists a long time to agree on what to call 'force', or 'energy', or 'heat'. Future teachers also learn how those struggles, even when resolved, leave the traces in our language (for example, we speak about heat as a substance) and even in the mathematical tools that we use to analyse physical phenomena. One example is the caloric model that is discussed in section 7.2 of this chapter. As the students learn the historical progressions through the ISLE lens, they also learn to use hypothetico-deductive reasoning. As we discussed above, this type of reasoning is

Table 7.1. An overview of two ISLE-based programs that develop high school physics teachers.

Overview of the program	Rutgers program	University of Ljubljana program
Where is the program housed?	Graduate School of Education	Faculty for Mathematics and Physics
Who are the students?	Those who have an undergraduate degree in physics or are in the last year of their undergraduate program.	Those who have an undergraduate degree in physics (completed first cycle in physics).
How long is the program?	Two academic years plus the summer in between.	Two academic years (students can study an additional year to complete a masters thesis).
How many physics specific courses do they take?	Six (each meets for 3 h a week), total of 270 h of physics methods instruction.	Five (from 4 to 6 h per week), total 390 h of physics methods instruction.
Do these courses prepare future teachers to teach every topic of a high school physics course (i.e. kinematics, dynamics, etc)?	Yes	Yes
Do all of these courses use the ISLE framework for learning physics?	Yes	No; 4 out of 5 use the ISLE framework.
In what semesters do PSTs engage in clinical practice (i.e. in teaching physics)?	In every semester of the program with a total of about 120 days.	Thirty days.
Does clinical practice engage PSTs in teaching in the ISLE-based environment?	Yes	Partly. Cooperating teachers where students do the clinical practice are using some ISLE ideas, but not all.
Is there a community of graduates	Yes	Yes
Do graduates of the program contribute to the development of the program and teaching in it?	Yes	They do contribute in the development of the program. They cannot teach in it, but they occasionally come as invitees to share their experiences with new students.

Table 7.2. Course content in physics teaching methods courses in the Rutgers program[a]. At Rutgers University, 1 credit stands for 1 h of in-person instruction. Most of the courses in the program have 3 credit-hours and involve 3 h a week in the classroom.

Year/ semester	Physics methods courses	Fundamental goals	Physics topics-specific goals
1/Autumn	Development of Ideas in Physical Science	To learn how physicists developed the ideas and laws that are a part of the high school physics curriculum.	Motion, forces, momentum, energy, first law of thermodynamics, kinetic molecular theory, electric charge, electric current, photon, atom, and nucleus.
1/Spring	Teaching and Assessment in Physical Science	To learn how to build student understanding of crucial concepts, and how to develop and implement curriculum units, lesson plans, to design assessment, and to learn how to respond to the students.	Motion, forces (including buoyant force), vibrations, energy, electric field, DC circuits, electric power.
	Technology in Physics Education	To learn how to use technology to develop and apply physics concepts and to assist in course management.	Topics depend on student choice but these are usually connected with available computer sensors (e.g. motion detector, force probe, microphone) and smartphone apps.
1/Summer	Engineering Education	To learn how to include engineering projects in a physics course.	Mechanics and DC circuits.
2/Autumn	Teaching Internship Seminar for physics students	To simultaneously support preservice teachers who are doing student teaching and to explore teaching approaches to the additional topics.	Motion, forces, momentum, statics, mechanical waves, and wave optics.
	Teaching Internship (counts as a 9-credit course)	To practice implementing the ISLE approach in a high school classroom continuously for 15 weeks.	Usually the topics are kinematics, dynamics, circular motion, momentum and energy.
2/Spring	Multiple Representations in Physical Science	To integrate different representations of physics knowledge into problem solving and to	Electric and magnetic fields, magnetic properties of materials, complex DC circuits,

| | connect learning of physics to what we know about the structure and function of the brain. | electromagnetic induction, e/m waves, geometrical optics, wave optics, and photoelectric effect. |

[a] These are not all the courses that the students take. General education courses such as Educational Psychology, Teaching Diverse Learners, etc, are not included in this table.

Table 7.3. Clinical practice in the Rutgers program.

Semester	Where the practice occurs	What responsibilities the PTSs have
1/Autumn	Local schools	*Observations*—4 h a week
	Physics teaching methods course	*Microteaching*—3 h
	Rutgers University, ISLE-based introductory physics course called Physics for the Sciences. The course coordinator has 20 years of ISLE experience.	*Sheltered teaching*—labs or problem solving sessions (3–6 h per week for 14 weeks), office hours, grading homework, and attendance of course meetings.
1/Spring	Local schools but only with cooperating teachers who graduated from the program and implement the ISLE approach.	*Observations*—1 full day per week of physics lessons for 15 weeks. PSTs plan and implement one 'real' lesson.
	Rutgers University, ISLE-based introductory physics course.	*Sheltered teaching*—(3–6 h per week for 14 weeks)
	Physics teaching methods course	*Microteaching*—3 h
2/Autumn	Local schools with the same cooperating teacher from 1/Spring	*Full-time teaching*—15 weeks
2/Spring	Rutgers University, ISLE-based introductory physics course.	*Sheltered teaching* (3–6 h per week for 14 weeks)
	Physics teaching methods course	*Microteaching*—3 h

very difficult (basing predictions not on your intuition but on the hypothesis under test) and it takes a long time for students to develop it. Therefore, this course is only the first step. In addition, in this first course, preservice teachers learn to notice the application of physics ideas in the surrounding world.

However, in order to see how the preservice teachers develop dispositions, knowledge, skills, habits, and routines consistent with the ISLE approach (and discussed in the previous chapters), we need not only talk about the structure of the courses but also how clinical practice is integrated into the program. Table 7.3 shows the types of clinical practice in which the preservice teachers (PSTs) participate semester by semester. Here we define several terms that will help you follow the table.

Observations—a PST is present in a high school physics classroom and observes, but does not teach.

Sheltered teaching—a PST has full responsibilities of a teacher while in the classroom but does not plan lessons, design assessments, or deal with classroom management and the interaction with parents or school administration. Sometimes PSTs plan and teach one or two lessons during a semester of observations.

Microteaching—a PST plans and enacts the lessons including the assessments. However, the students are their peers in the physics teaching methods course. Therefore, there is no classroom management and no interactions with parents or school administration.

Full-time teaching—a PST assumes full responsibilities of a high school physics teacher including lesson planning, assessment, enactment, classroom management, interactions with school administration and some interactions with parents.

From the above table, you can see the coherences and continuity of the program's clinical experiences. All of them—except the observations in the very first semester—occur in the ISLE-based classrooms with the mentors who are skilled in the ISLE approach. This aspect provides coherence. The PSTs engage in ISLE-based teaching with an increasing level of complexity over 2 years and *never* see examples of teaching that contradict the ISLE philosophy. This provides continuity of the experiences and gives the PSTs ample time to develop productive habits and even their own routines. Below, you see two quotes from current high school physics teachers who went through the program and are reflecting on the role of sheltered teaching in their development (Etkina 2015):

> I feel that teaching in Physics for the Sciences (PFTS) was incredibly useful in preparing me for both student teaching and my first teaching position. I definitely felt considerably more at ease in front of the class by the end of the PFTS year than I did the first time. Also, the format of [PFTS], where I was forced to work with students in small groups, who were working at different individual paces, prepared me for being a better teacher. Otherwise, I might have resorted to significantly more whole-class, teacher-led instruction because it seems easier, but the experience in PFTS showed me that managing small group work isn't as hard as it seems and has incredibly strong benefits for the students.
>
> I felt that being in charge of a classroom so early on was most helpful for me in two ways: (1) allowing me to experience the kinds of common ideas and questions students had and allowing me the opportunity to deal with those questions, and (2) giving me experience in dealing with the paperwork and non-teaching aspects of being in charge of a class. Those are two things that you can't fully understand unless you experience them and collaborate with others to deal with them.

Now that we have discussed *what* the preservice teachers learn and the kinds of clinical practice in which they engage, it is important to see *how* these two aspects—course work and clinical practice—work together to develop the habits. The structure of the courses and the clinical practice is based on the cognitive apprenticeship approach (Barab and Hay 2001). Cognitive apprenticeship follows the stages of traditional apprenticeship with additional activities that help make the master's

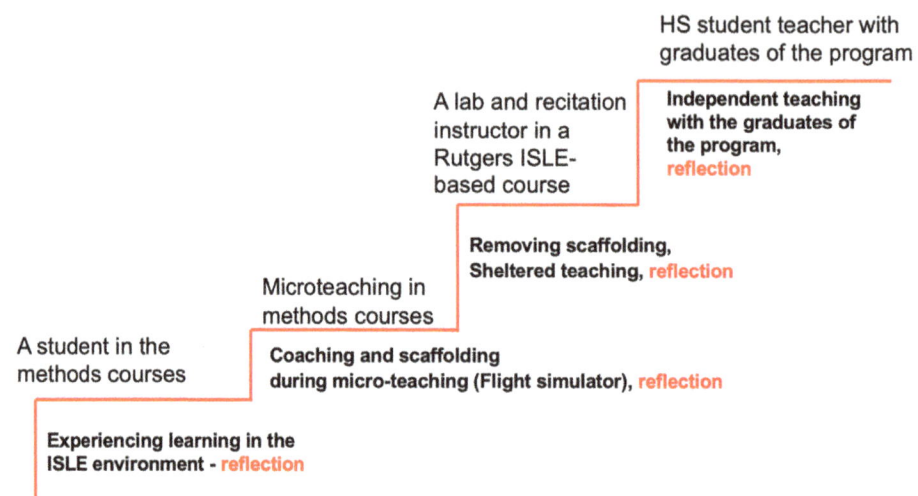

Figure 7.1. Cognitive apprenticeship approach to course work and clinical practice.

thinking visible. Namely, the master first models the desired skills and behaviors. Then, an apprentice tries some parts of the skills with coaching, scaffolding, and the master's feedback. The apprentice tries implementing the desired behavior in its whole with fading scaffolding and coaching. Finally, the apprentice has independent practice with feedback. The additional activities involve reflection on the part of the apprentices and on the part of the master. The progression is shown in figure 7.1.

During the first 5 or 6 weeks in every physics methods course, 130–140 min of each class meeting (out of 160 min of total class time) mimics a student-centered high school lesson in an ISLE-based classroom. During this time, preservice teachers play the role of high school students and the course instructor plays the role of a high school teacher. At the beginning of each class meeting, they take 10–15 min to respond to a quiz question. The quiz relates to the material of the previous class and assesses both their physics knowledge and their ability to respond to a high school student solving a similar problem. Then they explore physical phenomena in groups (e.g. collecting and analysing data, devising and testing explanations), presenting their findings on whiteboards, and participating in discussions, in the same way that high school students would. In the last 30–40 min of each class meeting, they reflect through whole-class discussions on the teaching methods used during the lesson, and how the same lesson can be implemented in high school. Following their reflections, the instructor reflects on the reasons for the decisions that they made during the lesson. At home preservice teachers continue working on the material related to the previous class meeting by solving additional problems and reading research papers. All student work (quizzes, homework, etc) can be revised and improved as many times as needed after the instructor provides feedback.

From week to week, these lessons follow the progression of an ISLE-based high school physics curriculum so that preservice teachers can see how ideas develop and build on each other over time. In the second part of the course (weeks 7 through 14), they microteach. In groups of two, they prepare and teach 140 min lessons to their peers, which represents roughly a week or two of possible high school instruction. (Usually, US students in high schools have four to five 45 min lessons per week in a one-year course and high school students in Europe have 2–3 lessons per week but study physics as a compulsory course over multiple years.) See more about preparation and enactment of microteaching in section 7.3, 'Clinical practice'.

The coaching and scaffolding provided by the instructor to the students preparing and enacting their microteaching in methods courses fades when they enact sheltered teaching in the ISLE-based university physics course. Finally, they only receive feedback on lesson planning and teaching when they have the semester of students teaching in the schools in the second year of the program. It is important to note that when preservice teachers return from full time student teaching to the last semester of the program and continue teaching in Physics for the Sciences, they become mentors to the students from the following cohort who are in their second semester of the program. Both cohorts interact with each other during course meetings and exchange weekly reflections on teaching. This interaction contributes to the development of the community of graduates which we will describe in section 7.4.

To summarize, the program establishes longevity (2 years) and coherence (the ISLE approach in all courses and in all clinical practice). Rutgers preservice physics teachers do not encounter any other teaching method during the program. They have multiple opportunities to develop dispositions, knowledge, and skills necessary to implement the ISLE approach. They observe and analyse their teachers' dispositions and develop their own. They practice all the tasks of teaching described in chapter 2 and do it multiple times. The purpose of this approach is to allow for time and continuity to develop productive teaching habits. However, as you remember from chapter 3, continuity and coherence are not enough to help people develop productive habits. Another condition is positive reinforcement—the person needs to feel good when doing something habitually so that the release of dopamine supports the development of the habit. This 'feeling good' moment is extremely important for habit development. How do we help our students feel good when they develop productive habits?

There are multiple ways we approach this issue. The first way is provided by the first intentionality of the ISLE approach—learning physics by practicing it. Conducting a testing experiment for an idea and getting a match with the prediction elicits very strong positive emotions. Interestingly, a mismatch also elicits positive emotions as the idea under test is never framed as personal intuition. Being able to go back to 'the need to know' and explain it, also elicits positive emotions. The second intentionality of the ISLE approach—motivating students and making them feel capable and belonging also provides multiple opportunities for the students to experience 'feeling good'. Seeing other groups come up with a solution similar to yours makes you feel good. However, seeing a new solution to the problem you have been working on also makes you feel appreciative (solving a problem leads to the release of dopamine in our brains). Revising your work multiple times and eventually reaching a 'perfect' level creates a very strong feeling of satisfaction. You feel that persistence and perseverance have produced results. One student commented on Ratemyprofessor.com about Eugenia's classes: 'I have never worked so hard in my life. But I never enjoyed any class more'. We will return to the ways of helping future physics teachers feel good during clinical practice later in the chapter.

The last condition for the development of habits is 'low friction' (see chapter 3 for the explanation). This condition is met when we simplify the actions that lead to the development of habits or remove possible obstacles. The nature of the ISLE approach yields to this condition perfectly as the teachers and students know exactly what comes after each step in the investigations. They do not need to wonder what to do after an observational experiment or after they have developed a mathematical model. Following the same approach in the sequence of activities allows for continuity and predictability. The fact that we have developed a large body of ISLE-consistent activities and problems for every topic of a physics course removes the need for the beginning teachers to design their own activities, which is a 'high friction task' but allows them to think about the goals of the existing activities, their sequence, and so forth. We will discuss how we reduce 'friction' in clinical practice in section 7.3 of this chapter.

One might ask: 'Everything you wrote above is about using the ISLE approach to prepare physics teachers. What about other approaches to teaching physics? Shouldn't beginning teachers be familiar with them and then make their own choice of what method is the best for them?' The answer is:

> Yes, there is Modeling Instruction, Active Physics, Physics by Inquiry and many other active learning, inquiry-oriented approaches to teaching physics. But knowing about them and being able to implement them day by day for every concept of a physics course takes time. Introducing future teachers to the myriad of existing approaches without them being able to practice those day after day prevents them from forming desired habits and does not allow for learning how to teach *every* concept of a physics course through this approach. When your students become teachers and start participating in professional development, they will learn about different methods and decide for themselves whether the ISLE approach is right for them. But in the meantime, starting their careers with the firm grasp and corresponding habits for one approach is better than knowing about many but having no skills to implement those.

In the following sections of this chapter, we focus on the details of the course work and clinical practice that help our students develop productive habits.

A final word is about professional development programs that help in-service teachers learn about and implement the ISLE approach. The same ideas apply here. Only when they have an opportunity to learn how to teach every topic of their curriculum through the ISLE approach and have multiple opportunities to practice it in the classroom and see that their students are responding positively to this approach will they develop robust-enough habits to sustain ISLE-based teaching in an adverse environment. To achieve this goal, multiple workshops are not enough. We need to supplement the workshops with in-class observations and coaching through team-teaching in the classroom so that the teachers new to ISLE can implement it successfully and receive positive reinforcement. It is a long process, but it is inevitable if you wish the people to develop new habits (Bologna 2023).

7.2 Course work

In this section, we provide the details of the course work in multiple physics methods courses (see table 7.2) and show how this course work helps future teachers develop productive habits and routines for implementing the ISLE environment in their classrooms. As we said above, we can split each course into two parts: 1. During the first 5–7 weeks, the course leader acts as a high school physics teacher, the PSTs act as high school students who are learning a specific concept or skill and at the end of the lesson both the students and the teacher reflect on the teaching methods. 2. During the rest of the semester (the remaining 9–7 weeks), the role of the teacher moves to the PSTs who, working in groups of 2–3, design and enact the lessons where their peers are the students. At the end of the lesson, all three groups ('teachers', 'students', and the course leader) reflect.

In both parts of the course, the students read research papers that are either relevant to the physics topic that they are learning or the pedagogical approach that they are implementing. The integration of the reading comes at the end of the lesson, which provides 'the need to know' for the discussion of the research papers. The focus of both parts of the course is on the development of productive habits as discussed in chapters 3–6 of this book.

Below we show examples of class activities in three sample courses, provide an overview of the rest of the activities, and discuss how those connect to the development of dispositions, knowledge, and skills, and build the foundation of the development of habits.

7.2.1 'Development of Ideas in Physical Science' course

Below you can see some of the activities that students in the course work on. The activities occur during the lesson when preservice teachers reconstruct the historical progression of experiments and reasoning that led to the development of the concept of specific heat, the creation of the caloric fluid model, and the rejections of the caloric fluid model.

When the students come to class, they are grouped into groups of 3–4. Each group has a whiteboard and a set of markers to record their observations and explanations. The groups receive one of the three assignments:

> *Group 1*: Obtain two exactly equal amounts of water (the mass does not matter as long as they are the same). Water #1 should be from the ice-cold container and water #2 should be from the hot water container. Record the temperature of #1 and #2. Mix them together in the Styrofoam container with the lid that has a thermometer and carefully observe what happens to the temperature of the mixture. Devise an explanation for your observation. Below is a sketch of what the students find:

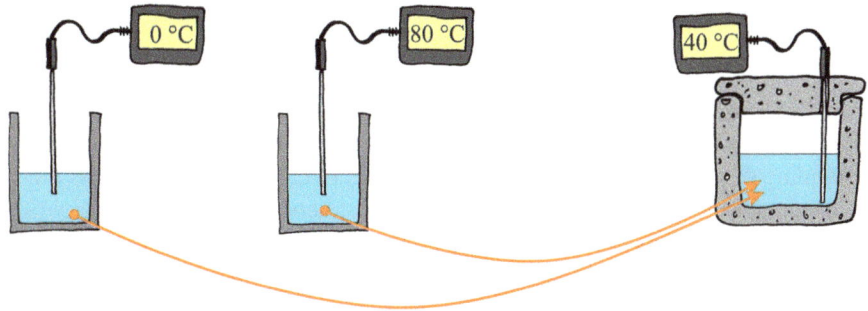

Group 2: Obtain two different amounts of water. Water #1 should be from the ice-cold container and should have mass m_1 and water #2 should be from the hot water container and should have mass m_2. Carefully measure the masses (volumes) so that $m_1 = 2m_2$. Record the temperature of #1 and #2. Mix them together in the Styrofoam container with the lid that has a thermometer and carefully observe what happens to the temperature of the mixture. Devise an explanation for your observation. Below is a sketch of what students find.

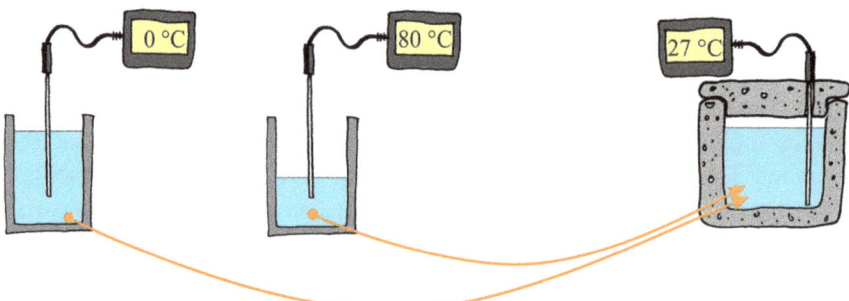

Group 3: Obtain a metal object from the cold reservoir and record its temperature. On your desk, you will have a container with hot water of exactly the same mass as the metal object. Record the initial temperature of the water and the temperature of the object. Place the hot water and the object in the Styrofoam container with the lid that has a thermometer and carefully observe what happens to the temperature inside the container. Devise an explanation for your observation. Below is a sketch of what students find.

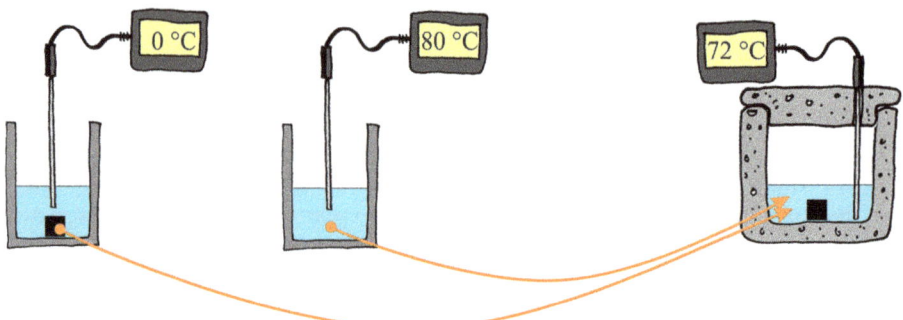

The goal of these observational experiments and the patterns that the students find is for them to invent two historically accurate ideas (but not necessarily accurate in our understanding of physics). The first idea is the model of caloric as a weightless fluid that resides inside objects and flows from warmer objects to cooler objects and is a conserved quantity (ideas brought into physics by Antoine Lavoisier). The second idea is the physical quantity of specific heat as a property of different materials to require different amounts of caloric to increase the temperature of 1 kg by 1 °C (historical contribution of Joseph Black). Although the model of caloric was later rejected experimentally, it still may appear to be a valid concept in physics. Such is the case with the relation: $Q_1 + Q_2 + Q_3 = 0$ in an isolated system, where Q_1, Q_2, and Q_3 represent energy transferred through the process of heating. Therefore, it is important to know how this mathematical statement relates to the ideas of energy conservation and why it is still valid.

Conducting the experiments and sharing their explanations involve the role of mass in thermal exchange and different specific heats of water and metals. However, the instructor tells them that the term 'specific heat' was not yet invented. Afterwards, the students read a passage from the original work of Joseph Black who developed the idea of specific heat and compare this idea to the explanations that they have just invented. Then, the course instructor shares the story of Antoine Lavoisier who came up with the model of caloric fluid to explain how 'heat' is transferred between warm and cold objects. After that, the students read the original writings of Benjamin Thompson, count Rumford, whose experiments rejected the model of caloric as an invisible weightless fluid. Both reading excerpts are taken from the *Source Book of Physics* (Magie 1963) where one can find original papers by Black and Thompson. At home, the students need to fill out the table shown below (see table 7.4; short possible answers are in italics) and they also read a research paper dedicated to how the term 'heat' being a noun obscures the meaning of heat in the first law of thermodynamics (Brookes and Etkina 2015).

Once a student submits the homework (online through Google docs), the course instructor provides feedback and the student revises the homework until it meets the requirements. This process of regular revision helps future teachers develop persistence and a growth mindset. At the same time, this process gives them an example of how to help their students improve their work. They start seeing the resubmission process as a natural part of learning, they experience the benefits of this process, and, therefore, start developing dispositions that will help them implement similar processes with their own students. We should never forget that teachers tend to teach the way they have been taught. For this reason, it is extremely important that in the physics methods courses where our students learn to how to teach, the teaching methods stay true to the dispositions that we wish our students to develop. If we want them to develop a disposition that everyone can learn physics and that it is their responsibility to create conditions for all students to learn, then they need to experience such conditions themselves and see how their own learning improves because of such conditions.

When the students come to class the next week, they respond to the following quiz at the beginning of the lesson.

Quiz: Describe how you understand the meaning of the physical quantities of temperature, internal energy, internal thermal energy, work, and heating. Give examples of each in real-life phenomena.

The course leader grades the quiz and returns it to the student the next week. While we try not to give formal grades, the feedback is usually given in the form of four signs (in descending order): +, ⊥, T, −. In case the grade is not the full plus, the students can resubmit their work using the following form (table 7.5). The same form is used for the quizzes in all other courses.

Table 7.4. Homework assignment.

Year	Name	What did he do?	Element of scientific exploration (ISLE process)
1780	Joseph Black	*Observed mixing of water of different temperatures and mixing of water and quicksilver of different temperatures.*	*Observational experiments, finding patterns, development of explanations.*
1787	Antoine Lavoisier	*Developed a model of caloric—a weightless odorless fluid surrounding particles of matters which follows the law of conservation.*	*Developing explanations/models.*
1799	Benjamin Thompson, count Rumford	*Accidental observations of the hotness of the cannon bores and deliberate experiments collecting data on how much the temperature of the bore rises and whether the specific heat of participating objects changes. He eventually ruled out the caloric model based on his testing experiments.*	*Accidental observations. Developing a hypothesis. Testing experiments. Arriving at a judgment.*
1799	Sir Humphry Davy	*Allegedly rubbed two ice cubes together at a temperature below freezing and fused them together due to melting. The experiment was designed to refute caloric model as the 'heat' was generated without changing the heat capacity of the objects. However, its attribution to Davy is questionable.*	*Testing experiment for a hypothesis and arriving at a judgement.*

Table 7.5. 'Resubmission' form.

Text of the problem/question.
Improved solution/response.
What did I do wrong in the first version and why did I do it?
How did I learn to solve the problem correctly?
What did I learn from this experience?

From this example, you see activities through which the students experience and reflect on the following tasks of teaching: II. Designing, selecting, and sequencing learning experiences and activities; IV. Scaffolding meaningful engagement in a science learning community; and VI. Using experiments to construct, test, and apply concepts. The homework helps develop a skill of reading and analysing original writings of scientists. The lesson and the homework contribute to the development of the following habits of mind and practice: 'Treating physics as a process not a set of rules' and 'Noticing physics everywhere'.

Additional types of activities in the course include:
  a. Reading and interpreting original writings by physicists describing the invention of physics ideas (Galilei, Newton, Leibnitz, Huygens, Joule, Ohm, etc).
  b. Analysing and synthesizing their work to trace the ISLE logical progressions.
  c. Designing lesson sequences that lead high school students through the historical development of a particular idea through the lens of the ISLE process with a lot of feedback from the course leader and enacting these lessons with their peers (see more on microteaching in section 7.3).
  d. Reading papers from journals for physics teachers (such as *The Physics Teacher* or *Physics Education*) that are related to the development of the same historical ideas and analysing those in relation to high school physics teaching,
  e. Interviewing novices and experts about their understanding of those ideas, analysing those interviews and interpreting them in light of student learning,
  f. Creating life stories about famous physicists (focus on women and representatives of under-represented groups, for example Marie Curie, Lise Meitner, Emmy Noether, Jocelyn Bell Burnell, and many more).
  g. Designing experiments that mirror fundamental historical experiments that led to the development or rejection of explanations/models.

From the above example and the list, you can see how this very first course contributes to the development of the dispositions, knowledge, and skills discussed in chapter 2. It also builds the foundation for the development of productive habits of mind (for both physicists and physics teachers). PSTs working in groups with whiteboards in every lesson experience the same context of the lesson every time they come to class. The same is true for the end of class meeting when they reflect on what they learned as physics students and as teachers. These repeated contexts are necessary for the formation of the habits of practice. But as it takes time and multiple exposures to develop a habit, one course is not enough. Other conditions needed for the formation of the habits are rewards and reduced 'friction'. Rewards or feelings of success are built into the ISLE progression of the material (observing outcomes of the testing experiments that match predictions and sharing your work with others are rewarding) and from the organization of the class (resubmission of work with clearly seen results is rewarding). Reduced friction comes from the prepared materials (the instructor has equipment necessary to run students' designed testing experiments ready; the instructor responds to students' submission of work quickly so that there is enough time to revise homework or resubmit the quiz before the next class meeting). It is important that there are no numerical grades for the assignments in the course, only feedback whether the work meets expectations or more effort is needed.

7.2.2 'Teaching and Assessment in Physical Science' course

In this example, the students in groups of 3–4 work on a series of activities that represent the ISLE process for constructing a qualitative concept of scalar components of vectors. The activities below are taken from the ALG. After the students complete these three activities and present their solutions on whiteboards, the course leader asks them to reflect: What did they learn? How did they learn it? What student ideas and difficulties do these activities anticipate and how do they help address those?

Activity 1: Observational experiment
The sketch below shows three strings pulling in different directions in a horizontal plane on a small ring (R) at the center. A force diagram for the ring is also shown on a grid.
    a. Based on what you see in the force diagram, explain why the ring does not accelerate in the positive or negative *x*-direction. Be explicit.
    b. Repeat for the *y*-direction.

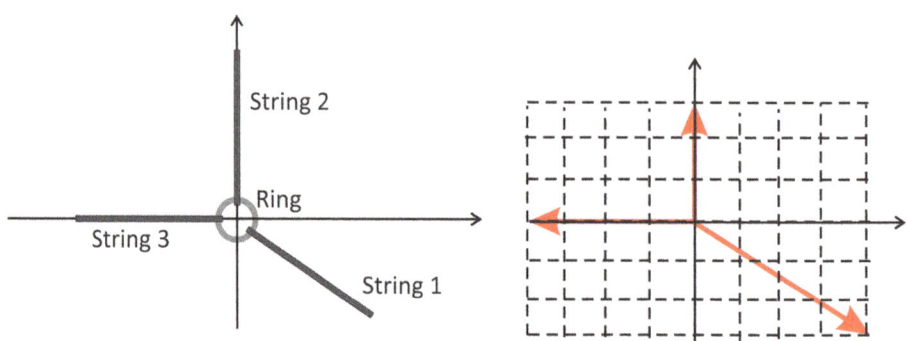

After the students find the components and share their whiteboards, the course leader summarizes what they have found (this is an example of what we call 'Time for telling'). The summary might look like the following:

> Notice that string 1 exerts a 4 N force toward the right, which balances the 4 N force exerted by string 3 toward the left. Similarly, string 2 exerts a 3-force upward, which is balanced by the 3 N downward pull exerted by string 1. If you don't see this, go back to the force diagram and try to visualize it. You should be able to realize that string 1 pulls in both the horizontal *x*-direction and the vertical *y*-direction. We say that $\vec{F}_{1 on R}$ has

an x-component $F_{1onRx} = +4N$, and a y-component $F_{1onRy} = -3N$. Normally, we don't have force diagrams on grids that allow us to visualize the components so explicitly in this way. In the next activity, we will do the same analysis using trigonometry.

Activity 2: Mathematical modeling

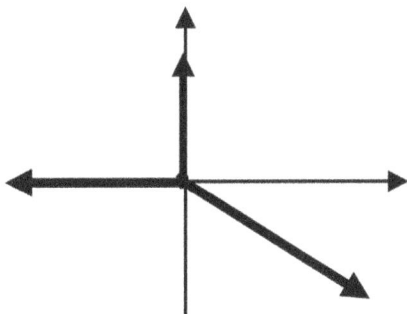

The sketch on the right shows the same three strings pulling on the ring as in the previous activity. However, an angle is now shown for the pulling direction of string 1 relative to the x-axis.
  a. How could you calculate the effect of string 1 pulling in the x-direction?
  b. How could you calculate string 1's effect pulling in the y-direction? That is, how could you calculate the x- and y-components of $\vec{F}_{1onR}$ if you know only the magnitude of the force (5 N) and the direction of the force relative to the x-axis (37° below the positive x-axis)? What are the magnitudes of the other two forces?

Activity 3: Testing experiment
Equipment per group: whiteboard and markers, metal ring, three spring scales (Alternative: Use a force table with ring, strings, pulleys, hangers and slotted objects)
  Work with your group members to recreate the situation in activity 2 and check whether the forces that you found keep the ring in equilibrium. (Alternative: watch the video of the experiment at https://mediaplayer.pearsoncmg.com/assets/_frames.true/sci-OALG-4-1-2)
  In this example, you can see how PSTs learn to proceed from a conceptual understanding of force components developed with the help of the grid tool to the trigonometric functions that help find the same components. Student difficulties with components arise from the disconnect of the physical situation and their knowledge of trigonometric functions. This example helps students to see the connection. At the same time, PSTs experience integration of all elements of the ISLE process into activities that take a very short time. Often teachers think that the whole ISLE process takes a long time, but this example shows that sometimes it takes only

10 min of class time. Finally, the last activity where the students make the prediction about the outcome of the experiment and then conduct the experiment to compare the outcome to the prediction elicits positive feedback when the outcome does match the prediction. This 'feeling good' moment is important for the formation of many physics habits of mind.

At home, the PSTs write a lesson plan for the lesson in which high school students invent the concept of a friction force as a component of the force that the surface exerts on an object (the other component is the normal force). While ISLE-based activities for such a lesson can be found in the ALG and the motivation and reasoning are in the IG, writing a lesson plan involves more than just planning activities. The availability of all the above materials reduces 'friction' in the development of a habit of writing lesson plans consistent with the ISLE approach. To write a lesson plan, future teachers need to formulate the goals (consulting National Standards, the IG, and the textbook CP: E&A), choose the activity for the 'need to know', decide how lesson activities connect to each other, how to motivate students, what difficulties students might have and how to address them, what questions/activities should be left for assessment, and what to assign for homework. They also need to think of what experiments the students will perform, what equipment they will need, and how to organize group work for these experiments. PSTs have an outline for a lesson plan provided to them in the syllabus of the course, which they follow for these assignments.

A part of the homework assignment is solving physics problems to develop the habits of expert approaches to problem solving (see figure 7.2). Finally, they read a paper by John Clement, 'Using bridging analogies and anchoring intuitions to deal with students' preconceptions in physics' (Clement 1993) that shows an example of how students develop the idea of normal force.

The course leader provides feedback on their lesson plan, which they can improve and resubmit again, but the solutions to the homework problems are assessed through peer learning. The students exchange their solutions and score each other using the same rubric that each of them used to make solutions (see IG, chapter 1). Then the authors of the solutions improve their work based on rubric feedback and resubmit them to the course leader. The rubric for problem solving is shown in table 7.6.

When they come to the next course meeting, they work on a quiz shown below.

If you place a book against the vertical wall that accelerates as shown in figure on the right, the book will not slide down. Explain how this is possible. Your friend claims that the necessary condition for the book not to slide down can be expressed as $a > \frac{mg}{\mu_s}$, where μ_s is the coefficient of static friction between the book and the wall and m is the mass of the book. Evaluate the expression (check units and limiting cases). If you think the expression is incorrect, derive and evaluate your own expression.

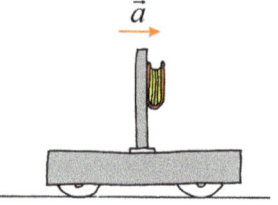

Figure 7.2. Example of problems that PSTs solve in their homework.

Table 7.6. Problem solving rubric.

Rubric item	Adequate	Needs improvement	Inadequate
…is able to clearly explain and justify the key steps of their reasoning process.	Student verbally explains *what* they are doing and *why*. Explanation is clear, sufficiently detailed, easy to follow, and shows physical and conceptual understanding.	Student explains *what* they are doing, but missing *why* they are doing it. And/or there is some difficulty in following their explanation.	Explanation is incoherent, confusing, or missing; and/or invokes incorrect/ irrelevant physics ideas; and/or is unrelated to that which is being explained.
…is able to create two or more *consistent* representations of the problem.	Two or more representations are constructed according to accepted standards learned in class, and the representations are consistent with each other.	Two or more different representations are present and they are consistent, but there are mistakes or missing elements in the representations.	There are major (key) mistakes/missing elements in representations or different representations are inconsistent with each other.
…is able to choose and apply productive mathematical procedures for solving the problem.	Mathematical procedure is productive for solving the problem. Implementation of procedure is free of major conceptual errors.	Productive mathematical procedures are chosen, but implementation reveals misunderstanding about how to implement them.	Mathematical procedures are unproductive/ inappropriate and will not lead to a physically reasonable answer to the problem, even if implemented correctly.
…is able to evaluate the reasonableness of final result.	Evaluates reasonableness of the result, correctly applying all the steps of one of the possible evaluation techniques listed below: a. limiting/ special case analysis, b. unit analysis, c. physical reasonableness of answer, d. two independent	An appropriate evaluation technique is used, but there are mistakes in the implementation of the technique (wrong units, misunderstanding of how reasonable the numbers are) and/or student neglects to draw a conclusion from their analysis.	There is no evaluation, or evaluation technique is implemented in an incoherent way, and/or an invalid conclusion is drawn, such as concluding the answer is reasonable when evaluation analysis shows it is not reasonable.

methods, e. cross substitution consistency, f. consistency of representations. A valid conclusion is drawn from the analysis.

Quiz during class meeting next week:

Your students are learning Newtonian dynamics and are arguing about the following problem: *A person is pulling a sled on a horizontal icy ground as shown in the figure below* (see part (a) of the figure). *Draw a force diagram for the sled. Indicate any assumptions that you made.* One of the group members suggests neglecting a friction force component of the force that the ice exerts on the sled. Another student draws and labels the diagram (see part (b) of the figure below).

(a)   (b)

You hear the third student say: 'There is a mistake in the diagram, the upward vertical force arrow should always be the same as the downward arrow.'
 1. Do you agree with the student? Explain your answer.
 2. Why do you think the student made this comment?
 3. How would you respond to this comment in class?

By analysing the force component activity (which is only a part of the lesson that day) and the follow-up homework, we see that the future teachers continue developing productive dispositions for teaching physics and engage in the following tasks of teaching in this course:
  I. Anticipating student ideas.
  II. Designing, selecting, and sequencing learning experiences and activities.

III. Monitoring, interpreting, and acting on student thinking.
IV. Explaining and using examples, models, representations, and arguments to support students' scientific understanding.
V. Using experiments to construct, test, and apply concepts.

By participating in the lessons as high school students, they experience themselves what it means to be members of a scientific community (task of teaching IV. Scaffolding meaningful engagement in a science learning community). They continue developing multiple habits of mind of physicist and physics teachers. They practice by means of using hypothetico-deductive reasoning, using mathematics in physics, responding to student ideas, and so forth.

Additional activities that future physics teachers engage in this course are:
a. Designing ISLE activity sequences (observational experiment–patterns–multiple explanations–designing testing experiments–making predictions based on different explanations–conducting testing experiments and comparing outcomes to predictions–modifying explanations in case of mismatches).
b. Creating ISLE-based lesson plans with a lot of feedback from the course leader and enacting them during microteaching experiences (see more in section 7.3 of this chapter).
c. Designing ISLE-based unit plans (e.g. circular motion unit, electric charge and force unit; DC circuits unit).
d. Designing ISLE-based formative and summative assessments for specific concepts and units.
e. Analysing student work and responding to student ideas.
f. Asking question and responding to student questions.
g. Monitoring group work.
h. Reflecting on their own teaching and the teaching of others.
i. Connecting research-based ideas to the design of ISLE-based curriculum materials and ISLE-based approaches to teaching.

7.2.3 'Multiple Representations in Physical Science' course

This is the last course in the sequence of physics teaching methods courses. The students in this course have gone through a whole semester of student teaching internships in high schools and they are in their third semester of teaching in physics for the sciences. They continue to explore how to engage their future students in learning physics through the ISLE approach. The focus in this course is on the reflection of the design of the ISLE-based curriculum materials and research of the brain structures that explain how approaches and activities in ISLE are based on our knowledge of the brain. While the students continue using the same textbook CP: E&A and other ISLE-based materials as before, each week they also read one or two chapters from the book *The Art of Changing the Brain* by James Zull (Zull 2002) (lovingly called 'The Brain Book' by the students). In class meetings, we discuss how each chapter is connected to the teaching and learning process in this particular class meeting. The example below shows one of the activities that addresses the above goals. Prior to this class meeting, the students read the chapter in The Brain Book called 'The natural relationship between brain structure and learning'. An important difference between the sequence of physics topics in this course and the other two courses described above is that the order of the topics in the course do not follow the logical progression of a high school curriculum. Instead, it is based on the connections between multiple representations that are used in different units.

Activity: Analysing multiple representations

The goals of this activity are to examine different representations used in chapter 20 of the textbook (magnetism) and to learn how to write a self-assessment rubric to help students master those representations. Work with your group to answer the following questions:

What are the representations used in the magnetism unit? How will students learn to make them? When will they use them? What difficulties do you expect that they will have?

Use both the textbook and the ALG file to find different representations and make a table with four columns: Name of the representation; what does it represent?; when is it used?; what are possible difficulties that students might experience with it?

After you finish, we will all work on making a rubric for assessing one of the representations and then you can add rubrics for other representations. See the outcome of this work in table 7.7.

Table 7.7. Example of a rubric for drawing $\vec{B}$ field lines for permanent magnets that emerged from the discussion.

Ability	Missing	Inadequate	Needs improvement	Adequate
To represent magnetic fields with $\vec{B}$ field lines.	There are no $\vec{B}$ field lines on the diagram.	The lines are present but are in the wrong direction, or direction is not marked and the density does not match the magnitude of the $\vec{B}$ field.	The direction and density of the lines are correct but the lines have a beginning and an end.	The lines have no beginning or end. Their direction is clearly marked with arrows and is either based on the right-hand rule for the fields created by electric currents or is correct for permanent magnets. The lines come out of the north pole and enter into the south pole. The density of lines represents the magnitude of the $\vec{B}$ field.

Reflect on the structure of the rubric. In what ways will the rubric help your students draw $\vec{B}$ field lines? What follows is the discussion on the roles of the rubrics in learning, how they help students, how to construct them, and so forth.

As homework, the students are asked to devise a rubric for electric field lines. As in the previous courses, the course leader provides immediate feedback on their work and the students improve their rubrics based on the feedback. In addition to reading the next chapter in The Brain Book, the students also read the paper describing the design of the rubrics for different scientific abilities (Etkina *et al* 2006) and explore the website with all existing rubrics (see https://sites.google.com/site/scientificabilities/). There they can find rubrics for several graphical representations, but not for electric and magnetic field lines. The existing rubrics serve as a model, but do not limit students' creativity. Finally, they are assigned complex problems on the topics of magnetic and electric fields. Those problems undergo peer assessment in the same way as was described for the course, Teaching and Assessment in Physical Science.

From the above example, we see that the students engage in the following tasks of teaching:
    I. Anticipating student thinking around science ideas.
    II. Monitoring, interpreting, and acting on student thinking.
    III. Explaining and using examples, models, representations, and arguments to support students' scientific understanding.

Additional types of activities in which the PSTs engage in this course are:
    a. Developing short lessons following the ISLE process for new topics without any help from the course leader and enacting them in the classroom.
    b. Solving non-traditional physics problems for different topics (see the classification in the first ISLE book (Etkina *et al* 2019a) and IG).
    c. Connecting solutions of paper and pencil problems to experiments or simulations to test their answers.
    d. Exploring applications of physics in modern devices.
    e. Analysing common scientific myths (cell phones cause cancer) and learning how to respond to those.
    f. Organizing and building equipment for the experiments during microteaching.
    g. Practicing responding to student ideas.
    h. Connecting teaching practices to research findings.

The above example of the lesson activity, homework, and additional activities in the course show that the students continue practicing all of the tasks of teaching in the ISLE environment while becoming increasingly independent in lesson design. Our observations show that, by this time, PSTs habitually design ISLE-based lessons not only by using available ISLE-based activities, but even inventing their own. They feel comfortable with the steps of the ISLE process and confident in using equipment for experiments. It is difficult to say that they have developed all of the previously discussed habits of mind and practice. However, it is very clear from their behavior in class meetings, their preparation of the lessons, and their responses to students during those lessons (see more about this aspect in the next section that many of the habits have been formed).

7.3 Clinical practice

When we use the term 'clinical practice', we mean all actions and activities in which the preservice teachers engage. This includes designing, enacting, and reflecting on the process of teaching and learning physics either with high school students, college students, or their peers in the teacher preparation program. As we said above, this process follows the cognitive apprenticeship approach with the goal of developing productive habits of mind and practice of physics teachers who help their students learn physics through the ISLE approach. In the Rutgers program, there are four types of clinical practice:

1. Observations of high school physics classes where students learn through the ISLE approach.
2. Teaching as lab or problem solving instructors in an ISLE-based introductory physics college course.
3. Microteaching in the physics methods courses where the students are peers in the program.
4. Full-time student teaching internship in a high school physics classroom.

It is helpful to see the progression of the difficulty of the activities in which the PSTs engage in these different types of clinical practice as well as the role of the 'master'. This progression is shown in table 7.8.

Table 7.8. Progression of clinical practice activities.

Type of clinical practice and when it takes place	What PSTs do	What the 'master' does	What dispositions (D), knowledge (K), and skills (S) develop
Observations of high school classes learning physics through the ISLE approach. (Semester 2)	Observe, discuss observations with the cooperating teacher and answer specific questions posed by the course leader.	There are two 'masters' here. The course leader poses reflection questions and the cooperating teacher helps make their thinking visible.	D: All students can learn physics. K: ToTs V and VI. S: Skills of observation and reflection.
Teaching as lab or problem solving instructors in an	Participate in training meetings, conduct class sessions, reflect	Conducts training meetings, plans class sessions, holds 'open	D: Every student brings valuable

(Continued)

Table 7.8. (*Continued*)

Type of clinical practice and when it takes place	What PSTs do	What the 'master' does	What dispositions (D), knowledge (K), and skills (S) develop
ISLE-based introductory physics college course. (Semesters 1, 3, and 4)	in writing on their teaching, hold office hours, grade lab reports and homework including resubmissions, and selected exam problems.	classroom' so that PSTs can come and observe him/her, observes PSTs and provides feedback, assigns homework, creates exams and exam problem scoring rubrics.	contributions to class discourse. K: ToTs III, VI, skills of managing group work. S: Skills of managing whiteboard sharing sessions.
Microteaching in the physics methods courses where the students are peers in the program. (Semesters 1–4)	Create lesson plans, collect and assemble equipment; teach the lessons for their peers, reflect on their teaching.	Provides feedback on the lesson plans, assists with equipment, coaches and scaffolds during the teaching, reflects on the teaching.	D: Preparing a lesson takes time and multiple revisions. K: ToTs—all. S: Creating lesson plans, creating formative assessments, managing equipment, reflection.
Full-time student teaching in a high school physics classroom. (Semester 3)	Create lesson plans, collect and assemble equipment; teach the lessons for high school students, design formative and summative assessments, reflect on their teaching.	Provides feedback on lesson plans and teaching, reflects on teaching.	D: It is not about me. K: ToT—all. S: Classroom management in addition to all of the skills above.

Below, we provide detailed descriptions of each type of clinical practice.

7.3.1 Observations of high school physics classes where students learn through the ISLE approach

As the goal of the program is to help PSTs develop productive habits for teaching physics through the ISLE approach, a fundamental aspect of clinical practice is observations of teachers teaching through ISLE and reflecting on different aspects of student learning through the ISLE approach. As many of the reasons for the teacher to make specific moves in the classroom are not evident from the moves, the PSTs need to discuss the lessons with the cooperating teacher to be able to benefit from the experience. As this clinical practice happens when the PSTs are taking the course Teaching and Assessment in Physical Science, the course instructor makes their reflection on observations a part of course work. Specifically, the course assignment is written in the course syllabus as follows:

> *Reflection on classroom observations:* Every week during your 10 weeks of observations in science classrooms, you will need to send me a report about your observations (the report should be submitted before Sunday of the week you did the observations). Each report should address specific questions that are listed at the end of the syllabus. During one of the observation visits, you will need to teach a lesson or a part of the lesson. Please make sure you arrange this with the cooperating teacher. Submit the lesson plan to them and to me in advance. After you teach the lesson, write a reflection.

Questions for classroom observations:
1. First, in detail, describe one of the lessons that you observed this week. Next, answer the question of the week with specific examples. In your reflection, paste the questions from the syllabus and answer them one by one.
2. *Week 1*: What were the goals of the lesson and how did the teacher make sure the goals were achieved? What did student understanding look like? Please provide at least three specific examples. What aspects of classroom management did you detect? List all examples.
3. *Week 2*: What evidence did you find of the teacher's awareness of student ideas (prior and current)? What did student difficulties look like? How did the teacher modify the lesson based on student feedback? Please provide at least three specific examples. What aspects of classroom management did you detect? List all examples.
4. *Week 3*. What forms of formative assessment did the teacher use? When did they use it? What kind of feedback did they provide? How did student questions, ideas, and difficulties affect the continuation of the lesson? What aspects of classroom management did you detect? List all examples.
5. *Week 4*: What questions did the teacher ask? How can you classify them according to Bloom's taxonomy? What answers did the teacher find satisfactory? Provide examples of both questions and answers. What aspects of classroom management did you detect? List all examples.
6. *Week 5*: How did the teacher start the lesson (please record this moment word to word)? Why did they do it this way? How much time were students actively engaged in the lesson? What did they do during these active engagement times? How much time did they spend sense-making? What did sense-making look like? How much time did they spend doing busy work? What did busy work look like? Why did they have to do it?
7. *Week 6*: What kinds of questions did the students ask? Were any of them essential questions? How did these questions shape the lesson? How did the teacher respond to the questions?

8. *Week 7*: What metacognitive strategies did students use? What were the examples of reflection?
9. *Weeks 8–10*. Provide three examples of student understanding. What did it look like when a student demonstrated understanding?

You probably noticed that in almost every reflection the PSTs had to describe classroom management techniques. Although by the second semester in the program PSTs have experience teaching physics in the ISLE environment, this teaching occurs in a college course and thus there are no (or almost no) issues of classroom management. A high school classroom is different. If the students are not occupied with meaningful work, or they are occupied with something else other than class work, issues of classroom management arise. Future teachers need to see how experienced teachers address or prevent such issues. These methods need to be observed and practiced many times to become habitual. While the best way to avoid the triggers for student misbehavior is to continuously engage them in exciting activities, some down time and even boring work is unavoidable. Skills of prevention and management of disruptive situations are best learned by observation of successful teachers. Therefore, the choice of cooperating teachers for clinical practice is crucial. And that is why the best way to do it is to place future teachers with the best graduates of the program.

7.3.2 Teaching as lab or problem solving sessions' instructor in an ISLE-based introductory physics college course

'Physics for the Sciences' is a large-enrollment (200 students) introductory algebra-based physics course for biology and health profession majors at Rutgers University. The course meets four times a week: two 55 min whole-class meetings, one 80 min recitation—a problem solving session for smaller sections (20–24 students per section), and one 3 h lab session for smaller sections (20–24 students per section). The course follows the ISLE approach and is led by a course coordinator professor, Michael (Mike) Gentile, who is an experienced ISLE developer and implementer. Mike teaches two weekly whole-class meetings and the first sections of lab and recitation every week. Mike also prepares all of the materials for all course meetings using ISLE resources (ALG activities for recitations, resources at https://sites.google.com/site/scientificabilities/isle-based-labs for the labs, and the textbook CP: E&A for running whole-class meetings and homework). He constructs exams (two midterms and one final per semester) and runs weekly training meetings for the recitation and lab instructors (PSTs in the Rutgers program).

All Rutgers PSTs teach recitations and labs in the course. This type of clinical practice is called 'sheltered teaching' as the PSTs do not design their own lesson plans for recitations or labs, and do not need to deal with classroom management issues. However, they are responsible for deeply understanding the material, the innovative teaching approaches inherent in ISLE and thorough conceptual understanding of the topics of class meetings. They also are responsible for knowing how to use equipment in the labs fluently, how to trouble-shoot problems, and how to quickly repair something when it is broken. Therefore, one of the benefits of this clinical practice is that the PSTs improve their knowledge of physics—both practical and theoretical.

The teaching materials for the students are provided to the PSTs by the course coordinator. The focus of this clinical practice of teaching in an ISLE-based classroom is on the crucial elements of the environment—active learning, collaboration, and experimental design. Therefore, the PSTs practice managing and monitoring group work, listening to the students and helping them solve problems or design experiments *without* telling them what to do. The PSTs also learn how to help individual students during office hours, and how to grade homework and lab reports using rubrics. In addition to developing teaching related skills, this clinical practice helps PSTs develop one of the most important skills for a teacher—the skill of time management. As they need to prepare for class meetings, grade labs and homework, hold office hours in addition to doing their own course work, such sheltered teaching requires very careful time management. It is during this sheltered teaching that the PSTs not only develop productive habits of running an ISLE-based classroom, but of their own time management. It is important to note that appointments of the PSTs are formal. They get paid for their work as university instructors and they hold full responsibility for their teaching duties. Such clinical practice is only possible due to the close collaboration of the Graduate School of Education where the physics teacher preparation is formally hosted and the Rutgers Department of Physics and Astronomy that offers physics for the sciences.

7.3.3 Microteaching in the physics methods courses where the students are peers in the program

As we noted above, the term microteaching refers to PSTs teaching lessons to their peers, not high school students. The goal of this clinical practice is for the PSTs to learn how to plan their own lessons and how to enact those lessons in the classroom. Microteaching is a part of physics teaching methods courses and all the coaching and scaffolding of this clinical practice is done by the corresponding course leader.

Microteaching is organized in the following way: at the beginning of the semester the course leader provides a table of topics to be taught by the PSTs and they sign up for the topics of their choice so that each lesson is taught by two students. Below is an example of such a table in the course Teaching and Assessment in Physical Science (see table 7.9). All lessons are expected to involve complete ISLE processes

Table 7.9. List of topics for microteaching and the names to be filled out by the students.

Week/date	Topic	Names
7 March 7	Circular motion	
10 April 4	Electric field	
11 April 11	Current electricity: Series and parallel circuits	
11 April 18	Current electricity: Electric power	
13 April 25	Static fluids	

for the development of the relevant concepts: qualitative observational experiments, patterns, explanations and testing experiments, quantitative observational experiments, patterns, relations among physical quantities, testing experiments, and application experiments. The lesson plans are expected to follow the template for the lesson plan that students receive in the syllabus (the abbreviated syllabus is shown in the appendix to this chapter).

7.3.3.1 Timeline for lesson preparation
After the students sign up for their chosen lessons, the course leader schedules several deadlines for them: the date for the 'content interview', the date for the submission of the lesson goals and formative assessments, the date for the first draft of the whole lesson plan, the date for equipment assembly, and the date for the final lesson plan and rehearsal. This list of dates shows the steps in the preparation of the lesson, the expectation of multiple revisions of the lesson plan, and the need to schedule special time for working with equipment for the lesson. An example of a timetable for the topic of microteaching 'Electric field' is shown below:

February 10—Content interview.
February 28—First draft of the lesson plan is due; feedback on 1 March.
March 14—Second draft of the lesson plan is due with the list of equipment; feedback on 16 March.
March 20–March 25—Equipment testing.
April 1—Final draft of the lesson plan is due; feedback on 2 April.
April 3—Final rehearsal.
April 4—Microteaching.

7.3.3.2 How to prepare a good lesson?
The key to a good lesson is preparation. That is why feedback from multiple instructors is needed.

How do students know how to write a lesson plan? Below, we note four stages of lesson planning:

1. During the very first course meeting in the Teaching and Assessment in Physical Science course, the students participate in the following activity: the instructor asks them to make a list of elements that are needed during the preparation of any lesson. The students work in groups for about 10 min and share their ideas on whiteboards. Afterwards, the instructor uses their whiteboards to focus on three important elements: learning goals of the lesson, formative assessments that would tell the teacher whether the goals were accomplished, and the sequence of activities that the students need to work through to achieve the goals. These three elements need to be thought of before one starts writing a lesson plan and especially before teaching the lesson.

2. As the course progresses and PSTs reflect on the lessons taught by the course leader, they need to focus their attention on those three elements. When they say what they learned—how did these things relate to the goals of the teacher? How did the activities build on each other to achieve the goals? How did the instructor know that the goals were achieved?

3. After two to three class meetings, the instructor needs to have another discussion about additional elements of a successful lesson. Those elements were empirically found by Eugenia's student, now a high school teacher, Heather Briggs. She worked with her cooperating teacher (Richard (Rich) Thekhorn, also Eugenia's former student) during her student teaching internship and noticed that while she did everything that they planned together, her lessons sometimes were good and exciting and sometimes they were not. Rich's lessons were always excellent. What was his secret? She started taking notes of everything that he did in every lesson and soon found that three important elements repeated all the time. Those were: *the need to know*, *the tools for success*, and *opportunities for success*. In other words, Rich always motivated each lesson. He told the students explicitly what tools they needed to use to be successful and he crafted the lessons in a way that the students could be and—most importantly—feel successful. After Heather found these elements and started incorporating them into her lessons, they improved dramatically. In the course that accompanied her teaching internship, Heather shared her findings with the class. Since then (more than ten years ago), all students in our teaching methods courses learn these three elements and incorporate them into their lesson planning.
4. Finally, the instructor provides the students with the formal lesson plan outline that they need to use when composing each of their lessons. It is important to remember that every country and, in the US, every school district has its own format for a lesson plan. Therefore, here we only provide general elements.

Outline of a lesson plan:
1. Title
2. What students should know before they start the lesson and why this knowledge is important.
3. *Goals* of the lesson:
    - Conceptual (what ideas or concepts will students construct during the lesson).
    - Quantitative (what mathematical relationships they will master).
    - Procedural (what skills they will learn and practice).
    - Epistemological (what they will learn about the nature of knowledge and the process of its construction—i.e. the elements of the ISLE process and connections between them).
4. List the *evidence* that you will collect that will convince you that students met the goals. Be specific.
5. Describe the most important ideas in terms of the subject area and make a list of real-life connections.
6. List student ideas that they will bring into the lesson. Focus on productive ideas and potential difficulties (what might cause trouble and resources—what you can build on). Make sure you explain *what you will do to address the difficulties*.

7. Make a list of equipment needed—group it into teacher use and student use. Think of what extra materials you need to have handy for possible testing experiments that the students might propose.
8. Write a script of the lesson
    - What is going to happen? What will you say? What questions will you ask? What will students do?
    - Reference all handouts that you plan to give to the students.
    - What elements of the ISLE process are present in the lesson? How are they connected?
    - Make sure you describe how you will start the lesson and how you will end it (to capture students' attention and to have some sort of closure).
    - Do not forget the need to know, tools for success, and opportunities for success for the students.
    - Finally, think of how you will motivate every activity and how you will 'string' them together so that the lesson does not look like a bunch of activities put together, but a coherent story that unfolds.
9. Create a timetable—who is going to be doing what and when during the lesson to make sure that students are actively engaged. An example is shown below.

Clock reading during the lesson	'Title of the activity' relation to the goals	Students doing	Me doing
0–6 min	Homework quiz, receive feedback on student understanding of motion diagrams—the foundation of today's lesson	Writing	Checking on equipment for the first activity

10. Describe the formative assessments that you plan to use, write what possible answers you expect from students, and how you will provide feedback (e.g. if these are problems—include solutions).
11. Describe modifications for different learners. Be specific as to how those modifications are related to this particular lesson.
12. Homework—make sure that it addresses two goals: strengthens this lesson and prepares students for the next lesson. Describe the guidance that you will provide to the students.

7.3.3.3 Details of student work

Having developed a great deal of knowledge about writing the lesson plan, having reflected on the actual lessons led by the course instructor, and having mastered the content of the lesson, future teachers are ready to submit their first draft of the lesson plan. They do it by using Google docs or some other file-sharing system. This approach allows the course instructor to provide feedback almost in real time, or at least very quickly, and the students revise, improve, etc. It is important to help the students see connections between the consecutive activities in the lesson, the motivation for each activity, and the importance of allowing enough time for 'time for telling'. By this time, all of them are familiar with the ISLE process and have experienced it as students in the Development of Ideas in Physical Science course and in the first half semester of the present course. They also implemented it during sheltered teaching in the university physics course and observed high school teachers implementing it in high school classrooms. However, even with all of this knowledge and experience, writing a good lesson plan takes a lot of time and effort and multiple rounds of feedback from the course instructor. This feedback serves simultaneously as the formative assessment of the process—it allows the students to improve their lesson plans and the course instructor to improve the practice of lesson plan preparation based on the strengths and difficulties that they detect in student work.

Another important issue needs attention here. After the preservice teachers decide on what experiments the students and themselves will do in the lesson, they need to prepare the necessary equipment. Usually, the course instructor has equipment available, but it needs to be organized and prepared before the lesson. If the equipment is needed for group work, they need to check every detail of every set-up. They might need to build or construct some new materials or order parts—that is why this work needs to be done in advance. If there is only one set-up for the teacher, the PSTs need to practice using it in front of the class to make sure that everyone sees the details. They need to think in advance as to where they would stand, whether the students will sit at their desks, or gather around the teacher table. All of these details may seem small and insignificant, but it is these details that 'make or break' the lesson. Therefore, working with the PSTs on the equipment is crucial for the success of future lessons.

7.3.3.4 Flight simulator
Now, let's imagine that all the preparation went well and the team of pre-service teachers is ready for implementing their lesson. We have developed a technique called a 'flight simulator' to use their teaching time productively for the development of skills and habits. A flight simulator is used in training aviation pilots. According to Wikipedia:

> A flight simulator is a device that artificially re-creates aircraft flight and the environment in which it flies, for pilot training, design, or other purposes. It includes replicating the equations that govern how aircraft fly, how they react to applications of flight controls, the effects of other aircraft systems, and how the aircraft reacts to external factors such as air density, turbulence, wind shear, cloud, precipitation, etc. Flight simulation is used for a variety of reasons, including flight training (mainly of pilots), the design and development of the aircraft itself, and research into aircraft characteristics and control handling qualities. (Wikipedia n.d.).

The metaphor of a flight simulator is extremely useful for teacher preparation. In the simulator, the trainee can learn to fly the plane in any kind of weather, crash the plane, and start again, can have an engine problem and learn how to still land the plane with it, etc, with no harm to the passengers. How can we create conditions similar to those in teacher preparation? Microteaching provides an excellent opportunity for such practice.

When future teachers teach their prepared lesson to their peers, there is no danger in repeating parts of the lesson or fixing problems that occur (something that does not happen in a real lesson). Therefore, we implement the strategy called 'stop and rewind'. Imagine that a lesson is going on and a student named Jasmine makes a comment that can potentially change the course of the lesson or indicates that she has an important misunderstanding. Andrew, who is the 'teacher' and is focused on his lesson plan, does not hear Jasmine's comment, or does not know how to respond to it and continues the lesson. The course instructor says: 'Stop and rewind. Let's go back for a second to Jasmine's comment. Andrew, did you hear what Jasmine just said?' Andrew responds negatively. The instructor asks the rest of the class and receives a negative response—they also missed Jasmine's comment. Then the instructor says: 'Jasmine, will you please repeat what you just said?' Jasmine repeats. The instructor turns to Andrew: 'What do you plan to do after Jasmine made that comment?' Andrew indicates that he does not know. Then the instructor turns to the class: 'What would you do?' While the class is thinking, Andrew comes up with a strategy, shares it with the class and changes the path of the lesson slightly to accommodate Jasmine. The lesson goes on. There can be several instances in the lesson when the course instructor utilizes the 'stop and rewind' strategy. When the lesson is over and all participants including Andrew (the 'teacher') reflect on the lesson, the instructor focuses special attention on the 'stop and rewind' episodes—how analysing them will make the lesson better.

You can see how this strategy is similar to the flight simulator. Future teachers have opportunities to 'redo' the elements of the lesson that needed to be corrected in real time. These instances help them solidify productive habits, instead of reflecting on the 'mistakes' at the end of the lesson when nothing can be done about them. Our experience shows that while in the first course, each team gets about 5–7 'stop and rewind' episodes per lesson (remember, the microteaching lessons are about 2.5 h long). In the last course of the program, there are hardly any 'stops and rewinds', at most one per lesson. The reasons for 'stop and rewind' can be grouped into several categories in which a future physics teacher:

1. Makes a serious physics mistake, and it needs to be corrected immediately so that the rest of the lesson goes in the 'right' direction.
2. Calls on the same people during whole group discussions while leaving the rest of the class inactive.
3. Does not give enough time for the students to ponder a question or does not put them into groups when the question is difficult.
4. Does not hear or understand a student's question or a comment and either does not respond or responds unproductively (e.g. a student asks a question, and the 'teacher' answers it straight away instead of engaging the class and/or the student who posed the question).

There are several pitfalls to consider when using the 'stop and rewind' strategy. First, the 'teachers' should not fear that those episodes will lead to a lower grade in microteaching. Therefore, microteaching is not graded. More, the course instructor informs the PSTs that the purpose of microteaching is to develop productive habits of mind and practice and not to produce a perfect performance. Therefore, the more repetitions of certain actions they experience (when writing the lesson plan and when enacting the lesson), the more habits they would form. Additionally, 'stop and rewind' episodes help the rest of the class focus on difficult moments in the lesson, which are usually connected to the complex tasks of teaching. When the experienced course instructor leads a lesson, those moments pass unnoticed and, therefore, are not utilized by future teachers as important instances and difficult situations. Finally, treating all PSTs with the utmost respect, providing supporting comments, and praising them sincerely for well-enacted tasks of teaching and ingenuity in lesson preparation makes them feel rewarded for their work.

Sometimes, a 'teacher' might lose their train of thought during the interruption, so it is very important to observe them carefully and help them get back on track without expressing any judgment if this occurs.

There is another important issue to consider here. During a lesson, there can be multiple moments when a teacher needs to make a split-second decision on how to proceed. If the decision is successful, the lesson improves. If the decision is not successful, the lesson might tank. How do we teach future teachers to make decisions? Eugenia introduced a metaphor of a 'double headed compass', which she shares with the PSTs after a microteaching lesson where the 'teachers' needed to make one of those split-second decisions and that decision was successful. She uses the example that occurred in class as 'the need to know' for the technique. The main idea of the compass is that it guides you when you need to make a decision. One 'head' of the compass arrow aligns with the direction of more learning (or less learning) and the other one aligns with the direction of better well-being for the students or worse well-being for the students. When making decisions one should think of whether the decision will lead to more learning and/or better well-being for the students or the opposite. For example, when a lesson is in progress, a student walks in late. What to do? Here, the compass tells you to not discuss the reasons of being late with the student in front of the entire class in order to not distract the rest of the students from learning (more learning). This also allows the student who was late to proceed to their seat and start working while helping them catch up (more learning). At an opportune moment or after class, talk to that student about the reasons for being late and provide help if needed (better well-being).

7.3.4 Full-time student teaching

Finally, the time comes for the PSTs to try teaching full time. This is what the student teaching internship is in the Rutgers program. PSTs are assigned to a school for the whole semester—15 weeks. In the USA, PSTs have the title of student-teachers while doing the internship (in other countries, this activity can be called teaching practice). There, they slowly assume the responsibilities of their cooperating teachers[2]—first observing their cooperating teachers for a few days, then team-teaching with them for another week, and, finally, taking over just one and then more classes until they almost do all the work of their cooperating teacher.

This time is incredibly important in the formation of a new teacher in any area. Therefore, if we wish our newly formed physics teachers to continue developing productive habits and simultaneously developing confidence in their abilities, we need to pair them with the cooperating teachers who would support them. This is exactly the reason why PSTs in the Rutgers program are always paired with the graduates of the program as cooperating teachers. These teachers are fluent in the ISLE approach and run their classroom exactly like the PSTs learned in their course work prior to full-time student teaching. This pairing is only possible if the leader of the teacher preparation program follows program graduates and makes sure that they indeed implement the ISLE approach in their classrooms. That is why the creation and maintenance of the community of the graduates (see section 7.5.2) is crucial for successful full-time student teaching of the PSTs.

In their full-time student teaching, the PSTs plan their own lessons, receive feedback from their cooperating teacher, revise the plans, implement the revisions in the classroom, reflect on this implementation, and receive feedback from the cooperating teacher. As the internship always occurs in the Fall semester of an academic year, the content of the lessons is familiar to the PSTs as they encountered this content three times in their program already. Such topics as kinematics, dynamics, circular motion, momentum, and energy are addressed in the first two physics teaching methods of the first year of the program and while teaching in Physics for the Sciences. And yet, independent lesson planning, attending to all aspects, including classroom management and their own time management is very difficult. This is where the supporting cooperating teacher and the leader of the program come into play. The role of the cooperating teacher is clear. They set the schedule for the PSTs to submit their lesson plans three days prior to teaching every lesson so that there is enough time to assemble equipment and revise the lesson based on feedback. During the lesson, they are in the classroom and when needed, pick up team-teaching mode if a true problem occurred or wait for the intern to figure out how to deal with the difficult situation. After the lesson, there is debriefing time. Here it is very important not to ruin the PST's confidence by focusing on the mistakes, but instead start by identifying strengths so that the intern can build on those in the future. We all make mistakes, they are unavoidable, but knowing our strong sides allows us to grow. This is the most important role of the cooperating teacher—to help the intern see what they are good at and to continue developing strong traits.

In addition to the cooperating teacher, the program leader pays an important role in full time student teaching. She does it in two ways: by observing the interns and by

---

[2] A cooperating teacher is an officially appointed high school physics teacher whose responsibility is to mentor a student-teacher in their own classroom. The responsibilities of the cooperating teacher involve mentoring lesson planning and enactment of a student-teacher and involving the student-teacher in other professional activities.

leading a course that occurs simultaneously with student teaching. This course, called 'Student teaching internship seminar', meets once a week and is dedicated to the support of the interns. As a part of the course, the program leader observes the PSTs four times during the semester. Before the observation, the student-teacher submits the lesson plan and the program leader provides feedback. After the lesson, both the cooperating teacher and the program leader debrief with the PST. The focus, again, is on their strengths and providing suggestions on how to fix problematic aspects. In addition, the program leader provides a rubric for the PSTs and to the cooperating teachers to plan and enact the lessons successfully. The rubric (see table 7.10), designed by Eugenia 15 years ago, is pasted below. Notice how the rows of the rubric help the interns develop productive habits of practice.

Table 7.10. Rubrics for self-assessment of teaching one lesson.

Ability	N/A	Well developed 3	Working towards it 2	Missed opportunity 1
To start a lesson in an organized productive way		Students start working from the first second, everything is planned and no time is wasted.	The first seconds are spent unproductively but the lesson got on track within the first 3 min.	The beginning of the lesson did not lead to organized, inspired work.
To create motivation for student learning		The content of the lesson is connected to student lives, or there is an interesting question, or motivation is created based on student success, students understand why they are doing what they are doing.	There is some attempt to motivate students but many do not know why they are doing what they are doing.	Motivation is based on 'need for the test' or is absent.
To keep track of what every student is doing		The teacher scans the classroom often and notices subtle details of student learning activities and behavior; most students participate in the lesson and speak.	The teacher follows most of the students but misses a few, the omissions do not lead to the disruption of the lesson.	The teacher does not notice a crucial moment/s that leads to the disruption of the whole lesson; few students participate.
To help students develop study habits		A great deal of attention is given to building study habits: taking notes, planning learning, metacognition, drawing sketches and	Some attention is given to building study habits but it is not systematic.	No attention is given to study habits.

(*Continued*)

Table 7.10. (*Continued*)

Ability	N/A	Well developed 3	Working towards it 2	Missed opportunity 1
		graphs, asking productive questions, time management.		
To use the board strategically		The board is a productive teaching tool that helps students organize their notes and follow the lesson, writing is clear, large letters, a ruler is used for the drawings and the whole lesson fits on one board.	The board is used but things are erased often, no ruler to draw graphs and other pictures, hard to follow.	The board is used randomly, it is clear that the teacher did not think it through.
To organize experimental work effectively		The experiments shown by the teacher are easy to see, students understand the point and either record and explain or predict, observe and reconcile. Experiments for the students are planned, who goes where and when is clear, no time is wasted, equipment is appropriate and works well	Experiments done by the teacher work well but student participation is minimal, the purpose is not clear. Student experiments are planned but student work is not well thought through beforehand, time is wasted.	Teacher experiments are hard to see, the discussion is limited. Student experiments are not thought through—either time is wasted, students are disorganized or the physics point is lost.
To organize whole-class discussion effectively		The teacher guides the discussion but does not dominate it, the summary is clear, lots of student–student talk, pauses for the students to take notes, main points are summarized on the board.	The discussion is two way mostly teacher–student–teacher, all summaries are done by the teacher, no time to take notes, the board is sketchy.	The teacher talks most of the time, students respond yes or no, the board is not used, no time or attention to notes.
To organize group work effectively		Students are used to working in groups, they arrange quickly, the teacher moves among the groups and group assignments are open-ended enough to promote fruitful discussions,	Students are used to working in groups but it takes some time to settle or group tasks are focused on one right answer, or whiteboards are not used productively, the teacher spends	Students are not accustomed to working in groups, many do not participate, no debriefing, the teacher does not attend to all groups.

Table 7.10. (*Continued*)

Ability	N/A	Well developed 3	Working towards it 2	Missed opportunity 1
		whiteboards are used and all students participate; at the end there is a debriefing.	too much time with one group.	
To manage time effectively		A productive *sense of urgency* is present, timing for activities is announced, the change of types of work occurs often but not too often.	The pace is either too slow or too fast.	The lesson drags.
To lead reflection effectively		All students participate, the reflection is focused on the important issues.	Few students participate, some comments are not useful.	Students reflect on non-important issues.
To assign homework effectively		The homework helps reinforce the past lesson or prepares for the future lesson, it is meaningful and instructions are clear.	The purpose of homework is unclear but the instructions are present.	No homework or no instructions.
To listen to the students		The teacher listens and responds to student comments productively.	The teacher listens but some responses are not productive.	Student comments are not noticed or ignored.
To use multiple representations		Multiple representations are used and are used productively.	Some representations are used productively.	Few representations are used and the purpose is unclear.
To use technology		Technology is used strategically.	Technology is used strategically sometimes.	Technology is used but is not really needed to improve learning.
To pose productive questions and to respond to students' questions		The questions are high level, responses to student questions are done through reflective toss technique, they lead to deep thinking, no wrong physics answers on the teacher's part.	The questions are mixed, students' questions are answered directly, the teacher's physics is correct.	The questions are mostly yes/no, students' questions are ignored, or teacher's responses have incorrect physics.
To encourage students to generate productive questions		There is a mechanism through which students learn to generate good questions, the teacher models how to ask	Students' questions are rare but are treated with respect	There are no students' questions.

(*Continued*)

Table 7.10. (*Continued*)

Ability	N/A	Well developed 3	Working towards it 2	Missed opportunity 1
		good questions, the atmosphere in class is conducive to students asking questions.		
To generate explanations		Students are continuously encouraged to explain and devise mechanisms for evidence; students, not the teacher, evaluate provided explanations, students are encouraged to argue their point of view and multiple points of view are tolerated as long as the explanations are logical; the explanations provided by the teacher are correct from the physics point of view.	Students sometimes are pressed for explanations but not always, the teacher evaluates explanations by saying good or ok, instead of tossing them back to students, the explanations provided by the teacher are OK but not really deep.	The teacher does not press for explanations, argumentation is not encouraged, phenomena are analysed macroscopically, mechanisms are missing, the explanations provided by the teacher have physics mistakes.
To build the lesson on students' ideas		The lesson plan takes into account student ideas documented in research and learned in course work and the lesson is continuously modified based on students' ideas emerging during the lesson.	The lesson plan takes into account student ideas documented in research and learned in course work and but during the lesson students' ideas largely go unnoticed.	Students' ideas are not taken into account during the planning stage and are not used productively during the lesson.

Once a week, student-teachers come to the university for a 3 h class (Teaching Internship Seminar). The goal of the course is to support the PSTs during their internship. Each class meeting starts with every PST sharing what they learned during the previous week. These reflections are focused on the positive aspects of their learning with the goal of having as few negative 'lessons' as possible. As we said above, helping the pre-service teachers see their strengths is more important than trying to eliminate their weaknesses.

One of traditional activities in the course is called 'Rehearsals'. A rehearsal means that 2–3 students individually prepare the beginning (5–10 min) of their next high school lesson and enact this beginning in the Internship Seminar. They act as teachers and the rest of the class assumes the role of their high school students. After each rehearsal is over, the whole class provides suggestions for improvement. The value of such rehearsal is tremendous. The 'teacher' receives valuable feedback for the most important part of the lesson and the participants receive an opportunity to observe a different variation of a lesson that they might have already taught or are about to teach.

Another activity in the course is working through difficult physics problems, analysing more complex experiments, and so forth. This part serves as a motivation for the pre-service teachers to keep pushing forward as the internship is always difficult for newly minted teachers. A huge number of responsibilities falls on them at once. Time management becomes an issue. Classroom organization, the smooth running of activities, and productive group work with teenagers are all very tough to master. The PSTs need continuous reminders of the positive aspects of their work and the fact that difficulties are common.

Observations show that the perception of their success as teachers fluctuates during the semester. At first, their confidence is low and then it increases slowly up to week 8. Then the fatigue sets in, frustration builds up, and some start doubting their ability to be teachers. It is at this moment in the internship that support and encouragement are crucial. Both the cooperating teacher and the program leader work together to help the pre-service teachers go over this emotional hump. After week 10, the mood goes up again and the lessons get better and better. By the end of the internship, most pre-service teachers establish excellent rapport with their high school students and their cooperating teacher. Both the students and the teacher feel sad to let the PST go. I (Eugenia) observed this many times and learned how important timely help and encouragement are for the growth of new teachers.

7.4 Assessment

After describing so many activities that take place in various methods courses and clinical practice, you might be asking yourselves: What assessments do you use to determine whether your pre-service teachers have developed desired habits of mind and practice? How do you know that they have developed the habits of maintenance and improvement? In this section, we provide examples of such assessments. It is important to note that we will not be discussing how we assign numerical grades to those assessments. Giving a numerical grade for learning was found to be detrimental to learning (Black and Wiliam 1998). Therefore, we will leave it to the readers to decide how they will grapple with the problem of how to structure their students' assessment experiences in the present educational system that requires a numerical grade. Eugenia solved this problem in the following way. On the first day of class in the first course that the students take with her, she says:

> I believe that all of you have come to this course and this program to become the best teachers you can possibly be. My role is to help you on this journey. Therefore, I will guide you on every assignment to the point

where the assignment is of high quality. If at some point you decide that you cannot work on it anymore and wish to stop learning, I will tell you what numerical grade it is worth. But if you keep working on improving it until it is 'perfect', then you will always get an A. To summarize, everyone gets an A in the course unless you do not want to work for it. I assume that everyone will want to work for it.

It is interesting that in 20 years of running the program and teaching four courses every year, no one ever told me (Eugenia) to stop pushing them to improve their work. Is it a problem that at the end of the course everyone gets an A? Traditionally, we would say that this is a terrible outcome. We need to have a bell curve! But does a bell curve in a course make sense? It means that some students learned a lot and some—almost nothing. Can you imagine a teacher who learned almost nothing in one of the courses and is still going to teach? Would you like your own child to be taught by a 'C' teacher? Or even a 'B' teacher? I think the answer is clear here.

But putting grades aside, how do we know at the end of a course that pre-service teachers developed some expertise in the goals being assessed by the course?

Below in table 7.10, we show examples of formative and summative assessments that are used during the course in our two programs. As a reminder, all the assessments can be improved after feedback is provided. We can think of many activities that future teachers engage in the program as formative assessments. We also included performance assessments. The official assessments of student teaching internship at Rutgers were conducted using the Danielson framework (Danielson 2013). However, we also use our own rubric used by the PSTs for self-assessment and us for formative assessment of microteaching and student teaching shown in table 7.9. We show this rubric in table 7.11. After the tables, we give examples of final exam questions.

Table 7.11. Summary of the types of assessments carried out in both programs.

Type of assessment	Rutgers program	University of Ljubljana program
Formative/individual	Quizzes in every class meeting	Quizzes in every class meeting
Formative/individual	Homework assignments	Homework (HW) assignments
Formative/individual	Article review; interviews	
Formative/group	Lesson plans	Lesson plans (typically in pairs)
Summative/group	Unit plans	Unit plans (whole class)
Summative/individual	Final course exams (oral or written)	One or two written tests per semester + points from HW and quizzes
Performance	Microteaching	Microteaching
Performance	Student teaching internship	Student teaching internship, sheltered teaching

Final exams

For many years, the Rutgers program practiced having oral exams at the end of the Teaching and Assessment Physical Science course and the Multiple Representations in Physical Science course. The PSTs would receive the questions for the exam about two weeks prior and were encouraged to work together to construct answers. They were allowed to bring the answers to the exam but only if they wrote those themselves (no borrowing notes during the exam). On the exam day, the course instructor prepared cards with the numbers of the questions and put those cards face down on the table. Every PST would take two cards while not knowing what questions they were choosing and would receive an additional problem chosen from the units that were a part of the course. During the next 25–50 min, each PST would prepare their answers to the questions and solution to the problem. They could write on a small whiteboard that we normally use for group work or on a big class board. After the prep time was over, the course instructor would first call on volunteers to present their answers and the rest of the class would listen to the answers. If a person had an issue, the course instructor would turn to the rest of the class for help. By the end of the exam, everyone would get an A for studying, trying, etc. The point of the exam was not to make the students produce perfect answers, but to start pondering and working with each other to try to answer the questions. As long as everyone participated in the development of the answers, the goal was reached. The final presentation was more to allow the future teachers to show off what they have learned, not so much to conduct a final assessment. Over the years, as the program grew, the oral exams would take too much time. They were replaced with written exams with similar questions while the group preparation, notes use, etc, remained the same. An example of an exam question set for the oral exam in the Teaching and Assessment of Physical Science course is shown below.

1. Show how to use NJ Science Standards while planning a lesson (using a specific lesson and specific standards, explain how the standards connect to the goal(s) of the lesson).
2. For the units that follow, make a list of conceptual goals, quantitative goals, epistemological and metacognitive goals, and explain why each goal is appropriate: kinematics, linear dynamics, circular motion, momentum, energy, fluid statics, vibrations, electric fields, DC currents, magnetism and electromagnetic induction.
3. Use any of the verbatim students' responses from your classroom observations (or from your own students) to show how this particular response will affect your instruction (you can use a made-up scenario).
4. Use student work from Physics for the Sciences or your classroom observations as evidence that students achieved a particular goal.
5. Use any lesson plan you wrote during this semester to show how you address all elements for the lesson plan.
6. Use the unit plan you wrote for your project to show how you address all elements of the unit plan.

7. Using an example of a lesson that you planned this semester, show that at the beginning of the lesson you build on student ideas and engage them in meaningful exploration of physics ideas.
8. What difficulties do students have with kinematics concepts? Make a list and suggest one strategy per difficulty that helps alleviate it.
9. For the question above, provide examples of formative assessments that will tell you whether a student has a particular difficulty.
10. What are the difficulties that students have with kinematics graphs? Provide examples of the difficulties and sequences of instructional moves that are considered helpful for those difficulties.
11. Describe what students will do in class to construct the equation, $\Delta x = v_{0x} t + \frac{1}{2} a_x t^2$.
12. Explain what concept of an index is and how you plan to use it to help your students construct the operational definitions of velocity, acceleration, density, $\vec{E}$ field, and electric resistance.
13. Explain how an operational definition of a physical quantity is different from a cause–effect relationship for the same quantity. Provide at least two examples of such relationships and explain how your students will learn the difference.
14. What should students know about *any* physical quantity and how will they learn it? Use two quantities as an example.
15. What are science practices? Give examples and describe examples of lessons where students can employ them. Be very specific. How are scientific abilities different from science practices?
16. What should students know about friction force? How will they learn it? What questions will you ask to find out if they mastered those ideas? What resources will you activate to help them learn about friction force?
17. Explain why one needs to draw a motion diagram first before drawing a force diagram while solving a dynamics problem. Provide an example of a problem to solve in which both representations are essential. Explain where in the dynamics unit students will solve the problem.
18. Describe a difference between a cook-book lab and a design lab. Provide an example of a design lab for energy and show how students will use science practices in this lab.
19. What are scientific abilities rubrics? Give an example and show how you can use them to help students develop scientific abilities.
20. Explain the relationship between normal and friction components of the force exerted by the surface on an object and demonstrate how to approach multiple-objects problems.
21. Explain why and under which assumption that the energy of a bound system is negative. Show how to derive the expression for gravitational potential energy and electric potential energy of two objects. Use bar charts for your derivation.

22. Show how to help students develop mathematical expressions for four types of energy used in mechanics and work. What resources will you activate during this derivation? What difficulties do you anticipate? What are helpful strategies?
23. Describe student difficulties that are documented in research in the area of work–energy.
24. Describe a curriculum sequence for teaching a work–energy unit.
25. Explain the difference between periodic motion and simple harmonic motion; describe and explain SHM; describe useful representations in this area.
26. Explain the difference between the concept of an electric field and the physical quantities that characterize and quantify it.
27. Use multiple representations to explain the behavior of conductors and dielectrics in an electric field.
28. Give an example of a DC circuit problem to solve in which one needs to reason through potential difference and not through current.
29. What should students know about power in electric circuits? How will they learn it? How will you know that they have learned it?
30. Give an example of a complete ISLE process for student learning for any of the concepts of your choice.
31. What is the difference between the two right-hand rules in magnetism? Describe how your students will learn those and how you will assess them.
32. Describe the essence of 'backwards design'.
33. What does 'an enduring understanding' and 'an essential question' mean? Give examples of both.
34. Outline the sequence of student learning for static electricity.
35. Outline the sequence of student learning for DC circuits.
36. Outline the sequence of student learning for magnetic fields.
37. Outline the sequence of student learning for electromagnetic induction.

7.5 The role of a community in the development of habits

In her book *Good Habits, Bad Habits* Wendy Wood (Wood 2019) talks about the importance of community support for the development of habits. We discussed this issue in chapter 3; here we return with practical recommendations on how to purposefully develop a community in a physics teacher preparation program, which extends beyond graduation. The longevity of the community is needed to help novice teachers sustain the habits developed in the program and build the habits of leadership and improvement.

In this section, we mostly focus on the approaches and practices of the Rutgers Physics Teacher Preparation Program as it has been developing a community for a much longer time (over 20 years, see Forbes *et al* (2019)) than the program at the University of Ljubljana.

The program at Rutgers is cohort-based. This means that during the two years of the program, future physics teachers take all courses in the teacher preparation program together. Every year, the program coordinator has two cohorts. The starting cohort is mostly doing course work with some clinical practice teaching in Physics for the Sciences and observing high school classrooms. The cohort that is in the first semester of the second year is doing full-time student teaching in schools and is then coming back for more course work in the second semester of the year. Additionally, every cohort interacts with the graduates of the program. The graduates teach some courses in the program, serve as cooperating teachers during student teaching internship, and later as mentors when the graduates start teaching. We can view the community in two dimensions—horizontal across a cohort and vertical—across cohorts, including the graduates. We can analyse the development of the community through the lenses of three different activities: course work, clinical practice, and professional development. To help the reader see the big picture, we present separately the horizontal and the vertical dimension (see table 7.12) of the activities necessary for the community building.

Horizontal dimension of the community development among students in the teacher preparation program

Course work:
- Participating in group work in every class meeting with different team members, whiteboard meetings, sharing with the rest of the class.
- Working on a long-term project with 2–3 team members, different partners in every course.
- Preparing for the final exam as a team.

Clinical practice:
- Teaching in Physics for the Sciences.
- Enacting microteaching in teams of 2–3 with different partners in every course.
- Sharing successes and problems during student teaching internship.

Professional development:
- Attending PD workshops with other physics teachers.

Table 7.12. **Vertical** dimension of the community development among students, graduates, and cooperating teachers who are graduates of the program.

Connections	Course work	Clinical practice	Professional development
Students–students		Teaching in Physics for the Sciences in the semesters when both cohorts are present.	
Students–graduates	Connecting with the graduates of the program who teach some of the courses in the program.		Participating in in-person meetings of the graduates.

| Students–
cooperating
teachers (who
are also program
graduates) | Reflecting on the
observations of
cooperating teachers—
graduates of the program. | Student teaching
internship with
graduates of the
program as
cooperating teachers. | Participating in
the online
community of
the graduates |
| --- | --- | --- | --- |

Above, we separated community building activities into categories. However, in real life, many of them occur simultaneously. What follows is the description of how these activities interact with each other.

7.5.1 In-the-program learning community

Preservice teachers start participating in the development of the learning community from the first day of starting the program. Since one of their clinical practice activities is teaching in Physics for the Sciences and the semester begins after Labor Day in early September, the students from the first cohort meet each other in the first training meeting for the course that occurs prior to the beginning of the school year. They all meet once a week during training meetings for the duration of the semester and share their teaching experiences on the course Google Discussion Group. Through this process, first-year students get to know each other. Simultaneously, in the Development of Ideas in Physical Science course, they work in groups on class activities. The group composition changes every week with the purpose being for the participants to get to know each other better. After three to four weeks pass, the students begin their work on a group project that culminates with their first microteaching. This activity lasts for about two months and becomes a real bonding experience. Such experience repeats in every physics teaching methods course with a different partner(s) as every course includes a big project which ends with micro-teaching. By the end of the first year, each pre-service teacher has worked with every other member of the cohort several times. This bonding solidifies during the preparation for the final exam in Teaching Physical Science as described above.

In the spring semester, the second, older cohort comes back from their student teaching internship and joins the first cohort in Physics for the Sciences. Two cohorts start bonding together in training meetings and through the Google Discussion Group where they post reflections about their experiences teaching a lab or a problem solving class every week.

It is important to establish a learning community of the first-year cohort prior to their semester of full-time student teaching internship. The student teaching internship is a very difficult time for beginning teachers. They receive a lot of responsibilities at once and if their time management skills are not fully developed, this experience is overwhelming. Having friends to share and to support and be supported is vital for the positive experience with the process of full-time student teaching internship.

The student teaching internship is also a step in the vertical development of the community. As the interns work with the graduates of the program as cooperating teachers, and many high schools have more than one graduate teaching physics, the interns join the local communities of the graduates and bond with their cooperating teacher and other teachers in the school as they all use the same approaches to learning physics. Having a cooperating teacher who is not only familiar with the ISLE approach but is also skilled in it is very important for the continuous development of productive habits. Not only do the interns not receive conflicting messages from their cooperating teacher, but they observe and learn from a practicing master of the ISLE approach. As one of the interns commented once to Eugenia: 'Rich, my cooperating teacher, does everything that you taught us to do, only he does it better than you do!' This was the best compliment that Eugenia could have received.

It is important to consider the benefits of interacting with interns for the cooperating teachers as well. As time goes by, new developments in education—and specifically in physics education research—lead to changes in the program and to the new approaches in teaching physics. Eugenia and Gorazd continuously develop new curriculum materials. Those who graduated from the program 10 years ago would lose the benefits of these new developments without contact with the interns.

7.5.2 After-graduation learning community

The importance of having a community after graduation cannot be overestimated. Watching first-year teachers struggle makes the need for support very clear. But not only first or second year teachers need a community to survive and to continue developing productive habits. Teachers who graduated 10 years ago need this as well. We all know how much energy teaching takes. How do we replenish this energy? How do we focus on the excitement of students learning our beloved physics instead of mundane bureaucratical tasks? Eugenia's observations of her community of graduates, which is over 20 years old, show that it is being together with like-minded people and working on cool physics problems that re-energizes her teachers and provides emotional support. Finally, without such community, it is impossible to place student-teachers with the cooperating teachers who are skilled in the ISLE approach. If student-teacher experiences do not support the development of productive habits during student teaching, these habits do not form.

However, the in-service teachers cannot meet and spend time together every day as they work far away from each other (although mostly in one state). Therefore, a long time ago (in 2004), the graduates of the program created an online discussion group (at first it was a Yahoo discussion group, about 14 years ago it migrated to Facebook). Preservice teachers join the group during their second year in the program. Today, the group has over 170 members. The group allows them to share new ideas, pose questions, and make announcements. Once a month, the community comes together either in person at Rutgers or virtually. Over the years, we have found that the best day for in-person meetings is Friday afternoon. The meeting usually starts at 5 pm and goes on until 8–9 pm. After that, many participants go out to dinner together. The meetings usually consist of dedicated presenters sharing new physics problems, modern physics ideas, or curriculum development and assessment ideas. During the meeting, the participants work in groups, use whiteboards, and discuss solutions as the whole class. This approach allows the teachers to immerse themselves into the ISLE environment at a higher level than that at which they are teaching. As one of the graduates said:

> The Friday meetings are helpful in terms of providing hints and specific activities I can use in the classroom, but more importantly, they provide a community where we feel free to share our difficulties and receive guidance and support in return. It is fun to spend time with a group of people who are as fascinated by physics as I am. (Etkina 2015 p 257)

Below we show a list of topics from the monthly meetings during the 2022–23 academic year. For each topic, you see the leader and, if it is a teacher, then how many years of experience of teaching physics they have.

September 6—Introductions and 'cool things' that we do at the beginning of the school year. Sharing, leaders Elana Resnick (program graduate, 9 years of experience) and Danielle Bugge (program graduate, 13 years of experience)

October 7—New problems in physics, leaders Eugenia and Gorazd

November 5—Rollerblading experience, leader Eugenia (see photos of this session in chapter 4)

December 2—Electromagnetic induction, leaders Daniel Lee and Oliver Islambouli (program graduates, 5 years of teaching experience)

January 20—Curved space time part I, leader Mike Gentile (over 20 years of teaching experience), the course coordinator for Physics for the Sciences

February 10—Energy situations involving living organisms, leader Eugenia

March 10—Teaching energy assessment, leader Eugenia

April 21—Report on the development of the ISLE approach for chemistry—leaders Julie Koft (program graduate, 3 years of experience), Samantha Strauss (5 years of experience), Ryan Berns (10 years of experience), Abigail Seo (program graduate, 1 year of experience)

May 12—Curved space time part II, leader Mike Gentile.

The teachers also bring their difficulties to the meetings. This is a tricky point as it is easy to turn those meetings into complaining sessions. Therefore, a long time ago, when the danger of such transformation became real, one of the community participants proposed a rule: in these meetings, one cannot complain about students, administration, or parents. One can state an issue and ask for solutions but cannot frame the question as a complaint. It is interesting how this rule transformed the meetings by making them more positive and productive.

Maintaining the life of the community requires time, energy, and the sacrifice of a Friday night once per month from the program coordinator. One might think that this is a heavy burden to carry. However, the meetings have such positive energy and provide such reinforcement for the program values that this emotional boost compensates for the invested time. Interestingly, since Eugenia retired from Rutgers in 2022, two of the program graduates (Danielle Bugge and Elana Resnick) took over organizing the meetings, doing the schedule, and inviting different people to run them. The Friday night program is continuing smoothly. Eugenia joins as a participant and runs the meetings about twice a year.

One last comment is in order. The teachers and the organizers of the meetings do not get paid for participating in and running these meetings. Those who need some documentation can receive a certificate showing how many hours of professional development they received from attending the meetings. This is the only tangible reward that the participants get. Why do they come?

This is what one of the teachers said:

> I look forward to these meetings. They fill me with energy and positive emotions. I meet my friends. I do new physics. I am reminded of how much fun it is. Teaching every day is draining. These meetings replenish my energy and enthusiasm. Every time I am about to start driving to Rutgers on a Friday night, I think of not going but driving home instead and relaxing. But then I come to the meeting, see all my people, and realize that this is the place where I need to be. (J L, graduate of 2012)

7.5.3 Community and habits

In their paper, Etkina, Gregorcic, and Vokos (Etkina *et al* 2017), introduced the framework of productive habits as a guide to teacher preparation and professional development. In addition to the physics teacher habits of mind and practice, they discussed the habits of maintenance and improvement. They wrote:

> …habits of maintenance and improvement are the habits that involve continuous learning on the part of the teacher as an individual and as a member of the community, as she organizes her professional life to give priority to maintaining the community, actively sharing new findings, and using the findings of other teachers. (Etkina *et al* 2017, 010107-6)

This statement underscores the importance of a time management habit as all the activities mentioned in the above sentence require significant time investment. We argue that it is much easier to engage in those activities if your friends are participating in them. Therefore, the community is crucial for the teachers implementing the ISLE approach to continue growing and learning new ideas. We would say that such a community is more important for ISLE-based teachers than for teachers implementing a traditional approach to learning physics as the latter probably constitutes the majority in a particular school. Being alone while implementing the ISLE approach and being continuously challenged for it makes having an outside community of like-minded people vital.

We recently noticed that, after hiring one graduate of our program, many schools seek more of our graduates. The reason is that their teaching is more in line with the demands of the twenty-first century and more engaging for the students. In the US, we have individual schools that have hired over five Rutgers program graduates in the last 15 years and some schools where all physics teachers are program graduates.

7.6 *Interlude* by Bor Gregorcic: AI and the learning of physics

This interlude is written by Bor Gregorcic, a Senior Lecturer/Associate Professor in the Department of Physics and Astronomy at the University of Uppsala (Sweden). He was a PhD student of Eugenia and Gorazd at the University of Ljubljana and in the past ten years has been working in Sweden. One of his duties is teaching physics methods courses for prospective physics teachers. In these courses, Bor uses the ISLE approach. In the past seven months since ChatGPT became available to the public, Bor has conducted several studies investigating the opportunities that AI provides for physics teachers. We invited Bor to write an interlude for our book and he gracefully agreed. Below is what Bor wrote; the references are included with the rest of the references for this chapter.

7.6.1 Preparing physics teachers for a life with artificial intelligence

Bor Gregorcic

In the last year, a new development had begun affecting physics education and, consequently, the preparation of physics teachers. About one year after the public release of ChatGPT, an artificial intelligence based chatbot, AI is already changing how we think about different aspects of our life, including education (Lo 2023). The user-friendly and intuitive design, along with the impressive writing capabilities of chatbots have driven their fast uptake by different groups, including students. According to some surveys (Ungdomsbarometern 2013), up to half of all high school students in Sweden had already used ChatGPT in 2023 to help them write homework assignments. It seems inevitable that AI-based tools will become even more capable, widespread, and easily accessible. Witnessing the fast uptake of ChatGPT, I would suggest, as many others have, that isolating the education process from AI technology is not only untenable, but also unproductive. The technology is here to stay, and teachers have the responsibility to prepare students for a life in a world *with* AI. Teachers should help students learn to use it in productive and responsible ways and to reflect on the implications of its use.

So, how should we prepare teachers for a world with AI? Here are some thoughts. If teachers are going to be helping students use AI productively, they themselves first need to have some understanding of how these tools work, including their strengths and their limitations, and be aware of the potential benefits and risks associated with their use (Polverini and Gregorcic 2023; Kasneci *et al* 2023). This is not an easy task. AI technology is developing at such a break-neck speed, that even scientific publications often cannot keep pace with it. Research papers on the topic can become outdated before they can make it through the several-month-long peer review process. So, teachers stand before a big challenge. While the physics that they teach in schools has remained more or less the same for decades or even centuries, the technological world in which they live is changing fast and in novel ways, ones which most of us did not anticipate as recently as the autumn of 2022. It would be great if teachers could learn about AI-based tools during their pre-service training, once and for all, and then use this knowledge in their profession. Unfortunately, this is not possible. AI technology is in rapid flux and teacher training can only be based on the current snapshot of the technological landscape. While some understanding of current AI tools is necessary, teacher training should also place emphasis on forming habits of improvement and life-long learning (Etkina *et al* 2017). I would also argue that making a habit of regularly interacting with AI tools is an important part of staying up-to-date on the topic. However, individual teachers cannot be expected to spend time scouring AI-research literature. It is the responsibility of the broader physics education and physics education research communities to curate relevant research on AI and generate research and educational materials on the topic. The challenge of preparing future generations for a life alongside AI is one that requires a communal approach, locally and globally.

7.6.2 What physics teachers should know about LLM-based technologies

While not sufficient for productive use, I believe there are some things physics teachers should know about ChatGPT and related technologies. This knowledge is needed in order to create solid footholds for their future learning on the topic (Polverini and Gregorcic 2023).

ChatGPT is a chatbot application that uses a large language model (LLM) as its 'engine'. A large language model is an artificial neural network created using a complicated machine learning procedure. The learning process consists of feeding huge amounts of text data to an algorithm that looks for patterns in the texts and capturing regularities. This way, LLM becomes a statistical machine that can predict which words are likely to appear together in a text, based on some given context. An analogy I like to use for the basic functioning of LLMs is that of an advanced auto-complete function that one nowadays often sees in online email services. When one begins to write an email, the function suggests a possible continuation of the sentence in order to save the user time. This is essentially what LLMs do, only on a more advanced level. Because the data LLMs are trained on are so extensive (essentially the size of the entire Internet), they can generate plausible-sounding sequences of words in almost any context. Generating such sequences does not, however, involve directly copying the data on which the LLM was trained. In fact, the LLM is not able to retrieve the data itself. It can only imitate, or mimic it, by generating novel text that 'sounds' like the data it was trained on. Sometimes, it is possible that certain parts of the data are reproduced verbatim, but that tends to happen in cases of very repetitive patterns, such as reciting well known physics laws or values of established physical constants, which appear with regularity and consistency in the training data.

The limitations of LLM-based chatbots
As a direct consequence of their architecture, LLM-based chatbots' output is often problematic in at least some of the following ways[3]: 1. They can produce false statements, containing confabulated information (Ji *et al* 2023). 2. They have difficulties with correctly executing mathematical operations (Wolfram 2023). 3, They can produce incoherent or contradictory chains of argumentation and often fail on common sense reasoning tasks (Forbes *et al* 2019, Polverini and Gregorcic 2023, Gregorcic and Pendrill 2023). 4. They are often biased in undesirable ways and can reinforce social stereotypes (Thakur 2023, Fang *et al* 2023, Khandelwal *et al* 2023). In more detail:

1. When prompted to provide specific information, especially such that does not appear repetitively in the training data, LLM-based chatbots are notorious for producing made-up facts, also referred to as AI hallucinations (Ji *et al* 2023). While this presents a major limitation for the use of LLM-based tools in fields such as history or medicine, reciting a lot of known facts is much less central in learning physics.

2. Another limitation, one that is more directly pertinent for LLM-based chatbot use in physics, is the LLM's limited ability to perform mathematical operations. If we understand how LLMs work, this is not too surprising. Indeed, imitating what a calculation typically looks like is not a good strategy for correctly calculating something. This limitation, however, can be compensated by allowing the LLM to outsource calculations to a more mathematically able tool (Wolfram 2023). While the freely available version of ChatGPT still does not allow this, the subscription-based version of it already allows the use of so-called plugins, which can significantly extend the range of ChatGPT's capabilities, especially in the domain of mathematical operations and programming.

3. The third limitation of ChatGPT and other similar LLM-based tools is their inconsistency in providing coherent chains of physics argumentation (Polverini and Gregorcic 2023, Gregorcic and Pendrill 2023). Because LLMs do not work by applying an underlying physics model (such as a Newtonian model of a point particle governed by laws of dynamics), their output can often be nonsensical from a physics point of view, while at the same time being sophisticated from a linguistic point of view.

4. The final limitation of LLM-based tools that I will discuss here is their tendency to display unwanted bias. While bias is an essential feature that makes LLMs work—they reflect and imitate subtle patterns in the training data—this often results also in unwanted bias. Unwanted bias, such as sexist (Thakur 2023, Kotek *et al* 2023), racist (Nadeem *et al* 2020), or casteist (Khandelwal *et al* 2023) tendencies are captured from the training data, just as any other pattern. While AI companies are working hard to mitigate such unwanted bias, this is difficult to do because bias lies at the very essence of how LLMs work.

---

[3] This is not an exhaustive list. For a more detailed analysis of opportunities and challenges of the use of LLM-based tools in education, see Kasneci *et al* (2023).

Teachers should be mindful of the above listed limitations of LLM-based chatbots when they consider using them, or letting students use them for tasks that require retrieval of information, mathematical tasks, or qualitative reasoning in physics and broader. I would suggest that the current unreliability of these tools and their tendency to display unwanted bias, makes them unsuitable to be used by students as AI tutors without the supervision of an experienced human teacher. Teachers of the future will need to spend time discussing such issues with their students and serve as a role-model for a critical attitude (Krupp *et al* 2023) toward the use of AI tools.

Strengths of LLM-based chatbots
Because of their ability to imitate existing writing styles and bring together different contexts in new ways, LLM-based chatbots can be great tools for creative work. This includes their use as brainstorming partners (Vasconcelos and Dos Santos 2023) and sounding boards for artistic, scientific. or pedagogical (which often contains elements of artistic and scientific) divergent thinking tasks. I believe we have only begun to scratch the surface of the many novel possibilities to use these tools to enrich the learning process in new, perhaps still unexpected ways.

Another common use of LLM-based chatbots is to receive fast feedback on technical aspects of one's writing, such as vocabulary, structure, grammar, voice, etc. The trend is clear here. Even academics are increasingly using these tools to assess their own writing (in terms of form and content) and use it to improve it before it is sent out for review (Nordling 2023).

Similarly to how calculators changed the way we think about manually doing calculations, AI will change how we think about writing[4]. The challenge here is that learning to write is often associated to learning to think. We need to further reflect on when the tool that is helping students do one thing is actually hampering their development. For example, to be able to critically assess chatbots' output, one needs domain specific knowledge. It may be possible for experts to recognize when a chatbot is generating fancy-sounding nonsense, but how can we make sure that students can also do it? Teachers need to help students become capable of using AI tools critically (Dahlkemper *et al* 2023; Küchemann *et al* 2023). This may require students not using them at first, and then gradually introducing them into the education process. But we still need to figure out when and how students should start using them to prevent them from using them *instead* of thinking in the first place.

7.6.3 What opportunities LLM-based chatbots bring to physics teacher education

Despite many unanswered questions and concerns, I believe that many of the aforementioned limitations and strengths of LLM-based chatbots open up new possibilities for physics education and the education of future physics teachers. Here, I propose some opportunities that I see arising, and what we can do in teacher

---

[4] The analogy to calculators should not be overextended, though. AI has a much broader potential for impacting education. See Lodge *et al* (2023) for a more in depth discussion of analogies for generative AI in education.

preparation programs to prepare teachers for the fruitful integration of AI into their own teaching. Note that many of the activities proposed below can also be used as classroom activities for students, not just pre-service teachers.

Addressing the discrepancy between the linguistic strength on one side, and the conceptual and mathematical difficulties of LLMs on the other, presents an opportunity to practice critical reading and assessment of written text with the focus on content, instead of form (Gregorcic and Pendrill 2023). Indeed, most text produced by chatbots has polished language, but often contains different types of errors, including mathematical, logical, or factual. Preservice teachers can read and assess AI-generated answers to physics tasks (Dahlkemper *et al* 2023, Bitzenbauer 2023) and identify productive and unproductive parts or aspects of answers. This does not actually require them to use the chatbot, just read what it has generated. This makes it a good introductory activity. In this case, the course leader must use the chatbot to generate interesting examples of answers to tasks, which can facilitate discussion and reflection. In my own experience, the unusual combination of polished language and odd reasoning errors invokes the need to know more about how chatbots work.

To build a body of experience in interacting with chatbots, another activity that I often use in my courses is asking pre-service teachers to find physics questions that the chatbot answers incorrectly. For readers who would like to give this exercise a try, I suggest starting with the freely available version of ChatGPT and asking it questions from the Force Concept Inventory or other conceptual physics questions, that do not require calculation. For example, I have found that non-traditional problems from the *College Physics: Explore and Apply* textbook (Etkina *et al* 2019b) are a great source of such tasks. One such task, adapted from problem 72 on p 214 that has turned out to generate interesting responses from ChatGPT is: A spherical street lamp accidentally explodes. Three equal pieces, A, B, and C fly off the lamp holder with equal speeds and from the same height. A flies vertically upwards, B flies horizontally, and C flies vertically downwards. Compare the speeds with which each piece hits the ground.

Using untraditional and conceptual tasks without much calculation is a good way to generate some intriguing responses, which will almost always trigger a lively discussion and generate opportunities to delve deeper into how LLM-based chatbots work.

Another fun task that I have tried in my class, which always had pre-service teachers completely engaged is a game of 'catch the bot'. In this game, actual student solutions to a given question are mixed in with chatbot-generated solutions. In my experience, pre-service teachers can develop a sense of when an answer is AI-generated, a skill that may be useful also in their future teaching careers. Following from the point above, analysing the errors in chatbot-generated text presents an opportunity to reflect on the source and nature of AI's difficulties and relate them to student difficulties. For example, interesting questions that often arise are: Can students' difficulties also be traced back to common patterns in language? Has the chatbot acquired these difficulties from being trained on incorrect answers and faulty reasoning texts?

Research on LLM has shown that the way we prompt these models is crucial for making their output useful. Preservice teachers need to develop skills for interacting with these chatbots to be able to use them productively. The field of 'prompt engineering' is a fast-evolving craft, which holds many interesting lessons about how LLMs function and what is required to make them behave in the way we want them to do so. Here, I would recommend some external reading on the topic of prompt engineering in the context of physics education (Polverini and Gregorcic 2023). In essence, giving the chatbot enough contextual information, giving it instructions on how it should behave, and even how it should 'think', is correlated with a better quality of responses. Knowing about prompt engineering techniques is also useful, when we want to get incorrect answers, for example, for use in critical reading activities discussed above.

Researching the field of prompt engineering together with Giulia Polverini, we have found that tasks which are written in everyday language and that do not use typical terms accompanying physics problems (e.g. the names of physics quantities and verbs such as derive, solve, and determine) tend to elicit responses that are less 'physics-y' and tend to go off on tangents that are often irrelevant from a physics perspective (Polverini and Gregorcic 2023). However, when one provides an explicit context of the task (e.g. forces in circular motion) and instructs the chatbot how to act (e.g. like a physics teacher), this tends to result in responses that are more relevant and of higher quality. For example, when asking ChatGPT the following question: 'A Nascar racer won the race in the last lap by grinding against the outer fence of the racetrack. Why did this trick work?' (Polverini and Gregorcic 2023, p 13), we got a response which was very racing-jargon-heavy and did not address the main point of the problem. When we added a couple of sentences to the prompt to better contextualize the problem and give instructions on how to act, the response became much better. The new prompt was: 'A Nascar racer won the race in the last lap by grinding against the outer fence of the racetrack. Why did this trick work? <u>Explain it like a physics teacher would from the perspective of forces in circular motion.</u>' (Polverini and Gregorcic 2023, p 15). For ChatGPT's responses to these prompts and a discussion of other prompt engineering techniques, see (Polverini and Gregorcic 2023).

When we know how to prompt them and when they are further augmented by external plugins, such as those that enhance their mathematical abilities, chatbots can become powerful tools, not only for students, but even for experts. The educational community needs to take this opportunity to rethink what knowledge and skills are and will remain essential for our students. While in the future we might experience a sharp decline in the need for certain abilities, some will likely increase in importance. I believe that the ability to think like a scientist is one of those abilities that will become more important. This is where the role of ISLE becomes extremely clear.

7.6.4 The role of ISLE in a world with AI

Just like in most aspects of life, AI will be, and in some contexts already is (Jumper *et al* 2021), playing an increasingly important role in science. It is increasingly being used in identifying patterns in scientific data, describing them, and generating potential hypotheses to explain them. To prevent mystification of science, we need to be able to see these steps as parts of the scientific process. ISLE allows us to see AI as a tool used in parts of the scientific process, instead of an entity 'taking over' science. For example, because physics is an experimental science, digital agents still cannot be doing all the work. Someone must perform the experiments, ensure the quality of collected data, as well as apply new knowledge in the real world. Furthermore, tasks that require the perception and physical manipulation of real-world objects are still very difficult to perform reliably by robots and AI (a phenomenon also referred to as Moravec's paradox (Moravec 1988)). It is thus very likely that such tasks will remain mostly in the human domain for some time to come. The teachers of the intermediate future should therefore focus on helping students learn about the interplay of experiment and models in science, and help them develop not only scientific habits of mind, but also a good understanding of the physical practicalities of scientific work, as well as their real-world implications.

Importantly, AI can also be abused, for example to generate fake data to support a hypothesis (Taloni *et al* 2023). Research ethics is thus another area of concern in science, that will be impacted by the development and uptake of AI tools.

I believe that a multifaceted and epistemologically refined understanding of the process of science is necessary for students to be able to recognize, critically assess, and use AI in responsible and ethical ways. I believe that ISLE, as an epistemological and pedagogical framework, holds great potential in this regard.

Appendix A. Teaching and Assessment in Physical Science course syllabus from Rutgers University

Learning goals

At the end of the course students will be able to answer the following questions:
1. What are the goals of learning physics/physical science in a high school? How do these goals relate to NGSS standards? How does one formulate assessable goals for instruction?
2. How is physics curriculum structured? What are the main topics and how do they relate to each other?
3. What are students' ideas about most important physics concepts and how do we build on them?
4. How do we make students active participants in the learning process and how do we make this process mirror scientific inquiry?
5. How do we help students develop scientific habits of mind—scientific abilities?
6. What is the difference between formative assessment, summative assessment, standards-based assessment and how do we implement each in each of the high school physics/physical science units?
7. How does one create a lesson and unit plan?
8. How do we create a positive, supportive, and caring classroom for all students?

Class materials

Make sure you are familiar with the content of the websites; browse them before the semester starts.
1. Provided and required: Etkina, Planinsic, and Van Heuvelen 2019 *College Physics: Explore and Apply* (New York: Pearson) (includes the *Active Learning Guide* and the *Instructor Guide*)
2. Required: New Generation Science Standards, available online at http://www.nextgenscience.org/next-generation-science-standards.
3. Required: The Danielson 2013 framework for teaching evaluation instrument—can be downloaded from the Internet and is posted on Google Classroom.
4. Required: Physics Union Mathematics (PUM) Curriculum modules, to download the modules and final assessment go to http://pum.islephysics.net/ then click on the Teacher login, and proceed to the download area. The website will ask you to input the information, after you do this, you will receive the access password.
5. Required: A set of papers from the reading list will be posted on the Google Classroom every week on Wednesday for the next class on Tuesday.
6. Strongly recommended (can be shared with a friend): Arons A 1997 *Teaching Introductory Physics* (New York: Wiley). ISBN 978-0471137078
7. Good to have Wiggins G P and McTighe J 2005 *Understanding by Design* (Alexandria, VA: ASCD).
8. PHET resource by the University of Colorado, Boulder: http://phet.colorado.edu

Websites with class activities:
http://islevideos.net/
https://sites.google.com/site/scientificabilities/
http://universeandmore.com

Grading and activities

Your course final grade will be based on how you meet the standards listed below. Each standard will be assessed *multiple times* according to the rubric—you *have* to convince me and your classmates that you meet the standard. If at any point you fail to meet the standard, you will have an opportunity to be assessed again. Each assignment can be improved. I encourage you to try as many times as you need to make the assignment perfect.

Rubric: 1. Working towards but is not meeting expectations yet. 2. Moving towards meeting expectations. 3. Meets expectations. 4. Exceeds expectations (I want to brag about you). I believe that every student in this course will work to exceed my expectations.

Content knowledge for teaching physics standards (broken down into content standards (CK) and pedagogical content knowledge standards (PCK))

Area of physics	CK	PCK
Kinematics	CK1—Can make connections between physical quantities used in kinematics and concrete and graphical representations, knows how derive $x(t)$ functions for different motions and is able to articulate the connections between the concepts and science practices in the unit, including what concepts yield better to specific practices.	PCK1—Is able to demonstrate an understanding of students' ideas in kinematics (productive and unproductive), is able to interpret student work on graphs and provide examples of formative and summative assessment in kinematics. Is able to use the concept of index to help students write linear functions.
Dynamics	CK2—Is able to articulate the relationships between Newton's laws and explain why particular representations are important.	PCK2—Is able to provide an example of how to set the goals for one lesson on Newton's laws and show the evidence that the goals are achieved.
Force laws	CK3—Is able to explain the relationship between normal and friction force and demonstrate how to approach multiple-objects problems.	PCK3—Is able to interpret student work on forces and suggest instructional sequences to address student difficulties.
Circular motion	CK4—Is able to identify productive and unproductive language in circular motion and derive the expression for centripetal acceleration without calculus.	PCK4—Is able to design a two hour laboratory for circular motion where students develop specific scientific abilities while constructing, testing or applying physics concepts.
Energy	CK5—Is able to demonstrate fluency with the system approach to energy and productive representations of work–energy processes. Is able to explain why the energy of a bound system is negative.	PCK5—Is able to show how to help students develop mathematical expressions for four types of energy used in mechanics and work. Can demonstrate understanding of student difficulties in this area. Can articulate the curriculum sequence for teaching work–energy unit.
Vibrations	CK7—Can explain the difference between periodic motion and simple harmonic motion; can describe and explain SHM; Is able to demonstrate familiarity with useful representations in this area.	PCK7—Is able to demonstrate an understanding of students' ideas in the area of vibrations, can design an instructional progression for the unit and final assessment.
Electric field	CK8—Can explain the difference between the concept of electric field and the physical quantities characterizing it. Is able to use	PCK8—Is able to address common difficulties that students have with the concept of electric potential in a lesson.

(Continued)

(*Continued*)

Area of physics	CK	PCK
	multiple representations to explain the behavior of conductors and dielectrics in electric field.	
DC current	CK9—Is able to reason through complex problems in electric circuits (including power) using the language of potential difference.	PCK9—Is able to design a lesson in which the students learn to reason through complex problems in electric circuits (including power) using the language of potential difference and connect this material to their everyday experience.

General standards

GS1: Is familiar with the New Generation Science Standards and can use them when planning and instruction and assessing student learning. Is able to connect NGSS to the ISLE process and its elements.

GS2: Is able to formulate the goals of the instructional unit that reflect the important ideas and practices in this unit and can be assessed.

GS3: Is able to interpret student responses (oral or written) and revise planned instruction based on the responses during microteaching.

GS4: Is able to collect (or to describe) evidence that will indicate that students achieved a proposed goal.

GS5: Is able to write an ISLE-based lesson plan that has all required elements and implement the lesson in practice.

GS6: Is able to write an ISLE-based unit plan that has all required elements.

GS7: Is able to devise a beginning of a lesson that builds on student ideas and engages them in meaningful exploration of physics ideas during microteaching.

GS8: Is able to solve (or explain why the solution is not possible) for any physics problem at the level of algebra-based physics in the areas that are addressed in the course using expert approach to problem solving.

Description of activities

Attendance and participation in class discussions: Attendance and participation in each class are essential for your learning in class.

Quizzes: Every class will start with a short quiz related to your knowledge of student difficulties in particular concepts and abilities (all quizzes are linked to the standards). To prepare for the quizzes make sure you can do relevant quizzes from the PUM modules. To do well on the quiz you need to do the readings and reflect on the learning in class. Each quiz can be improved. The number of attempts is not limited. The purpose is for you to learn, not for me to give you a grade. As the class time is limited, if you think that you need extra time for the quiz—please come early. We will be in the classroom at 4.30 pm.

Professional development activities: As a part of your preparation for being a physics teacher, you will need to join the American Association of Physics Teachers (AAPT at http://aapt.org) and/or the NJ section of the AAPT (http://njaapt.org)—check out the information on the websites, NJ AAPT is free for students. You will also need participate in two professional development activities of your choice and write a reflection about what you learned.

Danielson framework: This is the framework that will guide your growth as a teacher for years to come. Study it and use it for reflections on observations of lessons in class, lessons in schools, and your microteaching.

Homework: Every week after class you will
1. For the first eight weeks of classes, you will write a lesson plan or another assignment for one of the concepts discussed in class. Follow the outline at the end of the syllabus. The homework should be submitted by Thursday night on Google Classroom. I will read the reports on Friday, provide feedback typed in the document in a different color, and you will make revisions in the third color and send the revised document back to me. You can revise the homework as many times as you wish to improve your grade but you have to do it before the next class (on Tuesday). For the last six weeks of classes, you will write a reflective journal answering three questions as if you were a student in that class (answers relate to physics not teaching):

- What did I learn in class and how did I learn it?
- What remained unclear?
- If I were the teacher, what questions would I ask to find out whether my students understood the material?

Or you will complete another assignment if necessary.
2. Work with chapters/sections of the CP: E&A, IG, and the ALG to analyse the structure of the cycles and complete problem solving tasks—these are your responsibility. Every week you will be assigned problems from the textbook, ALG, or PUM modules to solve. To get help with the problems, attend a problem solving help session (we will schedule it to fit your schedules). If you need extra help, you can always stay after class.
3. Read chapters of the *Teaching Introductory Physics* book and *Understanding by Design* book and assigned articles and be prepared for class discussions.

Reading list: The papers for each class will be posted weekly on Google Classroom.

Reflection on classroom observations: Every week during your ten weeks of observations of science classrooms you will need to send me a report about your observations (the report should be submitted before Sunday of the week you did the observations). Each report should address specific questions that are listed at the end of the syllabus[5]. During one of the observational visits, you will need to teach a lesson or a part of the lesson, please make sure you arrange this with the cooperating teacher, submit the lesson plan to him/her and me in advance and then after you teach the lesson, you write a reflection.

Formative assessment activities: For the *kinematics* unit you will design five formative assessment activities that assess student construction of understanding of the concepts of velocity and acceleration and development of some scientific abilities. In the assignment you will specify: what the target understanding or ability look like when demonstrated by a student, how the activity assesses it, provide two possible student responses and describe how you will provide feedback to the student, and how you will modify your instruction based on the feedback from the student. Deadline: 15 February.

Summative assessment (a test): For the *dynamics* unit you will design a 45 min test. You will make a list of understandings and abilities that the test will assess, provide problems and tasks for the students, explain why you chose these problems and how they fit together in terms of difficulty. Tests will be discussed in class. Deadline 28 February.

Microteaching: During weeks 9–15 groups of students will teach lessons in class (the topic of the lesson is a part of the unit described below). A two-student group should work as a team. The length of a lesson is about 120–150 min. The goal of the lesson is that the students construct a physics concept through the ISLE approach. To prepare for classroom teaching each group has to meet with me for a content interview first, then for initial planning, the first draft, and the practice (about four meetings). Please make sure that you schedule your work accordingly during the semester. You are responsible for the materials (equipment) used during the lesson. Discuss them with me in advance and find time to check available equipment in the supply room and what you need to build/order.

Lesson plan: Before you teach the lesson, you will write a lesson plan, you will submit this lesson plan for feedback on Google Classroom after the planning meeting and then revise it before meeting for the first draft and then revise again. After teaching you will add a reflection and then submit both the plan and reflection on Google Classroom. The deadline for the lesson plan with the reflection is *one week after* you teach the lesson in class.

Unit plan: The lesson that you taught is a part of a unit for which you need to write a unit plan. You will submit the first draft ten days after microteaching and revise as many times as needed after feedback. Deadline: *ten days* after you teach the lesson in class. The elements of a unit plan are provided at the end of the syllabus.

Final exam: At the end of the course on there is an oral examination. You will receive a list of questions to prepare (about 50) in the middle of the semester. During the exam students will be randomly assigned two of the questions. You will present your answer in front of the class. In addition, each student will be given a problem to solve or a laboratory investigation to perform. The problems and laboratory investigations will be from assigned problems from the ALG, PUM, or the video website.

---

[5] We provided this list in the chapter in section 7.3, 'Clinical practice'.

Course schedule

Topics for discussions (by week, fill out appropriate NGSS standards)

Week	Topic
1, 2, 3	Backward design approach to curriculum design. Tools for teaching physics. Language and learning physics. Kinematics.
4, 5, 6	What does it mean to understand? NGSS. Newton's laws, mass and force. Dynamics. Force laws.
7	Asking the right question. Circular motion.
8	Evaluation of reformed teaching—Danielson framework. Energy. Multiple representations.
9	Experiments in physics instruction. Oscillations.
10, 11	What is pedagogical content knowledge and why should we worry about it? Static electricity—Electric field and electric potential.
12, 13	Teaching different students. Current electricity. Series and parallel circuits.
14	A meeting with an experienced teacher. Professional organizations for physical science teachers. Current electricity. Electric power.
15.	Final exam.

Unit plan
1. Title
2. NGSS standards addressed in the unit. Explain why you chose those.
3. Length total (days and periods).
4. What students should know and have done before the start of the unit. Explain why this is important.
5. Standards (goals) that you set for the students, e.g. conceptual (what ideas and concepts students should be able to apply), quantitative (what mathematical procedures they should be able to demonstrate, what quantitative problems to solve, etc), procedural (what science practices they should be able to demonstrate), and epistemological (what should they be able to do to show you that they understand how knowledge relevant to this unit was constructed).
6. What evidence will convince you that students achieved the standards? List and describe.
7. Most important ideas in terms of the subject matter—describe in detail. This is where I will look for your content knowledge, so make sure you go into detail. List cross-curricula links. Most appropriate science practices for this unit.
8. Student potential difficulties and helpful prior knowledge. How can you help with the former and build on the latter?

9. Lessons outline—list all lessons in the unit with the standards that they help achieve and brief descriptions. This should be very short to give a sense of the flow of the unit.
10. Relevance to students' lives.
11. Full text of a two-period lab. This should be the lab you designed, not the one you did in 193/194.
12. Final traditional (paper and pencil) and alternative (performance-based) summative assessment:
    a. Unit test (include expected high-quality responses to each assignment). Traditional does not mean multiple-choice, it means a limited time written test. Make sure that your state the unit standards that you can assess with different tasks. Describe the grading scheme for the assessment. Explain whether you are doing standards-based assessment or traditional grading.
    b. Performance summative assessment. Provide descriptions with brief guidelines for the students and expected outcomes. Describe relationships to the standards.
    c. Student projects. Describe what they are and how you will provide guidance to the students. Describe relationships to the standards.
    d. Out of classroom activities if appropriate (field trips, fun competitions, plays, etc). Describe relationships to the standards.
13. Modifications for different learners.
14. List equipment for the unit and resources for the students.
15. List complete references to all resources you use as a teacher.
16. Reflection on the implementation of the unit including commentary on obstacles in implementing it: how well does the unit meet the needs of diverse learners? How well did you teach the content and science practices and well did students learn those (the answer to this question should include deep analysis of the final assessments); what were the pedagogical strategies used and what needs improvement? How did you communicate the results of the unit to the parents?

Appendix B. Multiple Representations in Physical Science course syllabus from Rutgers University

Learning goals

At the end of the course students will be able to answer the following questions:
1. How can we connect learning of physics to the structure and function of the brain?
2. How do we apply the knowledge of brain science and cognitive science to problem solving in a physics class?
3. What are 'new types of physics problems' and how do we create those?

4. What are different ways to engage diverse students in meaningful problem solving in every unit of a high school physics/physical science course?
5. What is the role of multiple representations in problem solving?
6. How do we create rubrics for developing and assessing graphical representations?

The goal of the course is to acquaint prospective physical science teachers with the multiple representation method used in constructing concepts and problem solving in physical science. Multiple representations are powerful tools that aid the brain during concept acquisition and problem solving. Multiple representations enhance metacognition and epistemic cognition. Being familiar with the multiple representations used in a discipline is crucial for mastering and teaching it. In this course we will focus on such representations as pictorial representations, motion and force diagrams, graphs, energy bar charts, ray and wave front diagrams, and applications of these representations to problem solving. We will also learn how to help students construct and use analogies.

Class materials:
1. Zull J 2002 *The Art of Changing the Brain* (Sterling, VA: Stylus)
2. Provided by the instructor: Etkina P and Van Heuvelen A 2009 *College Physics: Explore and Apply* 2nd edn (San Francisco, CA: Pearson) (including the *Active Learning Guide* and *Instructor Guide*)
3. New Generation Science Standards, available online at http://www.nextgen-science.org/next-generation-science-standards.
4. Physics Union Mathematics (PUM) Curriculum modules, to download the modules and final assessment go to http://pum.rutgers.edu then click on Teacher login, and proceed to the download area. The website will ask you to input the information, after you do this, I will send you the access password. Please make sure you have all modules downloaded before the semester starts.
5. A set of papers from the reading list will be posted every week on Tuesday for the next class on Monday.
6. Websites with class activities:
   1. http://islevideos.net/
   2. https://sites.google.com/site/scientificabilities/
   3. http://universeandmore.com
PHET resource by University of Colorado, Boulder: http://phet.colorado.edu
7. Research-based physics teaching methods and assessments are at https://www.physport.org/ and https://www.compadre.org/.
   1. Make sure you are familiar with the content of the websites; browse them before the semester starts.
8. A great resource website: http://www.physics.usyd.edu.au/super/physics_tut/credits.html

Grading and activities

Your course final grade will be based on how you meet the standards listed below. Each standard will be assessed *multiple times* according to the rubric. If at any point you fail to meet the standard, you will have an opportunity to be assessed again. Each assignment can be improved. I encourage you to try as many times as you need to make the assignment perfect.

Rubric: 1. Working towards but is not meeting expectations yet. 2. Moving towards meeting expectations. 3. Meets expectations. 4. Exceeds expectations (I want to brag about you). I believe that every student in this course will work to exceed my expectations.

General standards

GS1: Is familiar with new types of problems and can devise them for specific topics and make self-assessment rubrics for formative feedback.

GS2: Is able to connect recommendations of brain research to physics instruction, specifically to interpret literature recommendations and apply to specific instructional moves.

GS3: Is able to interpret student responses (oral or written) and revise planned instruction based on the responses during microteaching.

GS4: Is able to collect (or to describe) evidence that will indicate that students achieved a proposed goal.

GS5: Is able to design rubrics to help students self-assess multiple representations used in a physics course.

GS6: Is able to devise a beginning of a problem solving lesson that builds on student ideas and engages them in meaningful application of physics ideas (problem solving) during microteaching.

GS7: Is able to write a unit plan that has all required elements.

GS8: Is able to solve (or explain why the solution is not possible) for any physics problem at the level of algebra-based physics in the areas that are addressed in the course.

GS9: Is able to describe at least ten available resources that she/he will use when planning physics lessons

Lesson specific standards (broken down into content standards (CKT-Physics) and pedagogical content knowledge standards (PCK))

Area of physics	CK	PCK
Magnetic fields	CKT1—Can use all graphical and mathematical representations of magnetic field to reason about magnetic processes in a vacuum, diamagnetics, paramagnetics, and ferromagnetics.	PCK1—Can show how help students write and use a rubric to self-assess their use of graphical representations of magnetic field (magnetic field vectors and magnetic field lines).

	CKT2—Can reason about electromagnetic induction using multiple representations.	PCK2—Can show how to engage students in solving non-traditional problems involving magnetic fields.
Electric field	CK3—Can use different graphical and mathematical representations of electric field to reason about electrostatic processes in a vacuum, dielectrics, conductors, and in capacitors.	PCK3—Can show how help students write and use a rubric to self-assess their use of graphical representations of electric field (electric field vectors, electric field lines and equipotential surfaces).
Geometrical optics	CKT4—Can use real optical systems and ray diagrams to locate an image produced by a reasonable optical system, including glasses.	PCK4—Can design a rubric for self-assessment of ray diagrams. Can show how to engage students in solving non-traditional problems involving geometrical optics.
Wave optics	CKT5—Can explain such light phenomena as reflection, refraction, double slit experiments, thin film colors, spectroscopy, etc, using the wave model of light.	PCK5—Can help students use Huygens principle and wave fronts representations to explain phenomena mentioned in CKT4. Can show how to engage students in solving non-traditional problems involving geometrical optics
DC current	CKT6—Can use analogies, microscopic models and physical quantities to reason about electric circuits including complex elements. CKT7—Can solve complex electric circuit problems involving multiple loops, multiple batteries and batteries having internal resistance.	PCK6—Can solve the same problem using analogies, microscopic models, and physical quantities and explain what students will benefit from which approach. PCK7—Can show a sequence of instructional moves to help students master the concepts of DC circuits using simulations.

Description of activities

Attendance, participation in class discussions: Attendance and participation in each class meeting are crucial for your learning. Discussions in class will focus on problem solving and research on student learning in a particular area.

Homework The goal of the homework is for you to improve your content knowledge, to read more research papers, to learn to compose new types of problems and to devise ways to assess them formatively.

1. For each lesson you will work with another group member to teach a 20 min segment of a lesson for a chosen topic focusing on MRs.

2. Each week you will be given a reading assignment. It is your responsibility to read the papers and the chapters from Zull's book.

The Physics Teacher article: You will need to find, read, analyse, and use a paper from *The Physics Teacher* for any of the topics outlines below, start looking now!

Final exam: At the end of the course there is a written examination. You will receive a list of exam questions in advance. The exam will consist of your responses to two of the questions (selected randomly) and problem solving.

Types of problems

Type of problem	Description
Ranking tasks	Students have to rank the values of a certain physical quantity.
Choose answer and explanation	Students have to choose the correct answer *and* the correct matching explanation.
Choose measuring procedure	Students have to choose (or propose) the correct (or the best) experimental procedure that will allow them to measure/determine a certain quantity.
Evaluate	Students have to critically evaluate the reasoning of some (imaginary) people or evaluate the suggested solution to a problem (words, graphs, diagrams, equation).
Make judgment (based on data)	Students have to make a judgment about one or more hypotheses, based on data or other forms of evidence that are given in the problem, sometimes with uncertainties
Linearization	First, students have to write an equation that describes the relevant situation. Then they have to rearrange the equation to obtain a linear function (note that the independent and dependent variables in this function can be any function of the data given in the problem).
Multiple possibility and tell all	Students have to list as many quantities as they can that can be determined based on data given in the problem, or tell everything they can about the physical attributes of the objects that appear in the text or the relations between them.
Jeopardy	Students have to convert a representation of a solution into a problem statement.
Design an experiment (or pose a problem)	Students have to design an experiment, an experimental procedure, or a device that will allow them to measure/determine certain physical quantities or that would meet specific requirements
Problem based on real data	Students have to solve problems that are based on real data, obtained in real-life situations, often using easily available equipment and/or equipment that is typically used in student labs.

Tentative list of topics for discussions and microteaching (by week; MR—multiple representations)

Week	Topic
1, 2, 3	Types of MRs and types of new problem. Magnetic field. Electromagnetic induction. Properties of magnetic materials.
4, 5, 6	Electric field and its representations. Physical quantities characterizing the field. Capacitors, Discussion of resources online (develop a rubric for resource analysis),
7, 8	MR in geometrical optics. Reflection and refraction. Mirrors and lenses.
9, 10, 11	MR in wave optics. Diffraction and interference. Coherent light and opportunities for interference. Thin films. Photoelectric effect.
12	Traditional physics assessments (FMCE, CSEM, BEMA).
13, 14	MR in gas laws and thermodynamics.
15	Final exam.

References

Barab S A and Hay K E 2001 Doing science at the elbows of experts: issues related to the science apprenticeship camp *J. Res. Sci. Teach.* **38** 70–102

Bitzenbauer P 2023 ChatGPT in physics education: a pilot study on easy-to-implement activities *Contemp. Educ. Technol.* **15** ep430

Black P and Wiliam D 1998 Assessment and classroom learning *Assess. Educ* **5** 7–68

Bologna V 2023 An early physics approach to improve students' scientific attitudes. The role of teacher habits *Doctoral Dissertation* University of Trieste

Brookes D T and Etkina E 2015 The importance of language in students' reasoning about heat in thermodynamic processes *Int. J. Sci. Educ.* **37** 659–779

Clement J 1993 Using bridging analogies and anchoring intuitions to deal with students' preconceptions in physics *J. Res. Sci. Teach.* **30** 1241–57

Dahlkemper M N, Lahme S Z and Klein P 2023 How do physics students evaluate artificial intelligence responses on comprehension questions? A study on the perceived scientific accuracy and linguistic quality of ChatGPT *Phys. Rev. Phys. Educ. Res.* **19** 010142

Danielson C 2013 The Framework for Teaching, Evaluation Instrument *Danielson Group* https://nctq.org/dmsView/2013_FfTEvalInstrument_Web_v1_2_20140825

Darling-Hammond L, Hammerness K, Grossman P, Rust F and Shulman L 2005 The design of teacher education programs *Preparing Teachers for a Changing World* ed L Darling-Hammond and J Bransford (San Francisco, CA: Jossey-Bass) pp 390–441

Etkina E, Brookes D and Planinsic G 2019a *Investigative Science Learning Environment: When Learning Physics Mirrors Doing Physics* (IOP Concise Physics) (San Rafael, CA: Morgan amd Claypool)

Etkina E 2015 Using early teaching experiences and a professional community to prepare pre-service teachers for every-day classroom challenges, to create habits of student-centered instruction and to prevent attrition ed C Sandifer and E Brewe *Recruiting and Educating Future Physics Teachers: Case Studies and Effective Practices* (College Park, MD: American Physical Society) pp 249–66

Etkina E, Gregorcic B and Vokos S 2017 Organizing physics teacher professional education around productive habit development: a way to meet reform challenges *Phys. Rev. Phys. Educ. Res.* **13** 010107

Etkina E, Planinsic G and Van Heuvelen A 2019b *College Physics: Explore and Apply* 2nd edn (New York: Pearson)

Etkina E, Van Heuvelen A, White-Brahmia S, Brookes D T, Gentile M, Murthy S, Rosengrant D and Warren A 2006 Developing and assessing student scientific abilities *Phys. Rev. Spec. Top. Phys. Educ. Res.* **2** 020103

Fang X, Che S, Mao M, Zhang H, Zhao M and Zhao X 2023 Bias of AI-generated content: an examination of news produced by large language models arXiv:2309.09825

Feiman-Nemser S 2001 From preparation to practice: designing a continuum to strengthen and sustain teaching *Teach. Coll. Rec.* **103** 1013

Forbes M, Holtzman A and Choi Y 2019 Do neural language representations learn physical commonsense? arXiv:1908.02899

Gregorcic B and Pendrill A-M 2023 ChatGPT and the frustrated Socrates *Phys. Educ.* **58** 035021

Hammerness L, Darling-Hammond L, Bransford J, Berliner D, Cochran-Smith M, McDonald M and Zeichner K 2005 *How Teachers Learn and Develop, in Preparing Teachers for a Changing World* ed L Darling-Hammond and J D Bransford (San Francisco, CA: Jossey-Bass) pp 358–89

Ji Z, Lee N, Frieske R, Yu T, Su D, Xu Y, Ishii E, Bang Y J, Madotto A and Fung P 2023 Survey of hallucination in natural language generation *ACM Comput. Surv.* **55** 248

Jumper J et al 2021 Highly accurate protein structure prediction with alphafold *Nature* **596** 7873

Kasneci E et al 2023 ChatGPT for good? On opportunities and challenges of large language models for education *Learn. Individ. Differ.* **103** 102274

Khandelwal K, Tonneau M, Bean A M, Kirk H R and Hale S A 2023 Casteist but not racist? Quantifying disparities in large language model bias between India and the West arXiv:2309.08573

Kotek H, Dockum R and Sun D Q 2023 Gender bias and stereotypes in large language models *Proc. of The ACM Collective Intelligence Conf.* pp 12–24

Krupp L, Steinert S, Kiefer-Emmanouilidis M, Avila K E, Lukowicz P, Kuhn J, Küchemann S and Karolus J 2023 Unreflected acceptance—investigating the negative consequences of ChatGPT-assisted problem solving in physics education arXiv:2309.03087

Küchemann S, Steinert S, Revenga N, Schweinberger M, Dinc Y, Avila K E and Kuhn J 2023 Can ChatGPT support prospective teachers in physics task development? *Phys. Rev. Phys. Educ. Res.* **19** 020128

Lo C K 2023 What is the impact of ChatGPT on education? A rapid review of the literature *Educ. Sci.* **13** 410

Lodge J M, Yang S, Furze L and Dawson P 2023 It's not like a calculator, so what is the relationship between learners and generative artificial intelligence? *Learn.: Res. Pract.* **9** 117

Magie F (ed) *A Sourcebook in Physics* (Cambridge, MA: Harvard University Press)

Moravec H 1988 *Mind Children: The Future of Robot and Human Intelligence* (Cambridge, MA: Harvard University Press)

Nadeem M, Bethke A and Reddy S 2020 StereoSet: measuring stereotypical bias in pretrained language models arXiv:2004.09456

Nordling L 2023 How ChatGPT is transforming the postdoc experience *Nature* **622** 655

Polverini G and Gregorcic B 2023 How understanding large language models can inform the use of ChatGPT in physics education arXiv:2309.12074

Taloni A, Scorcia V and Giannaccare G 2023 Large language model advanced data analysis abuse to create a fake data set in medical research *JAMA Ophthalmol.* **141** 1174–5

Thakur V 2023 Unveiling gender bias in terms of profession across LLMs: analyzing and addressing sociological implications arXiv:2307.09162

Ungdomsbarometern 2013 *Back2School 2023* https://ungdomsbarometern.se/rapportslapp-back2-school-2023/?fbclid=IwAR0qrPf9sYyiFzNy0c-f-lZ9APu8sSJPm3b9cOfGZuUfkg95DSCda9-Ob_w

Vasconcelos M A R and Dos Santos R P 2023 Enhancing STEM learning with ChatGPT and bing chat as objects to think with: a case study *EURASIA J. Math. Sci. Tech. Educ.* **19** em2296

Wikipedia n.d. Flight simulator *Wikipedia* https://en.wikipedia.org/wiki/Flight_simulator

Wolfram S 2023 ChatGPT Gets Its 'Wolfram Superpowers'! https://writings.stephenwolfram.com/2023/03/chatgpt-gets-its-wolfram-superpowers/?fbclid=IwAR1oyIMHMhH4Oi9yua7HkjOUGS3mznGDgewd9nZRnf1PyuHkflgMx2Cl4dg

Wood W 2019 *Good Habits, Bad Habits: The Science of Making Positive Changes that Stick* (London: Pan Macmillan)

Zull J E 2002 *The Art of Changing the Brain: Enriching Teaching by Exploring the Biology of Learning* 1st edn (Sterling, VA: Stylus)

Chapter 8
Success stories in the development of habits

In this chapter we share the stories of teachers who went through the ISLE physics teacher preparation program and now are teaching physics either at high schools or at universities. These teachers differ in their years of experience and their current professional roles.

8.1 Who are the authors of the stories?
June Lee

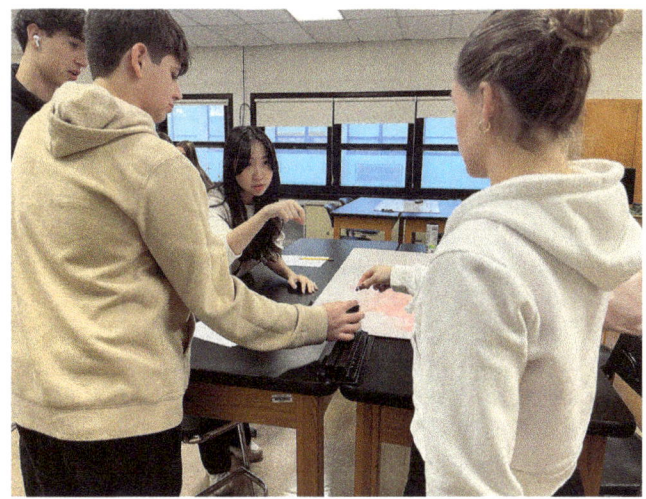

The first story is written by June Lee who is in her third year of teaching physics at Holmdel High school in New Jersey. She graduated from the Rutgers Physics

Teacher preparation program in 2021 and started teaching right away. Her story reflects how difficult it is to maintain the habits developed in the program and how sustaining them helps June enjoy her craft, inspire her students, and grow professionally. It is important to know that June's undergraduate degree is in chemistry and her physics teaching content knowledge was developed through 15 credits of physics courses, the physics methods courses in the teacher preparation program (see chapter 7) and teaching labs and problem-solving sessions in the Physics for the Sciences course. Her certification is in Physical Science—she is certified to teach both physics and chemistry.

Allison Daubert

The second story is written by Allison Daubert. She graduated from the Rutgers Physics Teacher preparation program in the spring of 2010 and taught physics for three years at Hopewell High School in New Jersey. She took a break for six years raising her three children and then went back to teaching in 2018, but this time at the university level at Bridgewater State University in Massachusetts. There, she helped develop a masters' level physics teacher preparation program and improved an undergraduate physics teacher preparation program that went from graduating 0–1 physics teachers a year to 5 teachers in 2023. She continued teaching physics ISLEizing algebra-based, calculus-based introductory physics courses, and bringing the ISLE approach to pre-service teachers and to the in-service professional development programs. In the Autumn of 2023 she was invited to teach physics at The Montrose School. She is now part time lecturer at Bridgewater State University and upper school physics teacher at The Montrose School. In 2022 she co-authored a paper in *The Physics Teacher* 'Refrigerator magnet investigation'. In the same year she received the 2022 Presidential Award for Distinguished Part Time Teaching at Bridgewater State University.

Danielle Buggé

The third story is by Danielle Buggé. Danielle has been teaching high school physics at the High School South for 14 years (since 2009) after she graduated from the Rutgers Physics Teacher Preparation program. After having taught for five years, in 2014, she applied for the PhD program in Physics Education at Rutgers University and finished her PhD work (with Eugenia as her advisor) in six years while being a full-time teacher. After finishing her PhD, Danielle did not stop teaching high school. She is still at West Windsor Plainsboro but she is engaged in physics education research at Rutgers and co-teaches a teaching internship seminar course in the Rutgers physics teacher preparation program. This school district now has seven(!) physics teachers who are graduates of the same program. In 2022 Danielle was chosen as the National Teacher of the Year by the Physics Teacher Education Coalition (PhysTEC). Danielle has published papers in *The Physics Teacher*, *Physics Education*, and *Physical Review Physics Education Research*. In 2022, after Eugenia retired from Rutgers, Danielle, with another graduate of the program (Elana Resnick), took over the maintenance of the Rutgers graduates' community, organizing meetings, keeping the calendar, and leading several meetings a year. In addition, together with Elana she is currently leading the efforts of the program graduates to develop ISLE-based curriculum materials for high school chemistry.

Josh Rutberg

The fourth story is by Joshua (Josh) Rutberg. Josh did not go through the Rutgers Physics Education Program as a pre-service teacher. He came to the Rutgers Graduate School of Education to get his PhD in physics education with Eugenia as his advisor in 2018 after having taught high school physics for five years. As a student in the PhD program, he not only took all physics teaching methods education courses described in chapter 7, but he also contributed to the development of experiments in those courses. He observed PSTs in the Physics for the Sciences course and attended all training meetings there. Therefore, Josh experienced all activities aimed at the development of productive habits as a pre-service physics teacher in the Rutgers Physics Teacher Preparation Program. In the second semester of his PhD studies, Josh started working with the team of faculty at Rutgers Newark first reforming and teaching one of their lab sections and then helping to ISLEize their algebra-based and calculus-based courses. He worked closely first with Eugenia and then independently running professional development for faculty and instructors. He not only trained all TAs in those courses but he also observed and documented how their behavior and teaching practices changed in the process. He came up with a list of important suggestions for those who run professional development for new ISLE faculty. This topic became one of the topics of his dissertation which he defended in February of 2022 and was recently described in his paper published in *Physical Review Physics Education Research*. He was immediately hired as a teaching faculty in the physics department at Rutgers Newark. While still a PhD student, Josh revamped all of the introductory physics curriculum during COVID lockdowns to preserve the spirit of ISLE. He continued this work after his PhD and now teaches an ISLE-based calculus-based physics course, develops curriculum materials, and trains all of the instructors in both algebra-based and calculus-based courses.

The stories that you will read are not edited. Every word is as the authors wrote it.

8.2 June Lee's story

Like many other new teachers, I look back on my work during student teaching and the first year with horror. There were so many mistakes and activities I would now do differently; I don't know what possessed me to even do it that way in the first place! But as I stumble through my amateur years, I recognize that the habits I develop now will be the habits that define the rest of my career—and luckily, I can confidently say that I am blessed to have been trained in some really great habits by the Rutgers Physics Teacher Preparation Program. Without the environment and experiences that this physics community consistently provides, I would not be able to feel successful or fulfilled in how I teach physics.

The first habit I developed was my dispositions towards students. As a student who experienced failure right before switching into teaching, I didn't enter as confident in myself as others who were majored in physics. But on the first day, and every day after that, Eugenia welcomed me into her physics family along with everyone else in the cohort that she taught. Knowing that she not only believed in me, but also expected excellent work was a significant driving force that let me believe I could be successful. Now when I interact with my students, who I know experience failure rather often, I make sure that they know I believe and expect them to be able to do physics with excellence—and they do! Michael Gentile (the course leader of Physics for the Sciences), under whose guidance I trained as a TA, taught me the habit of patience and understanding towards students. His forgiving policies have personally and emotionally impacted so many individuals, which create an environment of understanding for them to learn physics. Students are so grateful for taking his class, not because it is easy (actually far from it), but because they feel that they actually learn how to learn. With Eugenia and Mike as my first role models in the teacher prep program, I was able to solidify a disposition towards my students that balanced high expectations with generosity.

I cannot have high expectations for my students without extending that to myself in designing lessons that reflect my teaching philosophy. Just this past year, my supervisor asked me about my lesson on motion diagrams, how it connects to my school's curriculum, and if I really needed to spend that long on it. She knew that she was in for a long conversation when I took a very deep breath and said 'Okay—let me start at the beginning.' She patiently sat for 30 min while I convinced her that these diagrams were critical to understanding force diagrams and Newton's second law, which would extend to virtually every unit beyond just the one on forces! I would not have been able to have such a thorough conversation if I didn't habitually engage in knowing the 'full story' of my physics curriculum. I implement this 'story' habit on every level of teaching. Daily lesson plans either start or continue a story from the lesson before, which always include new phenomena or addressing phenomena we had seen before. Units always begin with the big Need-To-Know, and on the last day of the unit, I ask students 'Why did we learn all of this?' As the year progresses, students themselves start to remember why we learned physics in the first place, and feel immense satisfaction when they are able to explain how a cool phenomenon works on their own. And of course, each lesson is always heavily based on groupwork. Like many other ISLE teachers, my classroom is set up in groups of three to four students with no female students alone and a large whiteboard in the middle of the table. This grouping strategy also extends to my LGBTQ+ students so that they do not feel alone in the classroom. Due to this kind of environment, I have had my students burst into the classroom bragging about how they were thinking about circular motion taking an exit on the Garden State Parkway, or telling me how they were thinking about their change in internal energy while lifting weights at the gym.

My 'story' approach to designing lessons would not be possible without what is probably my strongest habit—time management. As a student teacher, my cooperating teacher Tina Lee (a graduate of the same program) sat with me over Google Meets for 2 h every Saturday morning and had me plan the upcoming week with her on her weekly and monthly planning templates. Even a few years later, I still use a version of these templates that I adapted for my own needs and I still plan weekly on Saturday mornings when I feel the most relaxed and creative. This let me weave the observational experiments, hypotheses, prediction-making, testing experiments, and application experiments in a way that made sense for my students consistently throughout the entire school year. Because I had such a strong habit of time management instilled in me through my student teaching experience with another ISLE teacher, I was able to tell the full 'story' in physics during my first year. Or in other words, complete the school's curriculum in one year—which I consider one of my proudest accomplishments as a first-year teacher.

The most controversial habit that I engage in each year is the resubmission habit. For two years, I was allowed to let my students resubmit and retake every single test that they took using the same method that I had learned from Matt Blackman (another graduate of the program, who taught the course Demonstrations and Technology in Science Education, a part of the program). My students understood that failure was a part of the learning process, much like how disproving hypotheses with testing experiments was a necessary part of ISLE. By the end of the year, my students had grown considerably because of these resubmissions that I allowed. What began as an extrinsic motivation over a higher grade turned into an intrinsic one, with my students reporting to me that they felt accomplished in actually understanding physics. Of course, allowing these kinds of multiple attempts came at a cost in both time and social connections.

While it was incredibly difficult at first, I believe it is a small price to pay after actually reaping its rewards at the end of the school year. Part of the value in these multiple attempts, is that it is inherently a mastery-oriented approach to learning. I teach a conceptual physics class that usually has students who have difficulty with memory, discipline, and basically every skill that any traditional class would require. So in June when it came time for their final review, I was nervous when I asked my students to draw a motion diagram for a problem about forces. They had not drawn a motion diagram in months, and I had planned for extra time dedicated to remembering how to draw diagrams from the first marking period. I cannot fully describe in words how proud I was when every single student (yes, every single one!) drew a perfect motion diagram, a perfect force diagram, and were able to relate the change in velocity arrow to the sum of the forces arrow. These were the same students who would come in every Monday truly needing to re-learn everything from the lesson last Friday, and here they were perfectly representing the motion of a paratrooper using skills they had learned nine months ago! The struggle that they went through to actually master those skills despite failure, and the struggle that I went through to continue the policy despite backlash was worth it. The students ended my course feeling successful. And for me, well, all teachers know that it is much easier to grade a final when the students know what they are doing!

Dispositions towards my students, lessons designed around the ISLE process, time management, and multiple attempts are all habits that I think let me be the best version of myself each day I walk into the classroom. Although they are all extremely important, I still believe that the most important habit is the habit of community. It is through the support and lifelong friendships of my cohort from the teacher prep program that I continue to love teaching with other like-minded teachers. Every habit I have is because of the mentors I had and still have through the program, and their habits became my habits. I find myself planning like Tina on the weekends. As I rotate through groups, I mimic Mike's (Michael Gentile is the course leader of Physics for the Sciences) patient line of questioning. I set student expectations firmly like Eugenia. My excitement for physics looks like Dan Lee's (Dan is a graduate of the program, a physics teacher who actively participates in the community of graduates). I try to be as thoughtful as Danielle Buggé (another graduate of the program who co-taught the seminar course accompanying Student Teaching Internship), as caring as Debbie Andres (yet another program graduate who taught the Engineering class). And in this amalgamation, my own personality as a physics teacher practicing good habits is able to grow. I would not be motivated to engage in these habits consistently without knowing that I have an entire community of ISLE teachers like the ones I mentioned above to lean on and learn from.

8.3 Allison Daubert's story

In the past 13 years since I graduated from the Rutgers Science Education program, I have taught in a variety of contexts. I have taught high school, algebra- and calculus-based university physics, pre-service physics teachers, in-service physical science teachers, and once again, I am back to teaching high school. I developed the broad habits that I bring to teaching every day in the Rutgers Science Education program. Additionally, I learned the dispositions and foundational ideas that have allowed me to develop my own set of habits over the years.

The first set of habits takes place in the days and weeks before I teach and defines my preparation for teaching. I carefully plan my lessons and look for clear beginnings, middle, and ends. I start by writing what the students will do and then structure what I will do around supporting the students in their work. I review my lesson plans and ask myself a set of questions:

- Is there a hook? Is there something that will get the kids excited?
- Do they have the tools to be successful in this lesson?

- Do they have opportunities to be successful in this lesson?
- Do I have a warm up? A clear task that they can begin working on before the bell rings.
- Do I have closure?
- Have I planned alternative back-ups in case equipment doesn't work as expected, such as videos of the experiments?

I provide closure in each class by asking students to tell me what they learned, what they feel better about, or what they have questions about. My students know that we always do this. Additionally, I never walk into the classroom without carefully having solved every physics problem that we will do in class this day. I never teach 'on the fly'. A practiced musician can walk on stage and appear to just jam with their band, but the audience hasn't seen the thousands of hours of rehearsal that happened before the performance. I remind myself of this.

The next category of habits includes things that I physically do while I am teaching. Many of these habits developed naturally in my years teaching but formed because of the bedrock laid in my graduate classes. These habits support a core idea: the experience and ideas of the student are centered before my own. To achieve this, I always doing the following:

- Stand in the back of the room, side of the room, or sit in a student seat as much as possible.
- Hand whiteboard markers to students and ask them to write at the board instead of me.
- Place my hands in my pockets while I work with small groups of students as a physical reminder of myself not to touch the whiteboard marker.
- Carry a cup of tea with me while I teach to resist the urge to take a whiteboard marker and to remind myself to slow down and give sufficient wait time to students so they can answer questions effectively.

I have habits surrounding what I say and write in my approach to both physics and teaching physics.

- I work each problem using the complete problem-solving strategy involving multiple representations with my students.
- I ask my students to always look for and identify consistencies between their multiple representations.
- I look for knowledge mobility within the room and point students to answer each other's questions before I do.
- I often ask my students to 'tell me in human words'; this is our class verbiage that serves as a reminder that we do not use fancy, specialized terminology until everyone fully understands it.
- I listen for what is correct in a student's comment, even if what they said contains errors. By helping them change the context or verbiage, together we make it correct.

- I ask 'How do you know?' often.
- I ask 'How could we test that idea?' often.

Additionally, I have habits around how I interact with my students. Building positive relationships with students and making sure that they feel known and loved in my physics class is integral towards their persistence in physics when the content becomes hard. I recognize that I represent what a physicist 'is' in the minds of my students. My habits of interaction with my students include:

- Greeting each student by name every day at the beginning of class.
- Learning something about each student—what do they love? What are they 'experts' in? What can they teach me?
- Celebrating successes, both in and out of the classroom, that my students achieve.
- Sitting next to students and working physics problems as if I were also student.

Lastly, I have my personal and professional habits of growth. I think of myself professionally within two lenses; that of a physicist and that of a person who guides the personal and intellectual growth of others. Many years ago, I developed a habit of always reading a book that is both enjoyable to me personally, but also pushes me and guides my professional growth within one of these two lenses. Over the years, I've worked through many books from *The Making of the Atomic* Bomb by Richard Rhodes, which helped advance my physics knowledge surrounding nuclear fission *to Failure is Not an Option* by Gene Kranz, which helped advance my understanding of how professional scientists work together and problem solve under pressure. I always share what I'm reading and what I'm learning about with my students. What I read keeps me continually excited as a physics learner.

8.4 Danielle Buggé's story

For the past 14 years, the theoretical framework of ISLE has been an integral part of my teaching practice. During my physics teacher preparation program at Rutgers University, I learned skills and dispositions that allowed me to develop and refine productive habits for teaching. Today, these habits are so engrained in my planning and preparation that I am able to spontaneously respond to situational cues in the classroom and obstacle removal for my students is seamlessly integrated into my instruction.

Before the start of every school year, I reflect on the previous year and write down new goals for both myself and the students. Having a clear direction helps with carefully planning my lessons, establishing the context for learning, and, later on, self-reflection. I explore not only my local community but also current global science news and ask myself is there a unique hook that can drive the learning throughout the entire school year? Recently, this hook has been space themed: be it the Perseverance rover landing or the launch and orbit of the James Webb Space telescope. I spend time at the end of the summer making sure my classroom is set up for collaborative learning: the tables are arranged to encourage groupwork and there are white boards and markers readily available.

As I plan individual lessons in the days and weeks leading up to each class, I double check that there are both additional hooks embedded in the plans to get the students excited about the topic and that each student has an opportunity to be successful in the classroom. I make sure there is time built in for reflection, asking 'how to do you know?' and 'what remains unclear?'. As I review activities, I carefully solve the problems, double check the links to videos, and make sure the necessary equipment is available prior to the lesson. These habits ensure my students will be able to focus their energy on doing physics, not troubleshooting because I was unprepared.

For me, the most important habit for my students is transparency in the classroom. If we are asking our students to learn in the way physicists work, there has to be transparency about the process—no tricks. This means making my physics lessons and assessments predictable. Students work in teams and the activities they complete are physically and cognitively demanding enough that everyone has to contribute. They come to expect that all new ideas begin with observational experiments and proposed explanations have to be tested prior to their use. There is also a sense of urgency in the classroom as they have a set amount of time to complete each activity. Students do homework and review their notes outside of class as they expect a check-in every few days. They also acknowledge that developing new ideas is difficult and they always have opportunities to demonstrate an improved understanding. Over time, their mindset shifts from viewing not understanding something as a negative to a welcome additional learning opportunity. These purposefully integrated habits lead to students establishing habits of their own that will help them in future courses and careers.

There also has to be a level of autonomy in the classroom where I put the ideas and questions of my students before my own. To achieve this, first, I always celebrate their questions. I display excitement when one of them asks a question that drives the learning forward. It doesn't take long for other students to instinctively answer their peers' questions since they know I'm going to bounce it back to the class. If they ask a question that we cannot yet answer due to needing to learn more physics knowledge, we write their question on an index card and display it on the bulletin board in the room. When we reach that point in the year, their question serves as the need to know for that lesson.

Furthermore, I try to minimize the amount of time I am in the front of the classroom. When students work in groups, I step back and give them space to grapple with each activity. I circulate the room and peek over their shoulders at their white boards. When I do engage in conversation, I ask them to explain their reasoning and pose fallback questions to help them work through their confusion. Teams who finish early are rewarded with more physics. Over time, these habits create a community of young physicists who are not afraid to be unsure, not afraid to ask their peers for assistance, and not afraid of challenging themselves as learners.

Finally, I would be remiss not to briefly mention my personal and professional habits of growth. My classroom community developed out of my experiences learning to be a physics teacher. As a student, I had no idea the impact my teacher preparation program was going to have on my life and career. I am fortunate to be part of a talented group of like-minded individuals, some of whom have become my closest friends, who share the same dispositions and fundamental knowledge. The Rutgers physics teacher preparation program alumni meet regularly to do physics and talk pedagogy. I also work with three alumni of the program, am an active member of the American Association of Physics Teachers, and stay up to date with current publications. These monthly meetings, conversations during planning periods, and engagement with the larger physics education community help me maintain and continue to develop habits integral to fostering growth mindsets in the classroom. I welcome these opportunities to connect with my colleagues and share with my students how I am constantly learning and growing in the profession. These actions and conversations help my students form lifelong habits of their own.

8.5 Josh Rutberg's story

As anyone who has spent even a little time in charge of their own classroom well knows, teaching involves making a lot of decisions. What questions to ask, when to press for more detail, how to group students, when to move on to the next activity, when to rephrase an important point, how to respond to students who are not engaged in your lesson, and on and on. Far too many decisions in far too little time. Decisions which must be made in a split second. Decisions which may shape how responsive your students will be to you for the rest of the year.

How are we supposed to make so many decisions without the time to properly consider the consequences or weigh our options?

The answer is: by knowing what we value as teachers and building productive habits to enact those values in our classrooms.

This may sound a bit ironic to my students, with whom I constantly emphasize the need for deliberate, careful, and intentional practice, but it is not so contradictory as it might first appear. The entire purpose of deliberate practice is to ingrain solid and productive habits. My students need to remind and force themselves to draw force diagrams when solving physics problems involving forces because they have not yet developed the habit of doing so automatically. But eventually they will, and those reminders will no longer be necessary.

Teaching is the same way. Things which required conscious effort on my part when I first started teaching around a decade ago now happen automatically. It took time and effort to train myself to ask questions in ways which invite discussion, to develop a sense of timing with my lessons, and to figure out how to establish the kind of supportive and learning-focused environment that I want in my classroom.

Providing professional development and training for our TAs and other faculty has made me realize just how ingrained many of my habits have become in the way I teach. I often find that there are things I forget to mention while preparing instructors to teach a particular lesson because they're not things I consciously think about anymore. To some extent I don't even realize I'm doing them. They just happen. Asking the rest of the class 'what do we think about that?' when a student answers a question rather than directly validating or rejecting it myself has become an automatic response, like saying 'bless you' when I hear someone sneeze.

And these habits don't only exist within the classroom. When watching YouTube, I often find myself thinking about how a video could be used in my classes ('this video of someone slipping on ice would be wonderful for helping students see the role friction plays when walking'). I also see physics in my everyday life and want to share that experience with my students. An incredibly vivid memory I have is of sitting at a table with a canned drink while waiting for a bus. I noticed that if I tipped the can up onto its edge and let it fall back down it would always slide a few centimeters before coming to a stop. For the 20 min it took for that bus to arrive, I kept doing this and trying to figure out exactly what factors controlled how much the can slid. More than one of the experiments we perform in my classes started out as these kinds of observations. When learning about rotational motion, for instance, we roll empty and full cans of soda down inclines instead of generic hollow and solid cylinders.

The more I try and help students see the ISLE approach to experimentation in the things that matter to them, the more I see and use it everywhere myself. This makes it that much easier to teach because I'm showing my students how I actually think, how I really approach physics and learning.

8.6 Reflection

As you can see from the above stories, we picked teachers with different levels of experience and different levels of physics courses that they teach, as well as different levels of responsibilities. June, who finished the program recently, focuses a great deal on specific people in the teacher preparation program who helped her grow, Josh, on the other hand does not mention the program at all but brings up the teachers who he trains. This spectrum of experience and responsibilities shows you that at any level of experience the habits are crucial.

At first, reading these stories we thought of analysing them, summarizing what they said, and commenting. But after reading the stories we realized that this step is not needed as the stories speak for themselves. And although they are very different in length, in tone, and in focus, all of them show all of the habits of mind and practice that were discussed in the previous chapters. Two of the reflections bring up the importance of the community for sustaining the habits. One discusses how to instill these habits in other teachers. All of them focus on the role of the ISLE approach in the development of their habits. While the stories have different format and different focus, they communicate the same idea that is the foundation of this book: if we wish to be successful as teachers implementing the ISLE approach, we need to develop productive habits.

Chapter 9

Summary

The goal of this book is to help teachers and teacher educators develop productive habits to engage their students in learning physics through the Investigative Science Learning Environment approach. To help achieve this goal, we have discussed the necessary knowledge, skills, and dispositions that ISLE teachers need to possess to implement the ISLE approach in their classrooms. We also discussed the processes through which teachers can develop habits of mind, practice, and improvement that will help them in the complex conditions of contemporary physics education.

The context of today's teaching is complex. On one hand, we want our student to experience physics as a process of inquiry, which takes time. On the other hand, the pressures of curriculum coverage (including the push to include new topics that are related to modern technologies, such as the Quantum Flagship Project) and standardized assessments dictate that students achieve numerous learning outcomes from this inquiry (normative knowledge). This only seems possible if students do not spend time experiencing how this knowledge came to be. We must consider that the teacher is in charge of what is happening in their classroom and makes all of the decisions that will benefit their students. However, these decisions are shaped by the school administration, parents, and the outside world. Additionally, the advent of digital technology, social media, and artificial intelligence seem to make the teacher's role obsolete. How do we stay true to our dispositions in this complicated environment?

While there is no one right answer to this question, we feel that a consistent implementation of the ISLE approach might help resolve the above controversies and mitigate the above problems. Let's start with the conflict between inquiry and coverage. Students learning physics through ISLE develop skills and abilities that scientists use in their work. One of those skills is critically reading scientific texts. If our students learn to read a textbook by using an elaborative interrogation technique, they can study several topics by themselves that we do not (or do not need to) touch upon in class. By using the ISLE approach, students are better prepared to learn (or deepen their understanding of) some topics on their own.

Next is the conflict between teacher academic freedom and standardized tests. The standardized assessments are more and more affected by the documents that focus on science practices (e.g. Next Generation Science Standards, Physics Advance Placement exams for AP Physics 1 and 2), which helps advocate for our approach. ISLE students continuously collect and analyse data and develop their own ideas.

What about the ubiquitous digital technology? Smart phone apps allow students to collect data in real time and thus simplify the teacher's process of preparing equipment and saving time on expensive interfaces for data collection. File sharing applications allow for student collaboration in real time and help the teacher follow and provide feedback on student work immediately.

Finally, teachers have worries about AI preventing our students to devise their own answers to problems. ISLE curriculum materials are full of new types of problems that do not have one right answer and thus cannot be successfully solved by large language models (LLMs; see chapter 7). Additionally, LLMs might help both teachers and students devise new problems, and engage our students in the evaluation techniques.

The ISLE approach calls for helping all students feel able to succeed in physics and provides tools for the teacher to assist their students in attaining this success. Isn't this what every parent and every administrator want?

All of the above shows that the ISLE approach helps resolve the existing tensions. The key here is that the ISLE approach is a holistic learning environment that has the intention of helping all students learn physics by practicing it every day. It is the change of the environment that is the focus of the approach, not of the students themselves. In this light, we can see the ISLE approach as an example of the Universal Design for education. The idea of Universal Design came to our language from architecture. It means that instead of 'fixing' people with physical impairments so that they can 'fit' into the existing environment, we need to think about fixing the environment so that it is suitable and accessible for people with different physical abilities. Architects did this by inventing ramps for wheelchairs, sidewalk bubbles for the canes of visually impaired people, and lots of other things. In physics, we can do this by making the learning environment hospitable to all students. We argue that the ISLE approach changes the learning environment to accommodate the needs of all students. It achieves this goal through the structure of *student learning experiences* and through the structure of *classroom organization*. Below is the summary of those structures.

Student learning experiences:
1. Students are not asked to make predictions about the outcomes of observational experiments before observing them. This step removes the fear of being 'wrong' with the prediction and the subsequent feeling of failed intuition when the outcome does not match the prediction. This approach is also consistent with how experts respond to questions that are completely out of their expertise (e.g. ask an astrophysicist to predict what a particular medicine will do to an ovarian cancer, etc).

2. Students are asked to describe what they observed in simple words, which removes the advantage of those who are familiar with the scientific vocabulary. Those who are familiar with the scientific vocabulary will rethink and ask themselves what is essential about the observed phenomenon. This step serves as an 'equalizer' among differently prepared groups of students. As everyone can say what they observed, the students experience the feeling of success.
3. Students working in groups develop their own hypotheses that explain the outcomes of the observational experiments. These hypotheses do not need to be correct; they only need to be experimentally testable. As these hypotheses are the result of group work, even those who have trouble coming up with new ideas can participate and contribute. This step helps develop confidence and appreciation for collaboration.
4. Students propose testing experiments for their hypotheses and make predictions of their outcomes using the hypotheses under test. Here, a mismatch of the outcome with the prediction does not mean personal failure but only the fact that the hypotheses under test might be incorrect. This step helps build student confidence in their ability to function as physicists as in physics, rejecting a hypothesis is a part of the process of knowledge development.
5. Students use graphical representations while reasoning about physical phenomena and making predictions about the outcomes of testing experiments. This step not only helps those who have difficulties with algebra or calculus, but also helps everyone create mental images, visualize, connect ideas, and build coherent knowledge.
6. Students are continuously engaged in experimental design. This step helps those who have practical/engineering interests. We often observe that those students who have trouble with traditional physics problems thrive when challenged with experimental design.

Classroom organization:
1. Systematic use of group work and whiteboards helps shy students voice their ideas in small groups while they might be hesitant to share them with the whole class. At the same time, group work helps students who come with an advantage see the value of their peers' contributions and gives them an opportunity to improve further, all while helping others (we all know how often it happens that when you try to explain something to others, you get better insight into the problem). Finally, team work is a foundation of success in the workplace. It is our duty to prepare our students for this aspect of their future lives.
2. Opportunities to resubmit their work for improvement without losing points for repeated attempts helps all students see the value of persistence and perseverance. It affects the development of a growth mindset, which in turn leads to more confidence in attempting difficult problems.

While these structures might seem natural and logical, the truth is that implementing those in practice is not easy. It requires specific dispositions of a teacher, i.e. believing that all students are able to learn physics and are willing to work hard to achieve it. It also requires deep knowledge of the subject matter and tools that will help students master it. Finally, skills in organizing group work, conducting experiments, listening to students' ideas and being ready to test them experimentally, and managing multiple resubmissions of student work are crucial in implementing these structures. But even with the development of the above dispositions, knowledge, and skills, it can be and is difficult to stay on course with the pressures of teaching in real time with all the complications of student psychological development, power structures inside students' groups, administration pressure, and so forth. This is where the idea of habits come into play. When we do something habitually, we do it even under pressure of time and events that we cannot control. The habits persist even in times of adversity. That is why this book is focused on the development of productive habits of teachers who help their students learn physics through the ISLE approach. We group these habits into three big categories: habits of mind, habits of practice, and habits of maintenance and improvement. We also argue that the key to the development of these habits is the presence of a strong learning community where all members feel safe to share their difficulties and help each other to face challenges.

In this book, we first discuss the dispositions, knowledge, and skills that are necessary to implement the ISLE approach (chapter 2) and then proceed to systematically describe the essence and the development of habits of mind of physicists and physics teachers implementing ISLE (chapters 3 and 4). We proceed to the discussion of habits of practice and maintenance and improvement in a community (chapter 5). We also separate habits and routines, which are necessary to develop and to sustain the habits (chapter 6). Finally, we share the structure and activities in the physics teacher preparation programs that focus on the development of the above habits (chapter 7) and produce teachers who use those habits every day (chapter 8). To conclude this book, we list free resources that are available to those who are interested in using the ISLE approach in their classrooms.

https://www.islephysics.net/ A website describing the essence of the ISLE approach and all available resources.

http://pum.islephysics.net/ ISLE-based modules for middle school and high school students.

https://sites.google.com/site/scientificabilities/ A website that describes scientific abilities and their development, provides a complete list of scientific abilities rubrics, and presents ISLE-based labs in which students design their own experiments to devise, test, and apply new ideas.

http://islevideos.net/ ISLE-based experiments and problems with supporting questions.

https://universeandmore.com/ ISLE-based video games designed by Matthew Blackman.

Finally, there is a Facebook group called Exploring and Applying Physics https://www.facebook.com/groups/320431092109343343. Over 2500 teachers of all levels from every continent of Earth except Antarctica are members of the group. Everyday posts provide continuous professional development to group members. The questions, comments, and discussions generated by group members span days. Stories of successful implementation of various ISLE activities ignite the biggest interest. Finally, Eugenia has been running monthly online professional development workshops, which are regularly attended by 20–30 teachers.

In the summer, teams of experienced ISLE teachers and professors run an eight hour introductory online workshop, another eight hour workshop at the summer meeting of the American Association of Physics Teachers, and a week-long workshop at Rutgers University. There are multiple ISLE workshops run in other US locations and in other countries throughout the year. There are plenty of opportunities to receive training in the ISLE methodology. We welcome you to our community!

www.ingramcontent.com/pod-product-compliance
Ingram Content Group UK Ltd.
Pitfield, Milton Keynes, MK11 3LW, UK
UKHW051341160426
5217IPUK00047B/119